SHELL THEORY

NORTH-HOLLAND SERIES IN

APPLIED MATHEMATICS AND MECHANICS

EDITORS:

J. D. ACHENBACH
Northwestern University

B. BUDIANSKY
Harvard University

W. T. KOITER
University of Technology, Delft

H. A. LAUWERIER
University of Amsterdam

L. VAN WIJNGAARDEN
Twente University of Technology

VOLUME 29

NORTH-HOLLAND – AMSTERDAM · NEW YORK · OXFORD

SHELL THEORY

Frithiof I. NIORDSON

Technical University of Denmark
Department of Solid Mechanics
Lyngby, Denmark

1985

NORTH-HOLLAND – AMSTERDAM · NEW YORK · OXFORD

© ELSEVIER SCIENCE PUBLISHERS B.V. — 1985

All rights reserved. No part of this publication may be reproduced, stored in a retrieval system, or transmitted, in any form or by any means, electronic, mechanical, photocopying, recording or otherwise, without the prior permission of the copyright owner.

ISBN: 0 444 87640 5

Publishers:
ELSEVIER SCIENCE PUBLISHERS B.V.
P.O. Box 1991
1000 BZ Amsterdam
The Netherlands

Sole distributors for the U.S.A. and Canada:
ELSEVIER SCIENCE PUBLISHING COMPANY, INC.
52 Vanderbilt Avenue
New York, N.Y. 10017
U.S.A.

Library of Congress Cataloging in Publication Data

Niordson, Frithiof I.
Shell theory.
(North-Holland series in applied mathematics and mechanics; v. 29)
Translated from Danish lecture notes, with changes and additions.
Includes bibliographies and index.
1. Shells (Engineering) I. Title. II. Series.
TA660.S5N56 1985   624.1'7762   84-18795
ISBN 0-444-87640-5 (Elsevier Science Pub. Co.)

PRINTED IN THE NETHERLANDS

To my wife
Hanne
and my children
Ulrika, Johan, Christian and Mathilde

PREFACE

This book originated as a set of lecture-notes to an introductory course on shell theory, given by the author to students of mechanical engineering at the Technical University of Denmark. The notes were originally written in Danish. Along with the translation into English several changes and additions were made.

In spite of the abundance of books on shell theory there seems to be no text-book available in English, suitable for an introductory course on modern shell theory on a graduate level.

It is my hope that this book will fill the gap and that others also might find these notes useful.

Several colleagues at our Department of Solid Mechanics have contributed with helpful suggestions. I gratefully acknowledge help from lektor Jes Christoffersen, lektor Jarl Jensen, and dr. Viggo Tvergaard.

My thanks are also due to Mrs. Bente Brask Andersen, who typed the manuscript and to Mr. Ebbe Dam Ravn, who draw the figures.

Lyngby   FRITHIOF NIORDSON
October 1983

TABLE OF CONTENTS

Preface vii

Table of contents ix

Introduction 1

Chapter 1. Tensor analysis 5

1. Introduction 7
2. Coordinate transformations 8
3. Contravariant and covariant vectors 9
4. Tensors 11
5. Algebra of tensors 13
6. The fundamental tensor 16
7. Associated tensors 19
8. Magnitude 20
9. Covariant differentiation 21
10. The Riemann-Christoffel tensor 25
11. The order of covariant differentiation for tensors . . . 27
12. Contravariant differentiation 28
13. Summary 28
    Bibliography 29

Chapter 2. The geometry of the middle-surface 31

1. Introduction 33
2. The first fundamental form of a surface 33
3. Normal coordinates 35
4. Second fundamental form of a surface 37
5. Invariants of curvature 41
6. The inverse problem 44

7. The equations of Codazzi 46
8. Surface and volume element 47
9. Coordinates of principal curvature 49
10. Summary 50
    Bibliography 51

Chapter 3. The kinematics of a surface 53

1. Introduction 55
2. Displacements 55
3. Stretching of the middle-surface 56
4. Rotation of an element of the shell 60
5. Bending of the middle-surface 64
6. Equations of compatibility 67
7. Summary 71

Chapter 4. The statics of a shell 73

1. Introduction 75
2. An element of the shell 75
3. The divergence theorem 77
4. Statically equivalent forces and moments . . . 79
5. External loads 82
5. Equations of equilibrium 83
7. Summary 87

Chapter 5. Effective forces and moments 89

1. Introduction 91
2. The principle of virtual work 91
3. Equations of equilibrium 99
4. Summary 99

Chapter 6. Boundary conditions 101

1. Introduction 103
2. External forces 103
3. Boundary conditions 105

4. Stress-distribution 109
5. Summary 110

Chapter 7. Stress–strain relations 113

1. Introduction 115
2. The elastic energy 115
3. The mathematical problem 122
4. Summary 123

Chapter 8. The static-geometric analogy 125

1. Introduction 127
2. Alternative measures of bending 127
3. Static-geometric analogy 129
4. Stress-functions 131
5. Summary and discussion 133
   Bibliography 133

Chapter 9. Plates 135

1. Introduction 137
2. Separation of the plate problem 137
3. In-plane loaded plates 138
4. Bending theory of plates 144
5. Bending of plates in rectangular coordinates . . . 144
6. Corners 145
7. Bending of a simply supported rectangular plate . . 148
8. The clamped and uniformly loaded elliptical plate . . 150
9. Vibrating plates 151
10. Energy methods 152
11. Bending of a clamped rectangular plate 160
12. Lowest natural frequency of a skew plate 161
13. Bending of plates in polar coordinates 165
14. Axisymmetrical bending of a circular plate 166
15. Free vibrations of a circular plate 168
16. Behaviour at a corner of a simply supported plate . . 172
    Bibliography 175

Chapter 10. The membrane state 177

1. Introduction 179
2. The membrane state in orthogonal coordinates 180
3. Axisymmetrical shells 181
4. Drop container 187
5. Domes 190
6. Conditions at the apex 192
7. Conditions at the base 193
8. Spherical dome of uniform thickness 193
9. Dome of uniform strength 194
10. Domes with given circumferential stresses 196
11. Deformation of axisymmetrical shells in the membrane state 197
12. General solution of the membrane state in axisymmetrical shells 199
13. Axisymmetrical loading 201
14. Wind load 201
15. The membrane state of a spherical shell 202
16. Membrane state of the toroidal shell 213
17. The membrane state of shells having zero Gaussian curvature 216
18. Discussion 221
    Bibliography 221

Chapter 11. Bending of circular cylindrical shells 223

1. Introduction 225
2. Basic equations 225
3. Boundary conditions 231
4. Solution of the mathematical problem 234
5. Bending due to axisymmetrical ring-loads 244
6. Non-axisymmetric bending of cylindrical shells 250
7. Free vibrations of cylindrical shells 259
8. Cylindrical panels 261
   Bibliography 262

Chapter 12. Axisymmetrical bending of shells 263

1. Introduction 265

2. Basic equations 265
3. Conical shells 272
4. Numerical example 276
5. Spherical shells 281
6. Toroidal shells 282
7. Approximate solution 283
   Bibliography 286

Chapter 13. Bending of spherical shells 287

1. Introduction 289
2. Basic relations 289
3. Solution of the mathematical problem 294
4. Boundary conditions 297
5. Spherical coordinates 298
6. Vibrations of spherical shells 312
7. Radially loaded spherical shell 334
8. Addendum 336
   Bibliography 336

Chapter 14. Bending of shells of revolution 337

1. Introduction 339
2. Shell equations 339
3. Boundary conditions 341
4. The differential equations for V_i and Q_i 343
5. Numerical solution 346
6. Bending of a toroidal shell under internal pressure . . . 349

Chapter 15. The Donnell–Mushtari–Vlasov theory . . . 353

1. Introduction 355
2. Simplified equations of equilibrium 356
3. Equilibrium in the deformed state 359
4. Summary 362
   Bibliography 362

Chapter 16. Nonlinear plate problems   363

1. Introduction   365
2. Simply supported rectangular plate   366
3. Plates with geometrical imperfections   368
4. Clamped quadratic membrane with a uniform transverse load   369
5. Circular membrane   372
6. Large amplitude vibrations of a clamped, circular plate .   375
   Bibliography   381

Chapter 17. Buckling of plates and shells   383

1. Introduction   385
2. The classical model of elastic stability   385
3. Buckling of axially compressed cylindrical shells   388
4. Inadequacy of the classical model   394
5. References and literature on the nonlinear buckling theories .   397
   Bibliography   398

Subject index   399

INTRODUCTION

This book deals with the foundations and applications of the theory of shells.

In books and papers on elasticity we meet concepts like column, disc, plate and shell. These terms refer partly to the shape of the bodies but, more important still, they also refer to the loading conditions, and most often indicate the kind of mathematical tool the author has in mind or is actually using. Nevertheless, a terminology referring to shape is useful.

If a body can be described by an inner surface and two outer surfaces such that a normal to the inner (middle) surface intersects the outer surfaces at the same distance $\pm\frac{1}{2}h$ (not necessarily constant), we call this body a shell. If the shell has no other boundary than the two outer surfaces, it is complete. If not complete, we always assume that it is bounded by a curve on the middle surface and a normal section along this curve.

Shells have many useful properties. When suitably designed, even very thin shells can support large loads, and shells are therefore utilized in structures like aircraft and ships, where light weight is essential. In other cases, the combined strength and enclosing properties of shells are utilized, for instance, in pressure-vessels, roofs and domes.

The mathematical formulation of problems in continuum mechanics leads as a rule to a system of partial differential equations and a set of boundary conditions. When the boundary is of a complicated geometrical shape like that of many shells, the numerical solution of the problem would defy even the most powerful computers available today, unless there were ways in which it could be simplified considerably.

The usual way by which a simplification is obtained consists in selecting coordinates in which the boundary conditions take a particularly simple form, usually at the cost of more complicated differential equations. But that is not enough. We need a more drastic and effective simplification—a *dimensional reduction*.

Dimensional reduction is the very nucleus of shell theory and is usually achieved by substituting averages and weighted averages (over the thickness of the shell) of the dependent variables (displacements, stresses etc.) for the variables themselves. In this process many details are necessarily lost but instead we gain a simplicity which in most cases is necessary for obtaining a solution to the problem and in addition we get a clearer survey of the essential aspects of the solution.

Shell theory originated historically with the special case of elastic plates. In early work by A. CAUCHY and S.D. POISSON in 1828–29 the dimensional reduction was based upon a power-series expansion of the dependent variables in the coordinate perpendicular to the middle-plane of the plate. For several reasons this approach was not very successful and fell into the background when G. KIRCHHOFF in 1876 introduced a theory for thin elastic plates,[1] based on the following assumptions,

(a) points which lie on one and the same normal to the un-deformed middle-surface also lie on one and the same normal to the deformed middle-surface;

(b) the displacements in the directions of the normal to the middle-surface are equal for all points on the same normal.

KIRCHHOFF's theory proved to be very fruitful and is widely used in mechanics and engineering today. It was applied to shells by A.E.H. LOVE in 1888.

The KIRCHHOFF assumptions lead an easy way to dimensional reduction, but since they cannot be fully reconciled with the three-dimensional theory of elasticity, there has always been some uneasiness connected with accepting them as a foundation for shell theory.

In an attempt to dispense with such contradictory hypotheses some authors have resorted into postulating a two-dimensional theory of elasticity, a theory of *elastic surfaces*. This approach, inspired by the

---

[1] For an account of the development of the theory of elastic plates and shells, see I. TODHUNTER and K. PEARSON (1960), *A History of the Theory of Elasticity*, Dover Publications, New York, in particular *Vol. 1*, pp. 241–276, pp. 336–357 and *Vol. 2*, Part II, pp. 83–91.

work of the COSSERAT brothers is quite analogous to the Euler theory of one-dimensional *elastica*. It has been demonstrated that there is no hindrance to the construction of such a two-dimensional theory, but the mere fact that a shell is a special case of a three-dimensional body should be a decisive argument against the introduction of any additional hypotheses in the theory. It is also contrary to the strive in science to unify theories.

The shell theory presented in this book is based on three-dimensional continuum theory without the burden of extra hypotheses, and brought as far as it seems reasonable before being specialized to thin elastic shells. Care has been taken to derive the fundamental equations without any restriction on the magnitude of the displacements, rotations, or strains. Thus, the complete nonlinear equations are derived and given in the text.

The dimensional reduction is achieved in three steps. In the first step (Chapter 4), two-dimensional stresses and moments are derived as averages and weighted averages of the three-dimensional stresses by formal integration over the thickness of the shell. In the second step (Chapter 5), the principle of virtual work is reduced to two dimensions using virtual displacements that are of the KIRCHHOFF type off the middle-surface. In the third step (Chapter 7), which applies only to thin elastic shells, the strain-energy density is reduced to a two-dimensional form. Introducing certain well-defined approximations the strain-energy density is given a form which is in the range permitted by the two-dimensional principle of virtual work (which due to the special type of virtual displacements is of limited generality) and the theory is completed.

In the eighth chapter, alternative formulations of the theory for thin elastic shells are discussed and the static-geometric analogy is shown, i.e., the analogy between the equations of equilibrium and compatibility. This also leads to the useful concept of stress-functions.

In Chapters 9–15, the general theory developed in the eight first chapters is applied to the linear case of thin elastic plates and shells. Different approximations and special geometries are treated.

Plane stress and bending of plates in rectangular, oblique and polar coordinates is discussed in Chapter 9. The efficiency of energy methods is demonstrated in some examples and the singular behaviour of solutions to the plate equation in obtuse corners is discussed.

The membrane theory of shells, which plays such an important role in engineering and design, is thoroughly treated in Chapter 10 with special emphasis on axisymmetrical shells, for which the general solution is given. Domes of different shapes are discussed and compared and the solution to spherical shells is given in the elegant and forceful complex form. The peculiarities of the membrane state of toroidal shells is discussed and the general solution for shells of zero Gaussian curvature is given.

To the bending theory of cylindrical shells Chapter 11 is devoted. The MORLEY–KOITER equations are derived and their general solution presented. A number of problems of interest to engineers are treated including the problem of free vibrations.

The REISSNER–MEISSNER method for axisymmetrical bending of shells of revolution is presented in Chapter 12. Specific results for conical and toroidal geometries are given and the important approximation connected with the name of J.W. GECKELER is derived and analysed.

The whole of Chapter 13 is devoted to spherical shells. The field equations are derived in an invariant form and then applied to spherical coordinates. The complete solution to these equations is presented, also for the case of free vibrations.

In Chapter 14 the general shell equations for non-axisymmetrical bending of shells of revolution are reduced to a set of eight ordinary first order equations and their numerical solution is discussed. The numerically stable 'field method' is described.

An often used approximation for solving problems in shell theory is the DONNELL–MUSHTARI–VLASOV (DMV) theory, especially suitable for shallow shells. This theory is given in Chapter 15 and the basis for all approximations involved is discussed. By considering equilibrium in the deformed state the linear theory is abandoned and the remaining part of this book deals with nonlinear problems.

The nonlinear DMV theory is applied in Chapter 16 to plates and shallow shells in which bending is affected by the membrane forces.

In the final Chapter 17 the classical model of elastic stability for shells in general is given and as an example specialized to axially compressed cylindrical shells. The inadequacy and the limitations of this theory are discussed.

CHAPTER 1

TENSOR ANALYSIS

1. Introduction	7
2. Coordinate transformations	8
3. Contravariant and covariant vectors	9
4. Tensors	11
5. Algebra of tensors	13
6. The fundamental tensor	16
7. Associated tensors	19
8. Magnitude	20
9. Covariant differentiation	21
10. The Riemann–Christoffel tensor	25
11. The order of covariant differentiation for tensors	27
12. Contravariant differentiation	28
13. Summary	28
Bibliography	29

CHAPTER 1

TENSOR ANALYSIS

1. Introduction

Tensors will be used throughout the text of this book and this chapter contains a short introduction to tensor analysis. Limitations of space make a complete or rigorous presentation of the subject impossible and therefore this chapter cannot substitute a textbook on tensor analysis. However, a student with an undergraduate background in calculus should be able to acquire here those elements of tensor analysis that are necessary for reading the text.

Why use tensors? The answer lies in the remarkable property of tensor equations: a tensor equation does not refer to any particular coordinate system, in fact, if it holds in just one system, it holds in all. Thus the validity of a tensor equation is independent of our arbitrary choice of coordinates. The importance of this fact for a shell theory is obvious.

The advantage of general tensor analysis will not be apparent if we restrict ourselves to *cartesian* (rectangular) coordinates only. But then such a restriction is not possible in shell theory where we have to use two-dimensional coordinate systems, even on surfaces where a cartesian reference system cannot be constructed.

Certain physical quantities like temperature or electric potential, for instance, require just one number at each point in space for their description. A change of coordinates does not affect these numbers. Such quantities are called *scalars*. Other entities like force or velocity need three numbers for their description. These numbers, the *components*, change according to certain rules when the coordinates are changed, in spite of the fact that they represent the same physical entity. We call those quantities *vectors*.

There are also quantities that require more than three numbers (in general 3^n numbers, where $n = 2, 3, 4, \ldots$) for their description. Such quantities are called *tensors*. The number n is the *order* of the tensor and consequently we call vectors and scalars tensors of order 1 and 0, respectively.

The concept of tensors is based on transformations between arbitrary coordinate systems. A starting point for tensor analysis is therefore the concept of *coordinate transformations*.

2. Coordinate transformations

To identify the points of three-dimensional space in a suitable manner, we attach three real numbers u^1, u^2, u^3, the *coordinates*, to each point and assume that there is a one-to-one relation between the coordinates u^i and the points of space.[1] In our notation, $i = 1, 2, 3$ is an *index* and not an exponent. Henceforward, we shall always use lower case Latin letters for indices that may assume the values 1, 2 and 3.[2]

If f^i are three independent, single-valued, continuously differentiable functions of u^i, the set of equations

$$x^i = f^i(u^1, u^2, u^3) \qquad (1.1)$$

defines other coordinates x^i and constitutes a *regular coordinate transformation*. The coordinates x^i might, for example, be cartesian coordinates and u^i, spherical coordinates, usually called r, ϕ, θ. This transformation would be regular for all $r \neq 0$.

The functions f^i are independent if and only if their *Jacobian determinant* is different from zero,

$$J = \det(f^i_{,j}) \neq 0, \qquad (1.2)$$

where the shorter notation $f^i_{,j}$ has been used for the partial derivative

---

[1] In some coordinate systems (e.g., polar coordinates) the one-to-one relation breaks down at certain *singular* points. This generally causes no trouble, but one should be aware of this.

[2] Later on we shall also use Greek letters for indices that assume the values 1 and 2 only.

$\partial f^i/\partial u^j$ of the function f^i with respect to the variable u^j. When inequality (1.2) holds, equations (1.1) may be solved for u^i, and we obtain

$$u^i = h^i(x^1, x^2, x^3), \qquad (1.3)$$

the *inverse transformation*. The functions h^i are then also independent, single-valued and continuously differentiable with respect to x^i.

Differentiating (1.1) and applying the chain-rule, the differential dx^i of x^i is found to be

$$dx^i = f^i_{,j} du^j, \qquad (1.4)$$

where we have utilized *Einstein's summation convention*. This convention implies the following. Whenever a letter appears twice as an index in the same term, a summation is implied over the range of the index. Such an index is called *dummy*. Clearly, the dummy indices may be changed in any term to any other letter (of the same range) that does not already appear in the term.[3] Letters appearing only once as indices in the same term (like i in equation (1.4)) are called *free*. We must take care never to let a letter appear more than twice in the same term as an index. With the Einstein convention, summations will occur in our investigations without waiting for our tardy approval. Thus the convention is not merely an abbrevation, but is of immense aid to the analysis.

3. Contravariant and covariant vectors

Equation (1.4) indicates how the differentials du^j transform to dx^i when the coordinate system is changed from u^i to x^i. We shall use this formula as a basis for our definition of contravariant vectors.

Any set of three quantities transforming according to (1.4) is called a *contravariant vector*. Thus, let A^i be the components of the vector in the coordinate system u^i. Then the components A^{*i} of the same vector in x^i are given by the linear transformation

$$A^{*i} = f^i_{,j} A^j, \qquad (1.5)$$

---

[3] Normally, an upper dummy index has its partner as a lower index, but this is not always so and it is not a requirement for the summation convention to apply.

in which the elements $f^i_{,j}$ of the *transformation matrix* are, in general, functions of the space coordinates.

We shall always use an upper index to indicate contravariant vectors. However, it must be remembered that not all quantities with an upper index are vectors. Thus, for instance, du^i is a contravariant vector, while u^i is not.

Let $T(u^1, u^2, u^3)$ be a continuously differentiable *scalar function* of the coordinates u^i. The same scalar is represented by another function $T^*(x^1, x^2, x^3)$ in the coordinate system x^i, i.e.,

$$T^*(x^1, x^2, x^3) = T(u^1, u^2, u^3),$$

whenever x^i and u^i refer to the same point in space. Thus, we have the identity

$$T^*(x^1, x^2, x^3) \equiv T(h^1(x^1, x^2, x^3), h^2(x^1, x^2, x^3), h^3(x^1, x^2, x^3))$$

and differentiation with respect to x^i yields

$$T^*_{,i} = T_{,j} h^j_{,i}.$$

This equation indicates how the *gradient* $T_{,j}$ is transformed when the coordinate system is changed from u^i to x^i. We shall use this rule to define another type of vectors. Any set of three quantities that transform according to this rule is called a *covariant vector*. Thus, if A_i are the components of the vector in the coordinate system u^i, and A^*_i its components in x^i, then

$$A^*_i = h^j_{,i} A_j. \tag{1.6}$$

A covariant vector is identified by a lower index. We have just seen that the gradient of any scalar function is a covariant vector.

Now, consider a sequence of successive coordinate transformations of either a contravariant or a covariant vector. It is easily shown that the transformation rules (1.5) and (1.6) have the property of making the result independent of the order in which the transformations occur. The rules are thus *self-consistent*. No other rules are known to have this remarkable property.[4] That is why contravariant and covariant vectors play such an important role in physics.

The reader may already be familiar with the concept of 'vectors' in rectangular coordinates. If we limit ourselves to transformations between rectangular coordinates, the rules (1.5) and (1.6) will coincide and we cannot discern between covariant and contravariant vectors.

4. Tensors

We shall now generalize the concept of vectors to quantities that require more than three components for their description.

Any set of $3^2 = 9$ quantities A^{ij} is called a *contravariant tensor of second order* if and only if it transforms according to the rule

$$A^{*ij} = f^i_{,k} f^j_{,l} A^{kl}. \qquad (1.7)$$

Analogously, a *covariant tensor of second order* is defined by the transformation law

$$A^*_{ij} = h^k_{,i} h^l_{,j} A_{kl}. \qquad (1.8)$$

There is still another possibility of defining tensors of second order. We shall call A^i_j a *mixed tensor of second order* if and only if it transforms according to the rule

$$A^{*i}_j = f^i_{,k} h^l_{,j} A^k_l. \qquad (1.9)$$

The *Kronecker delta* δ^i_j is defined by the following rule

$$\delta^i_j = \begin{cases} 1 & \text{if } i = j, \\ 0 & \text{if } i \neq j. \end{cases} \qquad (1.10)$$

In computations it often plays the role of a *substitutional operator*

$$\delta^i_j A(\ldots j \ldots) = A(\ldots i \ldots).$$

---

[4] Except for a rather trivial generalization, in which both sides of the equation are multiplied by a power of the Jacobian.

Note that index j here is substituted by i when $A(\ldots j \ldots)$ is multiplied by δ^i_j.

It is easily verified, as anticipated by the notation, that Kronecker's delta is a mixed tensor of second order by substituting δ^i_j into the right-hand side of (1.9). We get

$$\delta^{*i}_j = f^i_{,k} h^l_{,j} \delta^k_l = f^i_{,k} h^k_{,j}.$$

The product $f^i_{,k} h^k_{,j}$ can be computed as follows. Equations (1.1) and (1.3) yield the identity

$$x^i \equiv f^i(h^1(x^1, x^2, x^3), h^2(x^1, x^2, x^3), h^3(x^1, x^2, x^3))$$

and differentiation with respect to x^j gives

$$x^i_{,j} = f^i_{,k} h^k_{,j}.$$

But $x^i_{,j} = \delta^i_j$ and thus $\delta^{*i}_j = \delta^i_j$.

The Kronecker delta is therefore a mixed tensor of second order. It is, however, a very unusual tensor, having the same components in all coordinate systems.

In mechanics, tensors of second order play an important role. But there are also tensors of higher orders that are important. The rules (1.5) and (1.6) for vectors can be generalized to apply to tensors of arbitrary order in the following manner.

An entity $T^{ijk\cdots}_{pqr\cdots}$, that has u upper indices and v lower indices, thus having 3^n ($n = u + v$) components, is called a tensor of order (rank) n if and only if its components $T^{ijk\cdots}_{pqr\cdots}$ in the coordinate system u^i transform to $T^{*ijk\cdots}_{pqr\cdots}$ in the coordinate system x^i according to the relation

$$T^{*ijk\cdots}_{pqr\cdots} = f^i_{,l} f^j_{,m} f^k_{,n} \ldots h^s_{,p} h^t_{,q} h^u_{,r} \ldots T^{lmn\cdots}_{stu\cdots}, \tag{1.11}$$

where the functions f^i and h^i represent the transformation of coordinates and the inverse transformation according to (1.1) and (1.3). The tensor is said to be of contravariant order u and covariant order v. Two tensors are said to be of same *dimension* (*type*) when their contravariant and covariant orders match.

5. Algebra of tensors

In this section, algebraic rules of operation with tensors are established. Our notation for tensors is convenient for defining the sum, the difference and the (outer) product of tensors. For example,

$$C^{ij}_k = A^{ij}_k + B^{ij}_k$$

indicates that the elements of the entity C^{ij}_k are the sums of corresponding elements of A^{ij}_k and B^{ij}_k, so that, for instance, $C^{22}_3 = A^{22}_3 + B^{22}_3$. Likewise $C^{ijkl}_{pq} = A^k_{pq} B^{jil}$ denotes that $C^{3122}_{12} = A^2_{12} B^{132}$ etc.

Using the transformation law (1.11), it is easy to verify the following theorems.

Theorem I. *The sum of, or the difference between, two tensors at one and the same point and of the same dimension is a tensor of that dimension.*

It immediately follows that any linear combination of two tensors of the same dimension and at the same point is a tensor of that dimension.

Due to the fact that the transformation matrices $f^i_{.j}$ and $h^i_{.j}$ are functions of the coordinates, the sum of vectors (or tensors) at different points are not generally vectors (or tensors). This is in sharp contrast to the rules applying in elementary (cartesian) vector algebra, where, of course, $f^i_{.j}$ and $h^i_{.j}$ are constant.

The set of 3^{n+m} quantities consisting of the product of each component of a tensor of order n by each component of another tensor of order m defines the *outer product* of the two tensors.

Theorem II. *The outer product of two tensors at one and the same point is a tensor whose contravariant order is the sum of the contravariant orders of its factors and whose covariant order is the sum of the covariant orders of its factors.*

It is obvious that the outer product is distributive with respect to addition

$$(A^{\cdots}_{\cdots} + B^{\cdots}_{\cdots})C^{\cdots}_{\cdots} = A^{\cdots}_{\cdots}C^{\cdots}_{\cdots} + B^{\cdots}_{\cdots}C^{\cdots}_{\cdots}.$$

The components of a tensor in physics are represented by numbers that have physical dimensions like length, velocity, etc. Since the elements of the transformation matrix do not necessarily have the same dimensions, the components may not have the same dimensions either. Therefore, in general, no meaning is attached to any sums of components of one tensor. There is, however, a remarkable exception to this rule, called *contraction*.

If, in a mixed tensor, an upper index is put equal to a lower index, implying a summation of the components over the range of this index, the tensor is said to be contracted, resulting in an entity having $\frac{1}{9}$ of the components of the original tensor. Again, using the rule of transformation, (1.11), we may easily verify the following theorem.

Theorem III. *Any contracted tensor is a tensor of two orders less (one contravariant and one covariant).*

In particular, if a contravariant vector A^i is multiplied by a covariant vector B_j, the result will be the outer product $A^i B_j$, which is a mixed tensor of second order. Contraction results in a scalar $A^i B_i$, also called the *inner product* or *scalar product* of A^i and B_j.

A scalar is *invariant*, i.e., it has the same value in all coordinate systems. It is useful to learn that, in physics, only scalars can be measured directly. Thus, if F^i are the contravariant components of a force and n_i the covariant components of a unit normal in an invariantly defined direction, the scalar $F^i n_i$, the projection on, or the 'component' of, the force in that direction, will be a measurable quantity. $F^i n_i$ will have the physical dimension of force (Newton) while the components F^i may not.

Summation and multiplication of tensors formally follow the elementary rules of real numbers. But for division there is no corresponding simple rule. Instead, we have the so-called *rigorous quotient theorem*.

Theorem IV. *A quantity which, on inner multiplication by any covariant (alternatively any contravariant) vector, always gives a tensor or an invariant, is itself a tensor.*

The dimension of the tensor follows from Theorem II. We shall give the proof of this theorem in a special case.

Let us assume that the inner product of an unknown entity $A(jk)$ with two indices j and k and an arbitrary contravariant vector B^j is a covariant vector C_k, i.e.,

$$A(jk)B^j = C_k.$$

We shall show that $A(jk) = A_{jk}$ is a covariant tensor of second order (i.e., such that Theorem II is satisfied). Let an asterisk denote the components in the coordinate system x^i. Then

$$A^*(jk)B^{*j} = C_k^*.$$

Applying rule (1.5) for B^{*j} and (1.6) for C_k^*, we have

$$A^*(jk)f_{,m}^j B^m = h_{,k}^n C_n = h_{,k}^n A(jn)B^j.$$

Now we substitute m for the dummy index j on the right-hand side and obtain

$$\{A^*(jk)f_{,m}^j - h_{,k}^n A(mn)\}B^m = 0.$$

Since B^m is arbitrary, the quantity in braces must vanish. Multiplication with $h_{,p}^m$ yields

$$A^*(jk)f_{,m}^j h_{,p}^m = h_{,p}^m h_{,k}^n A(mn).$$

But $f_{,m}^j h_{,p}^m = \delta_p^j$ and hence

$$A^*(pk) = h_{,p}^m h_{,k}^n A(mn).$$

A comparison with (1.8) shows that $A(mn)$ is a covariant tensor of second order.

Using our rules for addition, subtraction, multiplication and contraction of tensors referring to a specific coordinate system and a specific point in space, we obtain a tensorial expression which is itself a tensor. If this tensor happens to vanish in one particular coordinate system, it vanishes in all. Putting the expression equal to zero, we obtain a *tensor equation*. That equation no longer refers to any particular coordinate system. We are accustomed to insisting (sometimes quite unnecessarily) that all equations in physics should be stated in a form independent of the units employed. Whatever additional insight into underlying causes is gained by this, still greater insight must be gained by stating them in a form altogether independent of the coordinate system.

6. The fundamental tensor

Let P and Q be two neighbouring points in space with the coordinates u^i and $u^i + du^i$, respectively. Let us also, for the moment, assume that the space is Euclidean and that x^i are cartesian coordinates in space, such that x^i and $x^i + dx^i$ are the coordinates of P and Q, respectively. The distance ds between the points follows from Pythagoras' theorem

$$ds^2 = dx^k\, dx^k = f^k_{,i}\, du^i f^k_{,j}\, du^j = f^k_{,i} f^k_{,j}\, du^i\, du^j.$$

With the notation

$$g_{ij} = f^k_{,i} f^k_{,j} \tag{1.12}$$

we have

$$ds^2 = g_{ij}\, du^i\, du^j, \tag{1.13}$$

where $g_{ij}(u^1, u^2, u^3)$ are nine functions that can be computed from (1.12) when the functions f^i connecting the coordinate system u^i with any cartesian coordinate system x^i are known. From (1.12) it follows that g_{ij} is symmetrical,

$$g_{ij} = g_{ji}, \tag{1.14}$$

and that, therefore, no more than six of the functions are independent. Clearly, for cartesian coordinates, $g_{ij} = \delta^i_j$.

Since ds^2 is invariant and du^j is arbitrary, $g_{ij}\, du^i$ must be a covariant vector. Again, since du^i is arbitrary, g_{ij} must be a covariant tensor of second order.

In deriving (1.12), we assumed the space to be Euclidean. Taking, instead, (1.13) as our starting point, we get the more general *Riemannian geometry*. It is then assumed that (1.13) is positive definite,[5] but the functions g_{ij} are otherwise arbitrary.

Any tensor of second order can be decomposed in a symmetrical and a skew-symmetrical part, viz.

$$g_{ij} = \tfrac{1}{2}(g_{ij} + g_{ji}) + \tfrac{1}{2}(g_{ij} - g_{ji}).$$

Multiplying by the symmetrical tensor $du^i\, du^j$, the product $\tfrac{1}{2}(g_{ij} - g_{ji})\, du^i\, du^j$ vanishes and thus only the symmetrical part of g_{ij} will contribute to ds^2. Thus, without loss of generality, we may always assume that g_{ij} is symmetrical.

The tensor g_{ij} is called the *fundamental* or *metric* tensor of the space. The term 'fundamental' is used because information can be retrieved from g_{ij} about some fundamental properties of the space such as, for instance, whether the space is Euclidean or not. The term 'metric' refers to the metrification of space by the distance relation (1.13).

By means of a suitable linear coordinate transformation, the metric tensor can always be put in the diagonal form $g^*_{ij} = \delta^i_j$ at any preselected point. Thus, writing

$$u^i = a^i_j x^j$$

where the a^i_j's are constants, we have

$$g^*_{ij} = a^k_i a^l_j g_{kl} = \delta^i_j.$$

These are only six equations to be satisfied by the nine constants a^i_j. There are thus many ways of making the reduction. The resulting coordinates x^i are then *cartesian at the point*.

Applying a well-known rule for determinants, we get

$$[\det(a^k_l)]^2 \det(g_{ij}) = \det(\delta^i_j) = 1.$$

---

[5] I.e., $g_{ij}\, du^i\, du^j > 0$ for all $du^i \neq 0$.

Thus, the determinant g of g_{ij} is always positive,

$$g = \det(g_{ij}) > 0. \tag{1.15}$$

In particular, if the coordinates x^i are cartesian at a point, then according to (1.12), which only holds if x^i are cartesian, at the point,

$$g = [\det(f^k_{,i})]^2 = J^2 \tag{1.16}$$

at the point.

The determinant $g(u^1, u^2, u^3)$ is, in general, a function of the coordinates, and since its value depends on the coordinate system selected, it is not a scalar.

The inverse of g_{ij} is given by

$$g^{ij} = \text{cofactor}(g_{ij})/g, \tag{1.17}$$

where the cofactor(g_{ij}) is the determinant obtained by excluding the ith row and the jth column from g_{ij} and multiplying the result by $(-1)^{i+j}$. From definition (1.17) it immediately follows that

$$g_{ij}g^{jk} = \delta^k_i. \tag{1.18}$$

To investigate the nature of g^{jk} we multiply (1.18) by a contravariant vector A^i,

$$(g_{ij}A^i)g^{jk} = \delta^k_i A^i = A^k.$$

Since $g_{ij}A^i = B_j$ can be solved for A^i given any B_j, we may consider B_j to be an arbitrary covariant vector. The rigorous quotient theorem then shows that g^{jk} is a contravariant tensor of second order, as indicated by the notation. g^{ij} is called the *fundamental contravariant tensor*. From the definition (1.17), it follows that g_{ij} is symmetrical

$$g^{ij} = g^{ji}. \tag{1.19}$$

Sometimes, it is convenient to write Kronecker's delta with the letter g instead of δ, i.e.,

$$g^i_j = \delta^i_j,$$

and g^i_j is then called the *fundamental mixed tensor*.

We have now introduced three fundamental tensors

$$g_{ij},\ g^i_j,\ g^{ij}$$

of covariant, mixed and contravariant dimension, respectively, all three being symmetrical.

7. Associated tensors

If A^i is a contravariant vector, then $g_{ij}A^i$ is a covariant vector. We shall denote this vector by A_j (indicating that it represents the same physical entity). Thus we write

$$A_j = g_{ij}A^i \qquad (1.20)$$

and regard it as an operation: *lowering* (and substitution) of an index. A_j is called an *associated* vector to A^i, and vice versa. If the contravariant components A^i are given, (1.20) gives the rule for computing the covariant components A_j. Conversely, if the A_j's are given, (1.20) may be solved (uniquely), since $g \neq 0$, to yield A^i, and due to (1.18), we get

$$A^i = g^{ij}A_j \qquad (1.21)$$

and regard this as an operation: *raising* (and substitution) of an index.

The convention may just as well be applied to tensors of any order. However, care must be taken to ensure consistency when dealing with mixed tensors. Thus, we write $g_{ij}A^{ik} = A^k_{\cdot j}$, leaving a unique space for j to return to upon raising. Of course, if A^{ik} is symmetrical, there will be no need for this precaution.

Elementary definitions of physical quantities like force or velocity refer to rectangular coordinates. For cartesian coordinates, $g_{ij} = g^{ij} = \delta^i_j$ and relations (1.20) and (1.21) leave the components unaltered. This leads to an enlarged view of a tensor as itself having no particular covariant or contravariant character, but having components of various degrees of covariance and contravariance as represented by the associated tensors. Raising or lowering of indices is thus not regarded as altering the individuality of a tensor; and reference to a tensor A_{ij} may (if the context permits) be taken to include the associated tensors $A^i{}_j$, $A_i{}^j$ and A^{ij}.

It is useful to note that dummy indices have a certain freedom of movement between the tensor-factors of an expression. For example,

$$A_{ij}B^{ij} = A^{ij}B_{ij} = A^i{}_j B_i{}^j = A_i{}^j B^i{}_j.$$

The vectors and tensors of physics are not contravariant and covariant by nature. Rather they can be represented in any coordinate system by their covariant and contravariant components. If, for example, force were defined by mass × acceleration, the force would naturally appear as a contravariant vector. However, if we were to assume that force is defined by work = force × displacement, the force would emerge as a covariant vector.

8. Magnitude

The *magnitude* A of a vector A^i is an invariant defined by the relation

$$A = (A^i A_i)^{1/2}, \tag{1.22}$$

where the positive square-root is taken. Thus

$$(A)^2 = A^i A_i = g_{ij} A^i A^j = g^{ij} A_i A_j. \tag{1.23}$$

In rectangular coordinates, the scalar product of the vectors A^i and B_i is

$$A^i B_i = AB \cos \alpha, \tag{1.24}$$

where α is the angle between them. But since it is a tensor equation, it holds for any coordinates.

Let α_{12} denote the angle between the coordinate lines u^1 and u^2. The tangent vector (not unit) to the coordinate line u^1 is $A^i = (du^1, 0, 0)$, and to the coordinate line u^2, $B^i = (0, du^2, 0)$. Then

$$A^i B_i = g_{ij} A^i B^j = g_{12} \, du^1 \, du^2,$$
$$(A)^2 = g_{ij} A^i A^j = g_{11} (du^1)^2,$$
$$(B)^2 = g_{ij} B^i B^j = g_{22} (du^2)^2$$

and hence

$$\cos \alpha_{12} = \frac{g_{12}}{(g_{11} g_{22})^{1/2}}, \quad \sin \alpha_{12} = \left(\frac{g_{11} g_{22} - g_{12} g_{21}}{g_{11} g_{22}} \right)^{1/2}. \quad (1.25)$$

If the coordinate lines intersect each other at right angles ($\alpha_{12} = \alpha_{23} = \alpha_{13} = \frac{1}{2}\pi$), then $g_{12} = g_{23} = g_{13} = 0$, i.e., $g_{ij} = 0$ for $i \neq j$. When this holds everywhere, the coordinates are said to be *orthogonal*. Simple examples of orthogonal coordinates, often used in mathematical physics, are rectangular, cylindrical and spherical coordinates.

9. Covariant differentiation

Tensors, like vectors, appear *isolated* or in *fields*. In this section we shall be dealing with continuously differentiable tensor-fields. We already know that the gradient of a scalar field (or using other words, the derivative of an invariant) is a covariant vector-field. But the derivative $A^i_{,j}$ of a vector-field A^i is not a tensor-field. There is, however, an operation in tensor calculus, called *covariant differentiation* that conserves the tensor character of the field.

Let $A^k(u^1, u^2, u^3)$ be a vector-field in the coordinates u^i. At a point P with coordinates u^i, the field is $A^i(P)$, and at a neighbouring point Q with coordinates $u^i + du^i$, the field is $A^i(Q) = A^i(P) + A^i_{,j}(P) \, du^j$.

However, the increment $A^i_{,j}(P) \, du^j$ does not necessarily represent an absolute change of the field, since $A^i(Q) - A^i(P)$ is not a vector, the points P and Q not being identical.

Let us again, for the moment, assume that the space is Euclidean, and let $A^{*i}(P)$ and $A^{*i}(Q)$ be the cartesian components of the field in P and Q, respectively. Then

$$A^{*k}(P) = f^k_{,m}(P) A^m(P)$$

and

$$A^{*k}(Q) = (f^k_{,m}(P) + f^k_{,ml}(P) \, du^l)(A^m(P) + A^m_{,j}(P) \, du^j),$$

where we have taken into account the fact that the functions $f^k_{,m}$ may change between P and Q.

The absolute change or increment of the field from P to Q is a vector δA^i, whose cartesian components are $A^{*k}(Q) - A^{*k}(P)$. Thus

$$f^k_{,i}\delta A^i = A^{*k}(Q) - A^{*k}(P) = (f^k_{,m}A^m_{,j} + f^k_{,mj}A^m)\, du^j,$$

where all arguments unless otherwise indicated, are evaluated at P.

Multiplication of both sides with $g^{ln}f^k_{,n}$ yields

$$\delta A^l = (A^l_{,j} + g^{ln}f^k_{,n}f^k_{,mj}A^m)\, du^j$$

using the relation $g^{ln}f^k_{,n}f^k_{,i} = g^{ln}g_{ni} = \delta^l_i$. The functions f^i may be eliminated from this expression using relation (1.12), which we differentiate with respect to the coordinates in the following combinations,

$$g_{jn,m} = (f^k_{,j}f^k_{,n})_{,m} = f^k_{,jm}f^k_{,n} + f^k_{,j}f^k_{,nm},$$
$$g_{mn,j} = (f^k_{,m}f^k_{,n})_{,j} = f^k_{,mj}f^k_{,n} + f^k_{,m}f^k_{,nj},$$
$$g_{jm,n} = (f^k_{,j}f^k_{,m})_{,n} = f^k_{,jn}f^k_{,m} + f^k_{,j}f^k_{,mn}.$$

Assuming the functions f^k to be twice continuously differentiable, we get, by adding the first two equations and subtracting the third,

$$g_{jn,m} + g_{mn,j} - g_{jm,n} = 2f^k_{,mj}f^k_{,n},$$

since the order of differentiation is immaterial.

For convenience we introduce the notation

$$\{^i_{j\ k}\} = \tfrac{1}{2}g^{il}(g_{jl,k} + g_{kl,j} - g_{jk,l}) = g^{il}f^n_{,l}f^n_{,jk}, \tag{1.26}$$

which is called the *Christoffel symbol*. In cartesian coordinates the $g_{ij} = \delta^i_j$ are constants and the Christoffel symbols vanish identically. But in spherical coodinates, for instance, they do not vanish. Hence, the Christoffel symbol is no tensor. However, it plays an important role in tensor calculus.

If the two first terms on the right-hand side are interchanged and the symmetry of g_{jk} in the third term is utilized, the result is the Christoffel symbol with the two lower indices interchanged. Thus, the symbol is symmetrical in the lower indices

$$\{^i_{j\,k}\} = \{^i_{k\,j}\}. \tag{1.27}$$

Using the notation (1.26), we can write

$$\delta A^l = (A^l_{,j} + \{^l_{m\,j}\} A^m)\, du^j. \tag{1.28}$$

Since du^j is arbitrary and δA^l a contravariant vector, the expression in parentheses must be a mixed tensor. This tensor is called the covariant derivative of A^l and is written [6]

$$D_j A^l = A^l_{,j} + \{^l_{m\,j}\} A^m. \tag{1.29}$$

Neither term on the right-hand side of (1.28) is a tensor but their sum is. The covariant derivative of A^l is also called the *gradient* of A^l. In Cartesian coordinates it reduces to the partial derivative.

The deduction of (1.29) was based upon the assumption that the space was Euclidean. This is, however, not necessary. Taking (1.29) as a starting point and using rule (1.9) for the transformation of a mixed tensor of second order, it can be verified that $D_j A^l$ is a mixed tensor making no assumptions about the space.

Had we started with a covariant vector-field $A_i(u^1, u^2, u^3)$ the result would have been

$$D_j A_l = A_{l,j} - \{^m_{l\,j}\} A_m, \tag{1.30}$$

which is the covariant derivative (gradient) of a covariant vector-field. It is useful to notice the minus sign and the position of the dummy index within the Christoffel symbol.

The covariant derivative can be generalized to tensor fields of arbitrary order in the following manner.

$$\begin{aligned} D_p T^{j\ldots k\ldots}_{q\ldots r\ldots} = {} & T^{j\ldots k\ldots}_{q\ldots r\ldots,p} + \{^j_{i\,p}\} T^{i\ldots k\ldots}_{q\ldots r\ldots} + \cdots + \\ & + \{^k_{i\,p}\} T^{j\ldots i\ldots}_{q\ldots r\ldots} + \cdots - \{^i_{q\,p}\} T^{j\ldots k\ldots}_{i\ldots r\ldots} \\ & - \cdots - \{^i_{r\,p}\} T^{j\ldots k\ldots}_{q\ldots i\ldots} - \cdots. \end{aligned} \tag{1.31}$$

The first term is the partial derivative. Then, for each contravariant

---

[6] In the literature greatly varying notations for the covariant derivative appear. Thus, we find $A^l_{;j}$, $A^l_{:j}$, $A^l_{\|j}$, $A^l_{/j}$, $A^l_{/j}$, and others.

index, a term with a Christoffel symbol is added, and for each covariant index, a term is subtracted. It can be verified that the result is a tensor one covariant order higher than the tensor $T^{j\ldots k\ldots}_{q\ldots r\ldots}$.

Let the tensor $T^{j\ldots k\ldots}_{q\ldots r\ldots}$ stand for the outer product of the tensors $A^{j\ldots}_{q\ldots}$ and $B^{k\ldots}_{r\ldots}$. Using (1.31), the following rule for the covariant derivative of a product is easily established,

$$D_p A^{j\ldots}_{q\ldots} B^{k\ldots}_{r\ldots} = A^{j\ldots}_{q\ldots} D_p B^{k\ldots}_{r\ldots} + B^{k\ldots}_{r\ldots} D_p A^{j\ldots}_{q\ldots} \qquad (1.32)$$

and we recognize the well-known distributive rule for the derivative of a product, which evidently also holds good for covariant differentiation.

A special case of (1.31) is the covariant derivative of a scalar

$$D_i S = S_{,i}. \qquad (1.33)$$

This was used as our first illustration of a covariant vector.

The usefulness of the covariant derivative arises largely from the fact that, in physics, the equations are stated in rectangular coordinates in which the g_{ij}'s are constants. If we replace the ordinary derivative in these equations by the covariant derivative, they will not be altered. This is a necessary step in reducing such equations to the tensor form, which holds true for all coordinate sytems.

Let us now compute the covariant derivative of the fundamental tensor. According to (1.31) we have

$$\begin{aligned} D_k g_{ij} &= g_{ij,k} - \{^{\ n}_{i\ k}\} g_{nj} - \{^{\ n}_{j\ k}\} g_{in} \\ &= g_{ij,k} - \tfrac{1}{2} g^{nm} g_{nj}(g_{im,k} + g_{km,i} - g_{ik,m}) \\ &\quad - \tfrac{1}{2} g^{nm} g_{in}(g_{jm,k} + g_{km,j} - g_{jk,m}). \end{aligned}$$

With the help of (1.18) we get

$$D_k g_{ij} = 0. \qquad (1.34)$$

Similarly, it is easy to show that

$$D_k g^{ij} = 0 \qquad (1.35)$$

and

$$D_k g^i_j = 0. \tag{1.36}$$

This proves that the fundamental tensor behaves like a constant under covariant differentiation. It is thus immaterial whether an index is raised before or after differentiation.

10. The Riemann–Christoffel tensor

Let $A_i(u^1, u^2, u^3)$ be a covariant vector-field and let $PQRS$ be an elementary coordinate mesh on $u^3 = \text{const}$ (see Fig. 1). Now, let us calculate the absolute change of A_i as we pass around the circuit $PQRS$. The absolute change:
(1) from P to Q is $D_1 A_i \, du^1$ evaluated at u^2.
(2) from Q to R is $D_2 A_i \, du^2$ evaluated at $u^1 + du^1$.
(3) from R to S is $-D_1 A_i \, du^1$ evaluated at $u^2 + du^2$.
(4) from S to P is $-D_2 A_i \, du^2$ evaluated at u^1.

Combining (1) and (3), the net result is the absolute difference between $D_1 A_i \, du^1$ evaluated at u^2 and at $u^2 + du^2$, i.e.,

$$-D_2(D_1 A_i \, du^1) \, du^2.$$

Similarly, (2) and (4) give

$$D_1(D_2 A_i \, du^2) \, du^1$$

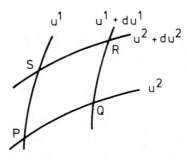

FIG. 1.

so that the absolute change around the circuit is

$$(D_1 D_2 A_i - D_2 D_1 A_i)\,du^1\,du^2\,.$$

We would expect that, on returning to our starting point P, the absolute change would be zero. Nevertheless, this is not always so. However, *if the order of covariant differentiation were permutable*, the absolute change around the circuit would vanish. This observation leads us to investigate the importance of the order of covariant differentiation.

Let us therefore calculate the tensor

$$D_i D_j A_k - D_j D_i A_k\,.$$

Using the rule for covariant differentiation, we get

$$\begin{aligned}
D_i D_j A_k &= D_i(A_{k,j} - \{{}_k{}^m{}_j\} A_m) \\
&= (A_{k,j} - \{{}_k{}^m{}_j\} A_m)_{,i} - \{{}_k{}^n{}_i\}(A_{n,j} - \{{}_n{}^m{}_j\} A_m) - \{{}_j{}^n{}_i\}(A_{k,n} - \{{}_k{}^m{}_n\} A_m) \\
&= A_{k,ji} - \{{}_k{}^m{}_j\} A_{m,i} - \{{}_k{}^m{}_j\}_{,i} A_m - \{{}_k{}^n{}_i\} A_{n,j} \\
&\quad + \{{}_k{}^n{}_i\}\{{}_n{}^m{}_j\} A_m - \{{}_j{}^n{}_i\} A_{k,n} + \{{}_j{}^n{}_i\}\{{}_k{}^m{}_n\} A_m
\end{aligned}$$

and the desired tensor is found when we have subtracted the corresponding expression with indices i and j permuted

$$\begin{aligned}
D_i D_j A_k - D_j D_i A_k &= \\
&= (\{{}_k{}^n{}_i\}\{{}_n{}^m{}_j\} - \{{}_k{}^n{}_j\}\{{}_n{}^m{}_i\} + \{{}_k{}^m{}_i\}_{,j} - \{{}_k{}^m{}_j\}_{,i}) A_m\,.
\end{aligned}$$

Since A_k is arbitrary, the expression in parentheses must be a tensor. We shall write

$$R^m_{kji} = \{{}_k{}^n{}_i\}\{{}_n{}^m{}_j\} - \{{}_k{}^n{}_j\}\{{}_n{}^m{}_i\} + \{{}_k{}^m{}_i\}_{,j} - \{{}_k{}^m{}_j\}_{,i} \tag{1.37}$$

for this tensor of order four, which is called the *Riemann–Christoffel tensor of second kind*. Multiplication by g_{ml} lowers the first index. After a few trivial calculations we get the *Riemann–Christoffel tensor of first kind*, a covariant tensor of fourth order

$$R_{ijkl} = \tfrac{1}{2}(g_{il,jk} + g_{jk,il} - g_{jl,ik} - g_{ik,jl}) + g_{mn}(\{{}^{m}_{jk}\}\{{}^{n}_{il}\} - \{{}^{m}_{jl}\}\{{}^{n}_{ik}\}). \quad (1.38)$$

In three dimensions, a tensor of order four has 81 components. It will be seen that R_{ijkl}, besides being skew-symmetrical in i and j, is also skew-symmetrical in l and k. Furthermore, it is symmetrical for the double interchange l and j, k and i. It has the additional cyclic property

$$R_{ijkl} + R_{iklj} + R_{iljk} = 0,$$

as is easily verified from (1.38). Hence, of the 81 components, only six are independent, e.g., the following combinations of indices,

$$ijkl = 1212, \ 1223, \ 1313, \ 2323, \ 1213, \ 1323.$$

All other components are either zero or equal to one of these six components with a plus or minus sign. In two dimensions, only one independent component remains, e.g., R_{1212}.

The Riemann–Christoffel tensor is derived solely from g_{ij} and therefore belongs to the class of fundamental tensors. In Euclidean space, cartesian coordinates may be constructed. In these, the g_{ij}'s are constants and the tensor vanishes. But as it is a tensor, it vanishes in all coordinate systems applicable to Euclidean space. In Riemannian space it does not generally vanish.

If the Riemann–Christoffel tensor vanishes, the order of covariant differentiation will be interchangeable, and it follows that *in Euclidean space the order of covariant differentiation can be interchanged.*

11. The order of covariant differentiation for tensors

In the previous section we found an expression for the difference between the two second-order covariant derivatives of a vector,

$$D_i D_j A_k - D_j D_i A_k = R^m{}_{kji} A_m. \quad (1.39)$$

Similarly, the difference between the second-order covariant derivatives of a covariant tensor of second order is

$$D_i D_j A_{kl} - D_j D_i A_{kl} = R^m{}_{kji} A_{ml} + R^m{}_{lji} A_{km}. \quad (1.40)$$

For a tensor of arbitrary order we find on the right-hand side one term for each index

$$D_i D_j T_{\ldots k \ldots} - D_j D_i T_{\ldots k \ldots} = \sum R^m_{kji} T_{\ldots m \ldots}. \qquad (1.41)$$

Generalization to mixed and contravariant tensors is trivial, since g^{ij} behaves as a constant under covariant differentiation.

12. Contravariant differentiation

For the sake of completeness, we also introduce the concept of *contravariant derivatives* by the operator

$$D^i = g^{ij} D_j. \qquad (1.42)$$

Using this notation, the *Laplacian*, for instance, will take the form

$$\Delta = D_i D^i. \qquad (1.43)$$

In working out this operator for a given coordinate system, we shall, however, rely on the relation

$$D_i D^i = g^{ij} D_i D_j.$$

It is useful to note that the Laplacian is an invariant operator.

13. Summary

Vectors and tensors are quantities obeying certain transformation laws related to changes of coordinate systems. Their importance lies in their property of *invariance*, which leads to the fact that, if a tensor equation is found to hold in one coordinate system, it will continue to hold when any transformation of coordinates is made. New tensors are recognized either by the use of the transformation law or by the property that the sum, difference, product or quotient of tensors, are tensors.

The principal operations of tensor calculus are addition, multiplication (outer and inner), contraction, covariant differentiation, substitution and raising or lowering of indices.

Of special interest are the fundamental tensors, of which we have introduced the metric tensor g_{ij} and the Riemann–Christoffel tensor R_{ijkl}. The latter has been expressed in terms of the former and its first and second derivatives.

The dimension of the space is indicated by the range of the indices. We have relied upon a three-dimensional space in which all indices take the values 1, 2 and 3. Generalization to other spaces with two, four or n dimensions is trivial. In deducing the shell theory in the following chapters, we shall make use of

(a) a three-dimensional Euclidean space in which the Riemann–Christoffel tensor vanishes identically, and embedded in this

(b) a two-dimensional space, the space of a surface in which the Riemann–Christoffel tensor does not necessarily vanish.

Bibliography

BRILLOUIN, L. (1964), *Tensors in Mechanics and Elasticity*, Academic Press, New York.
MCCONNELL, A.J. (1957), *Application of Tensor Analysis*, Dover Publications, New York.
MICHAEL, A.D. (1947), *Matrix and Tensor Calculus*, Wiley, New York.
SOKOLNIKOFF, I.S. (1951), *Tensor Analysis*, Applied Mathematics Series, Wiley, New York.
SPAIN, B. (1960), *Tensor Calculus*, Oliver and Boyd, London.
SPIEGEL, M.R. (1959), *Vector Analysis and an Introduction to Tensor Analysis*, Shaum's Outline Series, New York.

CHAPTER 2

THE GEOMETRY OF THE MIDDLE-SURFACE

1. Introduction 33
2. The first fundamental form of a surface 33
3. Normal coordinates 35
4. Second fundamental form of a surface 37
5. Invariants of curvature 41
6. The inverse problem 44
7. The equations of Codazzi 46
8. Surface and volume element 47
9. Coordinates of principal curvature 49
10. Summary 50
    Bibliography 51

CHAPTER 2

THE GEOMETRY OF THE MIDDLE-SURFACE

1. Introduction

The shape of a shell is usually described by the *middle-surface* of the shell and the shell-thickness h. The middle-surface is a surface such that any normal to it intersects the two free surfaces of the shell at the same distance $\pm\frac{1}{2}h$. We shall assume that the middle-surface is smooth but otherwise of arbitrary shape. In this chapter we shall study some properties of surfaces that are particularly important to shell theory.

2. The first fundamental form of a surface

From now on let x^i be a given fixed cartesian coordinate system in three-dimensional space in which the middle-surface is *embedded*.

We chose to describe the surface by means of a system of parametric relations

$$x^i = f^i(u^1, u^2), \qquad (2.1)$$

in which the parameters u^α ($\alpha = 1, 2$) serve as coordinates on the surface.[1] The form (2.1) is essentially due to C.F. Gauss (1827).

---

[1] Generally speaking, a surface, embedded in three-dimensional space, is given by an equation

$$F(x^1, x^2, x^3) = 0$$

such that when any two of the coordinates x^i are given, the third can be determined from that equation. It can be shown that if the determinant $\det(f^i_{,\alpha})$ is of rank two, the system (2.1) can be reduced to such an equation and thus determines a surface.

There are, of course, many ways of constructing coordinates on a surface, and we shall regard u^α as being an arbitrarily selected system.

We shall now apply the tools of vector and tensor analysis just developed to the two-dimensional space of a surface, and we shall use *Greek indices* for the range 1, 2.

Since the coordinates x^i of the given cartesian system are independent of our choice of coordinates u^α on the surface, the functions f^i of (2.1) are scalar.

The metric properties of the surface are easily established with reference to three-dimensional space. Thus, let u^α and $u^\alpha + du^\alpha$ be the coordinates of two neighbouring points on the surface. Their distance ds is determined by

$$ds^2 = dx^i\, dx^i = f^i_{,\alpha}\, du^\alpha f^i_{,\beta}\, du^\beta$$

or

$$ds^2 = a_{\alpha\beta}\, du^\alpha\, du^\beta, \qquad (2.2)$$

where

$$a_{\alpha\beta} = f^i_{,\alpha} f^i_{,\beta}. \qquad (2.3)$$

The metric relation (2.2) is also called *the first fundamental form of a surface* and $a_{\alpha\beta}$ is the metric or fundamental tensor of the surface.

We may take the three scalar functions f^i to be given. Equation (2.3) then defines the metric tensor $a_{\alpha\beta}$. It is useful to note that $f^1_{,\alpha}$ is the gradient of a scalar and thus a covariant vector. Thus, each term on the right-hand side of (2.3) (for instance $f^1_{,\alpha} f^1_{,\beta}$) is a covariant tensor and therefore also their sum.

The determinant a of the metric tensor $a_{\alpha\beta}$,

$$a = \det(a_{\alpha\beta}) = a_{11} a_{22} - (a_{12})^2, \qquad (2.4)$$

will prove to be of great importance in the analysis. It is not, of course, a scalar.

The parametric equations (2.1) of a surface can be obtained from the coordinate transformation (1.1) between cartesian coordinates x^i and a suitably selected three-dimensional coordinate system u^i in which u^3 is taken to be constant ($u^3 = $ const). Thus, for example, the parametric equations of a spherical surface appear when r is taken to be constant in a system of spherical coordinates r, ϕ, θ. The parameters u^α are then ϕ, θ. In this case, the elements of the fundamental tensor $a_{\alpha\beta}$ of the surface will coincide with the corresponding elements of g_{ij},

$$a_{\alpha\beta} = g_{\alpha\beta},$$

but a will, in general, be different from g. Also, the components of the Riemann–Christoffel tensor of the surface $B^\sigma_{\varepsilon\alpha\beta\gamma}$ will not, in general, coincide with corresponding elements of the Riemann–Christoffel tensor in three-dimensional space. Thus, according to (1.37), we have

$$R^\sigma_{\alpha\beta\gamma} = B^\sigma_{\alpha\beta\gamma} + \{{}^{\ 3}_{\alpha\ \gamma}\}\{{}^{\ \sigma}_{3\ \beta}\} - \{{}^{\ 3}_{\alpha\ \beta}\}\{{}^{\ \sigma}_{3\ \gamma}\},$$

since the range of the dummy indices is 1, 2, 3 in $R^m{}_{kji}$ but only 1, 2 in $B^\sigma_{\alpha\beta\gamma}$. However, $R^m{}_{kji} = 0$ in three-dimensional (Euclidean) space, and the Riemann–Christoffel tensor for the surface is therefore determined by

$$B^\sigma_{\alpha\beta\gamma} = \{{}^{\ 3}_{\alpha\ \beta}\}\{{}^{\ \sigma}_{3\ \gamma}\} - \{{}^{\ 3}_{\alpha\ \gamma}\}\{{}^{\ \sigma}_{3\ \beta}\}. \tag{2.5}$$

When, for a particular surface, $B^\sigma_{\alpha\beta\gamma}$ does not vanish, the space of the surface is non-Euclidean and no cartesian coordinates can be constructed on the surface. Nor is the order of covariant differentiation on the surface interchangeable.

3. Normal coordinates

The partial derivatives $f^i_{,1}$ of the three functions f^i with respect to u^1 are the cartesian components of a tangent vector to the coordinate line $u^2 = $ const, and thus the components of a tangent vector to the surface. Since the two cartesian vectors $f^i_{,1}$ and $f^i_{,2}$ are linearly independent, they uniquely determine a *tangent plane* to the surface.

A *normal* to the surface is perpendicular to the tangent plane and hence to both vectors $f^i_{,1}$ and $f^i_{,2}$. In cartesian coordinates, the normal X^i must then be proportional to the cross-product of $f^i_{,1}$ and $f^i_{,2}$, i.e.,

$$X^1 = c(f^2_{,1}f^3_{,2} - f^3_{,1}f^2_{,2}),$$
$$X^2 = c(f^3_{,1}f^1_{,2} - f^1_{,1}f^3_{,2}),$$
$$X^3 = c(f^1_{,1}f^2_{,2} - f^2_{,1}f^1_{,2}),$$

or, using a more compact way of writing,

$$X^i = ce_{ijk}f^j_{,1}f^k_{,2},$$

where the *alternating symbol* e_{ijk} is defined by [2]

$$e_{ijk} = \begin{cases} 1 & \text{if } ijk \text{ is a cyclic permutation of } 123, \\ -1 & \text{if } ijk \text{ is a cyclic permutation of } 132, \\ 0 & \text{in all other cases}. \end{cases} \quad (2.6)$$

The constant c may be chosen such that $X^i X^i = 1$, i.e., so that X^i becomes a *unit normal*. It is easily found that $X^i X^i = c^2 a$. It then follows that $c = \pm a^{-1/2}$, and

$$X^i = a^{-1/2} e_{ijk} f^j_{,1} f^k_{,2}, \quad (2.7)$$

where the positive square-root has been taken. *With this sign, the triad of vectors* $f^i_{,1} f^i_{,2}$ *and* X^i *will always be right-handed in that order* (see Fig. 2).

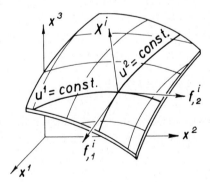

Fig. 2.

---

[2] The alternating symbol e_{ijk} is not a tensor.

A point in space (not too far from the surface) can be identified uniquely by its distance $z = u^3$ from the surface, measured along a normal to it (positively in the direction of the normal) and the coordinates u^α at the foot of the normal. Such coordinates u^i are called *normal coordinates* to the surface. Accordingly, the transformation between the normal coordinates u^i and the Cartesian coordinates x^i is given by [3]

$$x^i = f^i(u^1, u^2) + zX^i(u^1, u^2) = f^i(u^1, u^2, u^3). \qquad (2.8)$$

We have thus extended the two-dimensional coordinate system u^α into three-dimensional space and obtained coordinates u^i, such that $u^3 = $ const are surfaces parallel to the middle-surface $u^3 = 0$. In particular, the free surfaces of the shell are given by $u^3 = \pm\frac{1}{2}h$.

The components of the three-dimensional metric tensor are given by

$$g_{ij} = f^k_{,i} f^k_{,j} \qquad (2.9)$$

in the normal coordinate system. Substituting (2.8) into (2.9) and utilizing the fact that $X^k f^k_{,\alpha} = 0$, we find [4]

$$g_{33} = g^{33} = 1, \qquad g_{\alpha 3} = g^{\alpha 3} = 0 \quad \text{(for normal coordinates)}. \quad (2.10)$$

The Christoffel symbol $\{^{\ 3}_{\alpha\ \beta}\}$ takes the particularly simple form $-\frac{1}{2}g_{\alpha\beta,3}$ and from (2.5) we get

$$B_{1212} = \tfrac{1}{4}\det(g_{\alpha\beta,3}) \quad \text{(for normal coordinates and } z = 0). \quad (2.11)$$

4. Second fundamental form of a surface

When two surfaces are such that there exists a coordinate system on each, in terms of which the first fundamental forms of the two surfaces are identical, the surfaces are said to be *isometric*. Length of

---

[3] Cf. footnote 2 of Chapter 4.
[4] This is easily verified by multiplying (2.7) through by $f^i_{,\alpha}$ and working out the right-hand side. But it also follows from the fact that the tangent vector $f^k_{,\alpha}$ is perpendicular to the normal X^k.

arcs and angles between curves on the surface are expressible in terms of the metric tensor $a_{\alpha\beta}$; hence, any set of statements about such quantities (i.e., any geometry) that holds good on one of the two surfaces will hold good on the other. Two such surfaces are said to have the same *intrinsic geometry*. However, viewed from the embedding space, two such surfaces may be entirely different in form. It is this question of the geometry of a surface as viewed from the embedding space which is the subject of the present section.

Let C be a smooth curve on the surface (see Fig. 3) and consider a tangent plane at a point P with the coordinates $x^i(0)$ on the curve C. Let $Q(s)$ be an arbitrary point of the curve with coordinates $x^i(s)$, where s denotes the arc-length from P to Q. Expanding $x^i(s)$ in a Taylor series, we get

$$x^i(s) = x^i(0) + \left(\frac{dx^i}{ds}\right)_0 s + \frac{1}{2}\left(\frac{d^2x^i}{ds^2}\right)_0 s^2 + \cdots,$$

where index 0 indicates that the expressions are evaluated at P, i.e., at $s = 0$. Since

$$\frac{dx^i}{dx} = f^i_{,\alpha}\frac{du^\alpha}{ds}, \qquad \frac{d^2x^i}{ds^2} = f^i_{,\alpha}\frac{d^2u^\alpha}{ds^2} + f^i_{,\alpha\beta}\frac{du^\alpha}{ds}\frac{du^\beta}{ds},$$

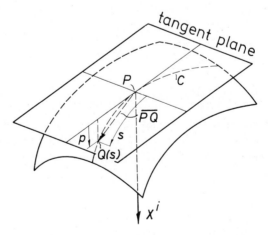

Fig. 3.

we have

$$x^i(s) = x^i + f^i_{,\alpha}\frac{du^\alpha}{ds}s + \tfrac{1}{2}f^i_{,\alpha}\frac{d^2u^\alpha}{ds^2}s^2 + \tfrac{1}{2}f^i_{,\alpha\beta}\frac{du^\alpha}{ds}\frac{du^\beta}{ds}s^2 + \cdots.$$

The distance p of $Q(s)$ from the tangent plane is the projection of $\overline{PQ}$ with the components $x^i(s) - x^i(0)$ on the normal X^i, i.e., $p = X^i[x^i(s) - x^i(0)]$. Thus

$$p = X^i\left\{f^i_{,\alpha}\left(\frac{du^\alpha}{ds}s + \frac{1}{2}\frac{d^2u^\alpha}{ds^2}s^2 + \cdots\right) + \tfrac{1}{2}f^i_{,\alpha\beta}\frac{du^\alpha}{ds}\frac{du^\beta}{ds}s^2 + \cdots\right\}.$$

It should be noted that p is positive in the direction of the normal. Since $X^i f^i_{,\alpha} = 0$, we get

$$p = \tfrac{1}{2}X^i f^i_{,\alpha\beta}\frac{du^\alpha}{ds}\frac{du^\beta}{ds}s^2 + \cdots$$

or, at the limit, when s approaches zero

$$p = \tfrac{1}{2}X^i f^i_{,\alpha\beta}\, du^\alpha\, du^\beta. \tag{2.12}$$

With the notation

$$d_{\alpha\beta} = X^i f^i_{,\alpha\beta}, \tag{2.13}$$

we have

$$p = \tfrac{1}{2}d_{\alpha\beta}\, du^\alpha\, du^\beta. \tag{2.14}$$

This equation is called the *second fundamental form* of a surface. Since p, the distance from the tangent plane, is invariant and du^α is arbitrary, $d_{\alpha\beta}$ must be a covariant tensor of second order. It is the *second fundamental tensor* or *curvature tensor* of the surface. Like $a_{\alpha\beta}$, it is symmetrical,

$$d_{\alpha\beta} = d_{\beta\alpha}, \tag{2.15}$$

which immediately follows from definition (2.13).

In normal coordinates we have particularly simple relations between the second fundamental tensor and the Christoffel symbols belonging to the embedding space. Thus, according to (1.26) we have

$$\{^{\ \ 3}_{\alpha\ \beta}\} = g^{3k} f^i_{,k} f^i_{,\alpha\beta} .$$

But, in normal coordinates,

$$g^{3k} f^i_{,k} = g^{33} f^i_{,3} = f^i_{,3} = X^i$$

and, hence,

$$\{^{\ \ 3}_{\alpha\ \beta}\} = d_{\alpha\beta} .$$

Similarly,

$$\{^{\ \ \alpha}_{3\ \beta}\} = g^{\alpha k} f^i_{,k} f^i_{,3\beta} = g^{\alpha\gamma} f^i_{,\gamma} f^i_{,3\beta} .$$

But

$$f^i_{,\gamma} f^i_{,3\beta} = (f^i_{,\gamma} f^i_{,3})_{,\beta} - f^i_{,3} f^i_{,\gamma\beta}$$
$$= - X^i f^i_{,\gamma\beta}$$
$$= - d_{\gamma\beta} - z X^i X^i_{,\gamma\beta} .$$

Hence,

$$\{^{\ \ \alpha}_{3\ \beta}\} = - d^\alpha_\beta$$

and substitution into (2.5) yields

$$B^\sigma_{\ \alpha\beta\gamma} = - d_{\alpha\beta} d^\sigma_\gamma + d_{\alpha\gamma} d^\sigma_\beta .$$

However, this is a tensor equation, and the condition of normal coordinates can thus be omitted. By lowering index σ we obtain

$$B_{\sigma\alpha\beta\gamma} = d_{\alpha\gamma} d_{\sigma\beta} - d_{\alpha\beta} d_{\sigma\gamma} . \qquad (2.16)$$

Of these sixteen equations, only one is significant, viz.,

$$B_{1212} = d_{11}d_{22} - (d_{12})^2 = \det(d_{\alpha\beta}) = d, \qquad (2.17)$$

the others being either repetitions of this or else identities.

The relation (2.17), $B_{1212} = d$, is called *the equation of Gauss*. From this is follows that the Riemann–Christoffel tensor for a surface vanishes if and only if the determinant of the curvature tensor vanishes.

5. Invariants of curvature

We wish to calculate the radius of curvature of the surface in a plane containing the normal X^i at a point P. Let Q be a point on the curve C, the intersection between the surface and the plane, at a distance s from P measured along the arc from P to Q (see Fig. 4). The normal to C at Q in the plane intersects the normal at P in O. When s tends to zero, $\overline{OP}$ tends to the radius of curvature R and thus

$$\frac{1}{R} = \lim_{s \to 0} \frac{1}{s} \frac{dp}{ds},$$

where p is the distance of Q from the tangent plane, i.e.,

$$p = \tfrac{1}{2} X^i f^i_{,\alpha\beta} \frac{du^\alpha}{ds} \frac{du^\beta}{ds} s^2 + \cdots .$$

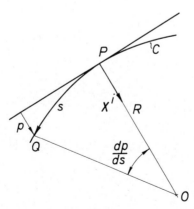

Fig. 4.

Hence,

$$\frac{1}{R} = \left[d_{\alpha\beta} \frac{du^\alpha}{ds} \frac{du^\beta}{ds} \right]_{s=0}$$

is the *curvature* of the surface in the direction du^α.

Introducing ds^2 from (2.2) we get

$$\frac{1}{R} = \frac{d_{\alpha\beta}\, du^\alpha\, du^\beta}{a_{\varepsilon\gamma}\, du^\varepsilon\, du^\gamma}. \tag{2.18}$$

The curvature is a function of the direction (determined by du^α) at any given point. The extremal or *principal curvatures* are determined from the condition that $1/R$ is stationary in the corresponding principal directions of curvature. We find the variation $\delta(1/R)$ by giving du^α an increment δu^α.[5]

Putting $\delta(1/R) = 0$, we get

$$ds^2 d_{\alpha\beta}(du^\alpha\, \delta u^\beta + \delta u^\alpha\, du^\beta) - \frac{1}{R} ds^2\, a_{\varepsilon\gamma}(du^\varepsilon\, \delta u^\gamma + \delta u^\varepsilon\, du^\gamma) = 0,$$

or, after changing dummy indices,

$$\left(d_{\alpha\beta} - \frac{1}{R} a_{\alpha\beta} \right) du^\alpha\, \delta u^\beta = 0.$$

This must hold good for any δu^β, and hence

$$\left(d_{\alpha\beta} - \frac{1}{R} a_{\alpha\beta} \right) du^\alpha = 0. \tag{2.19}$$

The result is a linear, homogeneous system of equations for the unknown principal directions du^α. The requirement for a meaningful solution is that the determinant vanishes,

---

[5] In (2.18) the magnitude of du^α cancels out and du^α may be taken to represent a contravariant vector of arbitrary magnitude.

$$\left(d_{11} - \frac{1}{R} a_{11}\right)\left(d_{22} - \frac{1}{R} a_{22}\right) - \left(d_{12} - \frac{1}{R} a_{12}\right)^2 = 0.$$

After working out the products and collecting terms in powers of $1/R$, we get

$$d - (d_{11}a_{22} + d_{22}a_{11} - 2d_{12}a_{12})\frac{1}{R} + a\left(\frac{1}{R}\right)^2 = 0$$

or

$$\left(\frac{1}{R}\right)^2 - d_{\alpha\beta}a^{\alpha\beta}\frac{1}{R} + \frac{d}{a} = 0. \tag{2.20}$$

The two roots to this equation, $1/R_1$ and $1/R_2$, are the principal curvatures, and we find

$$\frac{1}{R_1 R_2} = \frac{d}{a} = d_1^1 d_2^2 - d_1^2 d_2^1 = K \tag{2.21}$$

and

$$\frac{1}{R_1} + \frac{1}{R_2} = d_\alpha^\alpha = 2H, \tag{2.22}$$

where K and H are two characteristic invariants of curvature of the surface. The scalar K is called the *Gaussian curvature* and H the *mean curvature*. The following relation holds good between the Gaussian curvature and the Riemann–Christoffel tensor of the surface,

$$K = B_{1212}/a, \tag{2.23}$$

which is evident from a comparison of (2.21) and (2.17).

The principal curvatures $1/R_1$ and $1/R_2$ are found to be

$$\left.\begin{matrix}1/R_1\\1/R_2\end{matrix}\right\} = H \pm \sqrt{H^2 - K} \tag{2.24}$$

in terms of H and K.

For any surface isometric with the plane, the Riemann–Christoffel tensor vanishes and thus also its Gaussian curvature. At least one of the principal curvatures vanishes everywhere. Such surfaces are called *developable*, since any such surface can be developed upon a plane, that is, rolled out without stretching or contracting any part of it. It is evident that this can be done with for instance a cylinder or a cone.

6. The inverse problem

The first and second fundamental tensors of a surface, six functions in all, are determined by the three parametric equations (2.1). In this section we shall be concerned with the *inverse* problem: to determine the functions f^i and X^i when the fundamental tensors of a surface are given.

Since X^i and $f^i_{,\alpha}$ are orthogonal, their scalar product $X^i f^i_{,\alpha}$ vanishes. By differentiating this product with respect to u^β we get

$$X^i_{,\beta} f^i_{,\alpha} + X^i f^i_{,\alpha\beta} = 0$$

or, by (2.13),

$$X^i_{,\beta} f^i_{,\alpha} = -d_{\alpha\beta}. \tag{2.25}$$

However, as f^i and X^i are scalar functions, their partial and covariant derivatives coincide. Hence, we have

$$D_\beta(X^i D_\alpha f^i) = X^i D_\beta D_\alpha f^i + X^i_{,\beta} f^i_{,\alpha} = 0$$

and, together with (2.25), this yields an alternative expression for the second fundamental tensor

$$d_{\alpha\beta} = X^i D_\beta D_\alpha f^i. \tag{2.26}$$

It is interesting to compare this equation with (2.13). Writing

$$X^i D_\beta D_\alpha f^i = X^i D_\beta f^i_{,\alpha} = X^i (f^i_{,\alpha\beta} - \{^{\gamma}_{\alpha\,\beta}\} f^i_{,\gamma}) = X^i f^i_{,\alpha\beta}$$

we see that the term containing the Christoffel symbol vanishes due to the orthogonality of X^i and $f^i_{,\gamma}$.

Differentiating $X^i X^i = 1$, we get $2X^i X^i_{,\alpha} = 0$, which shows that $X^i_{,\alpha}$ is perpendicular to X^i and can therefore be resolved into $f^i_{,1}$ and $f^i_{,2}$. We write

$$X^i_{,\alpha} = t^\beta_\alpha f^i_{,\beta},$$

where t^β_α is a matrix of coefficients. The notation anticipates the tensor character of t^β_α. Multiplying through by $f^i_{,\gamma}$ yields

$$f^i_{,\gamma} X^i_{,\alpha} = t^\beta_\alpha a_{\beta\gamma}.$$

But the left-hand side is equal to $-d_{\gamma\alpha}$ according to (2.25), so that $t^\beta_\alpha = -d^\beta_\alpha$. It follows that

$$X^i_{,\alpha} = -d^\beta_\alpha f^i_{,\beta}. \tag{2.27}$$

This is the first part of the equations needed for solving the inverse problem. To obtain the remaining part we start with the expression

$$f^i_{,\gamma} D_\alpha D_\beta f^i = f^i_{,\gamma} D_\alpha f^i_{,\beta} = f^i_{,\gamma} (f^i_{,\alpha\beta} - \{^\varepsilon_{\alpha\,\beta}\} f^i_{,\varepsilon})$$

$$= f^i_{,\gamma} f^i_{,\alpha\beta} - a_{\gamma\varepsilon} \{^\varepsilon_{\alpha\,\beta}\}.$$

But according to (1.26) and the equation just above, we have

$$f^i_{,\gamma} f^i_{,\alpha\beta} = a_{\gamma\varepsilon} \{^\varepsilon_{\alpha\,\beta}\}$$

and hence

$$f^i_{,\gamma} D_\alpha D_\beta f^i = 0. \tag{2.28}$$

This equation shows that $D_\alpha D_\beta f^i$ is perpendicular to $f^i_{,\gamma}$ and must therefore have the direction of the normal X^i,

$$D_\alpha D_\beta f^i = h_{\alpha\beta} X^i.$$

Again, the tensorial character of $h_{\alpha\beta}$ has been anticipated by the notation and is verified when the equation is multiplied through by X^i,

$$X^i D_\alpha D_\beta f^i = h_{\alpha\beta} X^i X^i = h_{\alpha\beta}.$$

However, the left-hand side is equal to $d_{\alpha\beta}$ according to (2.26) and therefore $h_{\alpha\beta} = d_{\alpha\beta}$. Thus, we get the *Gaussian equations*

$$D_\alpha D_\beta f^i = d_{\alpha\beta} X^i, \qquad (2.29)$$

which are the remaining equations to be solved for the inverse problem. Introducing the six functions $p_\alpha^i = f^i_{,\alpha}$, we can write the complete system (2.27) and (2.29) as follows,

$$(p_\beta^i)_{,\alpha} = d_{\alpha\beta} X^i + \{{}^{\;\varepsilon}_{\alpha\;\beta}\} p_\varepsilon^i, \qquad X^i_{,\alpha} = -d_\alpha^\beta p_\beta^i, \qquad (2.30)$$

where $d_{\alpha\beta}$, d_α^β and $\{{}^{\;\varepsilon}_{\alpha\;\beta}\}$ are given. The solution to (2.30) must, however, also satisfy the following conditions,

$$p_\alpha^i p_\beta^i = a_{\alpha\beta}, \qquad X^i X^i = 1, \qquad X^i p_\alpha^i = 0. \qquad (2.31)$$

This is a so-called *mixed system*. In the next section we shall find necessary conditions for the integrability of this system.

7. The equations of Codazzi

Since $D_\gamma f^i$ are (three) covariant vector-fields, we have, according to equation (1.39),

$$D_\alpha D_\beta D_\gamma f^i - D_\beta D_\alpha D_\gamma f^i = f^i_{,\varepsilon} B^\varepsilon_{\;\gamma\beta\alpha}. \qquad (2.32)$$

But due to Gauss's equations (2.29), this is equivalent to

$$D_\alpha(d_{\beta\gamma} X^i) - D_\beta(d_{\alpha\gamma} X^i) = f^i_{,\varepsilon} a^{\varepsilon\sigma} B_{\sigma\gamma\beta\alpha}.$$

After the covariant differentiation has been performed, we have

$$d_{\beta\gamma} X^i_{,\alpha} + X^i D_\alpha d_{\beta\gamma} - d_{\alpha\gamma} X^i_{,\beta} - X^i D_\beta d_{\alpha\gamma} - f^i_{,\varepsilon} a^{\varepsilon\sigma} B_{\sigma\gamma\beta\alpha} = 0.$$

According to equation (2.27),

$$X^i_{,\alpha} = -d_\alpha^\varepsilon f^i_{,\varepsilon} = -a^{\sigma\varepsilon} d_{\alpha\sigma} f^i_{,\varepsilon}.$$

After substitution and collecting terms, we get

$$(d_{\alpha\gamma}d_{\beta\sigma} - d_{\beta\gamma}d_{\alpha\sigma} - B_{\sigma\gamma\beta\alpha})a^{\varepsilon\sigma}f^i_{,\varepsilon} + (D_\alpha d_{\beta\gamma} - D_\beta d_{\alpha\gamma})X^i = 0.$$

The first parenthesis vanishes[6] due to (2.16), and multiplying through by X^i yields

$$D_\alpha d_{\beta\gamma} - D_\beta d_{\alpha\gamma} = 0. \tag{2.33}$$

These are the *equations of Codazzi*. Together with Gauss's equation (2.17), they are conditions that are always fulfilled by the fundamental tensors of a surface and are thus necessary conditions for the integrability of the mixed system (2.30), (2.31). It can be proved[7] that these conditions are also sufficient, i.e., any six functions $a_{\alpha\beta}$, $d_{\alpha\beta}$ fulfilling the conditions $a > 0$, Gauss's equation (2.17) and Codazzi's equations (2.33) are the fundamental tensors of a surface.

8. Surface and volume element

On the surface $u^3 = $ const in any three-dimensional coordinates system u^i an elementary mesh *PQRS* (see Fig. 1) has the sides $(g_{11})^{1/2}du^1$ and $(g_{22})^{1/2}du^2$. The area of the elementary mesh is thus

$$(g_{11}g_{22})^{1/2} du^1 du^2 \sin \alpha_{12},$$

where α_{12} is the angle *QPS*, which is given by (1.25). Eliminating α_{12}, we get

$$dA = (g_{11}g_{22} - (g_{12})^2)^{1/2} du^1 du^2. \tag{2.34}$$

For a coordinate mesh on the middle-surface we get

$$dA = a^{1/2} du^1 du^2. \tag{2.35}$$

In normal coordinates, equation (2.34) takes the form

$$dA = g^{1/2} du^1 du^2.$$

---

[6] Alternatively, multiplying through by $f^i_{,\alpha}$ shows that this parenthesis vanishes and this gives an independent derivation of (2.16).
[7] See, for example, EISENHART (1947) pp. 114–121.

Due to (2.8) and (2.9) we have

$$g_{\alpha\beta} = f^i_{,\alpha} f^i_{,\beta} = (f^i_{,\alpha} + X^i_{,\alpha} z)(f^i_{,\beta} + X^i_{,\beta} z)$$
$$= a_{\alpha\beta} + (f^i_{,\alpha} X^i_{,\beta} + f^i_{,\beta} X^i_{,\alpha}) z + X^i_{,\alpha} X^i_{,\beta} z^2.$$

With $f^i_{,\alpha} X^i_{,\beta}$ from (2.25) and $X^i_{,\alpha}$ from (2.27) we get

$$g_{\alpha\beta} = a_{\alpha\beta} - 2 d_{\alpha\beta} z + d^\gamma_\alpha d_{\gamma\beta} z^2. \tag{2.36}$$

The right-hand side of (2.36) can be factorized in the following manner,

$$g_{\alpha\beta} = a^{\xi\eta}(a_{\alpha\xi} - d_{\alpha\xi} z)(a_{\beta\eta} - d_{\beta\eta} z)$$

so that

$$g = \frac{1}{a}[\det(a_{\alpha\beta} - d_{\alpha\beta} z)]^2.$$

The determinant entering this expression can be simplified by using the relations (2.21) and (2.22),

$$\det(a_{\alpha\beta} - d_{\alpha\beta} z) = (a_{11} - d_{11} z)(a_{22} - d_{22} z) - (a_{12} - d_{12} z)^2$$
$$= a(1 - 2Hz + Kz^2).$$

Hence,

$$(g/a)^{1/2} = 1 - 2Hz + Kz^2. \tag{2.37}$$

The coordinate mesh du^1, du^2 on any surface $u^3 = $ const has an area that depends on the distance u^3 from the middle-surface. In normal coordinates the dependence is given by (2.37).

The volume dV of the three-dimensional coordinate element $du^1 du^2 du^3$ is found to be

$$dV = (g_{11} g_{22} g_{33})^{1/2} \sin\alpha_{12} \sin\alpha_{23} \sin\alpha_{31} \, du^1 du^2 du^3$$

from elementary considerations. Calculating $\sin\alpha_{12}$ from (1.24) and sin α_{23}, sin α_{31} correspondingly, we get

$$dV = g^{1/2}\, du^1 du^2 du^3 . \tag{2.38}$$

This is equivalent to the form $dV = J\, du^1 du^2 du^3$, where J is the Jacobian determinant relating the coordinates u^i to cartesian coordinates x^i.

9. Coordinates of principal curvature

Returning to equation (2.19) we recall that for each of the roots $1/R_1$ and $1/R_2$ of the characteristic equation (2.20) we have a nontrivial solution, defining two directions λ_1^α and λ_2^α: the *directions of principal curvature*. Thus,

$$\left(d_{\alpha\beta} - \frac{1}{R_1} a_{\alpha\beta}\right)\lambda_1^\alpha = 0, \quad \left(d_{\alpha\beta} - \frac{1}{R_2} a_{\alpha\beta}\right)\lambda_2^\alpha = 0 . \tag{2.39}$$

Multiplying the first equation (2.39) by λ_2^β and the second one by λ_1^β and subtracting, we get

$$\left(\frac{1}{R_2} - \frac{1}{R_1}\right) a_{\alpha\beta} \lambda_1^\alpha \lambda_2^\beta = 0 . \tag{2.40}$$

If the radii of principal curvature R_1 and R_2 are different, equation (2.40) can only be satisfied if λ_1^α and λ_2^β are orthogonal. It is therefore always possible to construct mutually orthogonal coordinates, following the directions of principal curvature. Such coordinates are called *coordinates of principal curvature*.

In these coordinates, both fundamental tensors reduce to diagonal form, and in particular, we have

$$d_1^1 = \frac{1}{R_1}, \quad d_2^1 = d_1^2 = 0, \quad d_2^2 = \frac{1}{R_2} .$$

Equation (2.20) can be written with the help of (2.21) and (2.22) in the form

$$\left(\frac{1}{R}\right)^2 = 2H\frac{1}{R} - K, \tag{2.41}$$

which holds good for $R = R_1$ as well as for $R = R_2$.

Now, consider the equation

$$d^\alpha_\beta d^\beta_\gamma = 2H d^\alpha_\gamma - K a^\alpha_\gamma. \tag{2.42}$$

In coordinates of principal curvature this equation reduces to (2.41) for $\alpha = \gamma$ and to the identity $0 = 0$ for $\alpha \neq \gamma$. Thus it holds good for all α and γ in coordinates of principal curvature. On the other hand it is a tensor equation and therefore it holds good in all coordinate systems.

The equation is a special case of *Hamilton–Cayley's theorem* for the second-order tensor $d_{\alpha\beta}$. The formula will prove to be useful in Chapter 7.

10. Summary

Surfaces are two-dimensional spaces characterized primarily by their intrinsic geometry, represented by the first fundamental form or the metric tensor $a_{\alpha\beta}$. In terms of this tensor, length of arcs, angles between curves and surface areas are expressed, in fact, everything that has to do with plane (two-dimensional) geometry. But, as seen from the embedding space, the intrinsic geometry does not determine the surface uniquely. That is determined[8] by the behaviour of the surface in the neighbourhood of a characteristic point to the tangent plane at that point and is given by the second fundamental form, where the second fundamental tensor $d_{\alpha\beta}$ of a surface enters. However, the six functions $a_{\alpha\beta}$ and $d_{\alpha\beta}$ are not independent. There are four conditions to be satisfied by these functions. Firstly, the determinant a of $a_{\alpha\beta}$ must be positive; secondly, Gauss's equation relating the determinant of $d_{\alpha\beta}$ to the Riemann–Christoffel tensor of the surface (i.e., with the $a_{\alpha\beta}$ and their first and second derivatives) must be satisfied, and thirdly, the two equations of Codazzi, relating the covariant derivatives of $d_{\alpha\beta}$ must hold good. These conditions will be seen to play an important role in shell theory.

---

[8] Apart from a rigid-body motion.

The strength and elegance of general tensor analysis is apparent in the formulation of the differential geometry of surfaces.

Bibliography

EISENHART, L.P. (1947), *An Introduction to Differential Geometry with Use of the Tensor Calculus*, Princeton University Press.
SOKOLNIKOFF, I.S. (1951), *Tensor Analysis*, Applied Mathematics Series, Wiley, New York.
STOKER, J.J. (1969), *Differential Geometry*, Wiley–Interscience, New York.
THOMAS, T.Y. (1961), *Concepts from Tensor Analysis and Differential Geometry*, Academic Press, New York.

CHAPTER 3

THE KINEMATICS OF A SURFACE

1. Introduction 55
2. Displacements 55
3. Stretching of the middle-surface 56
4. Rotation of an element of the shell 60
5. Bending of the middle-surface 64
6. Equations of compatibility 67
7. Summary 71

CHAPTER 3

THE KINEMATICS OF A SURFACE

1. Introduction

In this chapter we shall deal with the deformation of the middle-surface. There are two different modes of describing the deformation of a continuum, the *Eulerian* and the *Lagrangian*. The Eulerian mode employs as independent variables the coordinates of a material point in the deformed state, while the Lagrangian description uses the coordinates of a typical point in the initial state.

We shall employ Lagrangian coordinates. This means that a given material point on the middle-surface will always have the same coordinates u^1, u^2 irrespective of the state of deformation. The reader may visualize the coordinate system u^1, u^2 as engraved on the middle-surface once and for all. During deformation this coordinate system deforms and when we refer to 'the deformed coordinate system' we shall have this picture in mind. However, the coordinate system x^i is fixed in space and the cartesian coordinates x^i of a typical material point will, of course, change during deformation.

2. Displacements

A point on the middle-surface with the coordinates u^α has the cartesian coordinates $f^i(u^1, u^2)$ in the initial state. After deformation the same point has the cartesian coordinates $f^i + \bar{v}^i$, where $\bar{v}^i$ are the cartesian components of the *displacement vector*.

Let $v^i(u^1, u^2)$ denote the components of the displacement vector in normal coordinates. Then, according to (1.5) and (2.8) we get

$$\bar{v}^i = f^i_{,j} v^j = f^i_{,\alpha} v^\alpha + X^i v^3$$

or

$$\bar{v}^i = f^i_{,\alpha} v^\alpha + X^i w, \tag{3.1}$$

where $w(u^1, u^2) = v^3$ is a scalar field, representing the *normal displacement* and $v^\alpha(u^1, u^2)$ is a vector-field representing the *tangential displacements*.

The three functions v^α and w give complete information about the deformation of the middle-surface. In fact, they also give information about the displacement of the surface as a rigid body, its translation and rotation. In order to relate the stresses in the shell to its deformation we need a *local measure* to describe the deformation, leaving irrelevant motion out of the picture. We shall find that, for such a purpose, the deformation of the shell is more adequately given by two tensor-fields, describing the *stretching* and the *bending* of the middle-surface.

3. Stretching of the middle-surface

The distance ds between two neighbouring points on the middle-surface in the initial state will change to ds^* in the deformed state. Thus

$$ds^{*2} = a^*_{\alpha\beta} \, du^\alpha \, du^\beta, \tag{3.2}$$

where $a^*_{\alpha\beta}$ is the metric tensor of the *deformed coordinate system*. We shall use an asterisk to distinguish quantities in the deformed state from the corresponding quantities in the initial state. In the Lagrangian mode of description, which we have adopted,

$$du^{*\alpha} = du^\alpha. \tag{3.3}$$

Since the fundamental tensor $a^*_{\alpha\beta}$ of the deformed system may differ from the fundamental tensor $a_{\alpha\beta}$ of the undeformed system, it is not unimportant which one is used for raising or lowering indices. We shall therefore adopt the following convention: Raising and lowering indices of tensors denoted by an asterisk is always performed with the help of

$a^*_{\alpha\beta}$ and $a^{*\alpha\beta}$. Thus, for example,

$$du^*_\alpha = a^*_{\alpha\beta} du^\beta = du_\alpha + (a^*_{\alpha\beta} - a_{\alpha\beta}) du^\beta,$$

and therefore du^*_α may differ from du_α.

The cartesian coordinates of a point on the middle-surface change from f^i to $f^i + \bar{v}^i$ during deformation. Hence, and according to (2.3), the metric tensor of the deformed surface is

$$a^*_{\alpha\beta} = (f^i + \bar{v}^i)_{,\alpha}(f^i + \bar{v}^i)_{,\beta} = a_{\alpha\beta} + f^i_{,\alpha}\bar{v}^i_{,\beta} + \bar{v}^i_{,\alpha}f^i_{,\beta} + \bar{v}^i_{,\alpha}\bar{v}^i_{,\beta}.$$

The first three terms constitute the linear part of $a^*_{\alpha\beta}$ in terms of the displacements. For many purposes the linear theory is fully adequate; however, we retain the quadratic term since it is important in certain problems with which we shall deal.

From (3.1) we get

$$\bar{v}^i_{,\alpha} = f^i_{,\alpha\beta} v^\beta + f^i_{,\beta} v^\beta_{,\alpha} + X^i_{,\alpha} w + X^i w_{,\alpha}.$$

The second derivative is given by (2.29),

$$f^i_{,\alpha\beta} = \{{}^\gamma_{\alpha\beta}\} f^i_{,\gamma} + d_{\alpha\beta} X^i$$

and substitution yields

$$\bar{v}^i_{,\alpha} = f^i_{,\gamma}(D_\alpha v^\gamma - d^\gamma_\alpha w) + X^i(d_{\alpha\gamma} v^\gamma + w_{,\alpha}),$$

where (2.27) has also been utilized. Introducing the notation

$$p_{\alpha\beta} = D_\alpha v_\beta - d_{\alpha\beta} w, \tag{3.4}$$

which is the generalized two-dimensional *displacement gradient*, and

$$q_\alpha = d_{\alpha\beta} v^\beta + w_{,\alpha}, \tag{3.5}$$

which is the *rotation*, we get

$$\bar{v}^i_{,\alpha} = f^i_{,\gamma} p_\alpha{}^\gamma + X^i q_\alpha. \tag{3.6}$$

Substitution into $a^*_{\alpha\beta}$ yields

$$a^*_{\alpha\beta} = a_{\alpha\beta} + p_{\alpha\beta} + p_{\beta\alpha} + p_\alpha{}^\gamma p_{\beta\gamma} + q_\alpha q_\beta. \qquad (3.7)$$

The quantity

$$(\mathrm{d}s^*)^2 - \mathrm{d}s^2 = (a^*_{\alpha\beta} - a_{\alpha\beta})\,\mathrm{d}u^\alpha\,\mathrm{d}u^\beta$$

is a measure of the change of the distance between two neighbouring points on the middle-surface during deformation. We shall express this measure with the *strain tensor* $E_{\alpha\beta}$, defined by[1]

$$E_{\alpha\beta} = \tfrac{1}{2}(a^*_{\alpha\beta} - a_{\alpha\beta}). \qquad (3.8)$$

Clearly, $E_{\alpha\beta}$ is symmetrical,

$$E_{\alpha\beta} = E_{\beta\alpha}.$$

With $a^*_{\alpha\beta}$ from (3.7) this yields

$$E_{\alpha\beta} = \tfrac{1}{2}(p_{\alpha\beta} + p_{\beta\alpha} + p_\alpha{}^\gamma p_{\beta\gamma} + q_\alpha q_\beta) \qquad (3.9)$$

or

$$E_{\alpha\beta} = \tfrac{1}{2}(D_\alpha v_\beta + D_\beta v_\alpha) - d_{\alpha\beta} w + \tfrac{1}{2}[D_\alpha v^\gamma D_\beta v_\gamma - d_{\beta\gamma} w D_\alpha v^\gamma - d_{\alpha\gamma} w D_\beta v^\gamma \\
+ d_{\alpha\gamma} d^\gamma_\beta (w)^2 + d_{\alpha\gamma} d_{\beta\delta} v^\gamma v^\delta + d_{\alpha\gamma} v^\gamma w_{,\beta} + d_{\beta\gamma} v^\gamma w_{,\alpha} + w_{,\alpha} w_{,\beta}], \qquad (3.10)$$

where the displacements v_α and w have been reinstalled. If all quadratic terms are omitted, we get the linear approximation

$$E_{\alpha\beta} \simeq \tfrac{1}{2}(D_\alpha v_\beta + D_\beta v_\alpha) - d_{\alpha\beta} w. \qquad (3.11)$$

Due to the symmetry of $E_{\alpha\beta}$, the stretching of the middle-surface is described by three functions that relate the metric tensor of the deformed surface to the metric tensor of the original, undeformed surface.

---

[1] The factor $\tfrac{1}{2}$ is conventional and serves the purpose of making our measure of strain conforming with the traditionally accepted measure.

STRETCHING OF THE MIDDLE-SURFACE

The strain in an arbitrary direction du^α is given by

$$\frac{1}{2}\frac{(ds^*)^2 - ds^2}{ds^2} = E_{\alpha\beta}\frac{du^\alpha}{ds}\frac{du^\beta}{ds}.$$

We may conclude from an argument very similar to the one used in Chapter 2 on the principal radii of curvature that the extreme values e of the strain are found in the directions du^α, given by the equations

$$(E_{\alpha\beta} - ea_{\alpha\beta})du^\alpha = 0,$$

and that e is determined from the condition that

$$\det(E_{\alpha\beta} - ea_{\alpha\beta}) = 0.$$

But

$$\det(E_{\alpha\beta} - ea_{\alpha\beta}) = \det a_{\alpha\gamma}(E^\gamma_\beta - e\delta^\gamma_\beta) = a \det(E^\gamma_\beta - e\delta^\gamma_\beta).$$

Since $a \neq 0$ we have

$$\det(E^\gamma_{\beta,} - e\delta^\gamma_\beta) = 0,$$

which yields the following equation of second degree for e,

$$e^2 - eE^\gamma_\gamma + \det(E^\gamma_\beta) = 0.$$

Thus, if e_1 and e_2 are the *principal strains*, we have

$$e_1 + e_2 = E^\gamma_\gamma = \zeta_1, \tag{3.12}$$

$$e_1 e_2 = \det(E^\gamma_\beta) = E^1_1 E^2_2 - E^1_2 E^2_1 = \zeta_2, \tag{3.13}$$

where we have used the symbols ζ_1 and ζ_2 for the first and the second *strain invariant*, respectively.

Since

$$a^* = \det(a^*_{\alpha\beta}) = \det(a_{\alpha\beta} + 2E_{\alpha\beta})$$

and

$$\det(a_{\alpha\beta} + 2E_{\alpha\beta}) = \det a_{\alpha\beta}(\delta^\beta_\gamma + 2E^\beta_\gamma) = a \det(\delta^\beta_\gamma + 2E^\beta_\gamma),$$

we get

$$a^* = a(1 + 2\zeta_1 + 4\zeta_2). \tag{3.14}$$

With this expression we may calculate the area-element $(a^*)^{1/2}du^1du^2$ of the deformed middle-surface. For small strains, we get the following *linearized* equation for the relative change in area,

$$\left(\frac{a^*}{a}\right)^{1/2} \simeq 1 + \zeta_1, \tag{3.15}$$

where ζ_1 due to (3.12), and (3.11) may be written as

$$\zeta_1 \simeq D_\alpha v^\alpha - d^\alpha_\alpha w = p_\alpha{}^\alpha. \tag{3.16}$$

4. Rotation of an element of the shell

The vector q_α was called the rotation and in this section we shall discuss its physical significance and give a reason for the term rotation. To do so, we shall assume throughout this section that all displacements and their derivatives are small. The reader should consider this section as a parenthesis in this respect.

The motion of a rigid body is completely described by two vectors, the *translation* vector $\bar{v}^i$ and the *rotation* vector $\bar{r}^i$. The rotation of a small element of the shell can be resolved into a rotation around the normal X^i and one perpendicular to this. For sufficiently small rotations we can write (see Fig. 5)

$$\bar{r}^i = f^i_{,\alpha} r^\alpha + X^i \Theta \tag{3.17}$$

analogously to (3.1). The displacements v^α, w uniquely determine the deformation, translation and rotation of the middle-surface.

Since the (unit) normal X^i retains its length during deformation, its increment δX^i must be perpendicular to X^i and lies therefore in the tangent plane,

$$\delta X^i = c^\alpha f^i_{,\alpha}, \tag{3.18}$$

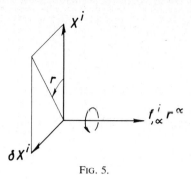

Fig. 5.

where the coefficients c^α form a contravariant vector, as we shall see.

The rotated normal $X^i + \delta X^i$ is perpendicular to the tangent plane to the deformed middle-surface, i.e.,

$$(X^i + \delta X^i)(f^i + \bar{v}^i)_{,\alpha} = 0$$

or, neglecting second-order terms,

$$\delta X^i f^i_{,\alpha} + X^i \bar{v}^i_{,\alpha} = 0 \,.$$

From this relation we find after substituting $\bar{v}^i_{,\alpha}$ from (3.6),

$$\delta X^i f^i_{,\alpha} = -q_\alpha \,. \tag{3.19}$$

Multiplication of (3.18) by $f^i_{,\gamma}$ yields, with (3.19),

$$c^\alpha a_{\alpha\gamma} = -q_\gamma$$

and therefore

$$\delta X^i = -q^\alpha f^i_{,\alpha} \,. \tag{3.20}$$

The cartesian vector $f^i_{,\alpha} r^\alpha$ (see Fig. 5) is the cross-product between the normal X^i and its increment δX^i,

$$f^i_{,\alpha} r^\alpha = e_{ijk} X^j \delta X^k \,.$$

Multiplying through by $f^i_{,\beta}$ and substituting δX^i from (3.20) we get

$$r_\beta = -q^\alpha e_{ijk} f^i_{,\beta} X^j f^k_{,\alpha}.$$

Due to (2.7) these two equations can be written as

$$r_1 = a^{1/2} q^2 \quad \text{and} \quad r_2 = -a^{1/2} q^1$$

or using the two-dimensional *alternating symbol* $e_{\alpha\beta}$, defined by

$$e_{12} = -e_{12} = 1, \qquad e_{11} = e_{22} = 0, \qquad (3.21)$$

we can summarize the equations in the formula

$$r_\alpha = a^{1/2} e_{\alpha\beta} q^\beta$$

or

$$r_\alpha = \varepsilon_{\alpha\beta} q^\beta, \qquad (3.22)$$

where $\varepsilon_{\alpha\beta}$ is the *alternating tensor* in two dimensions defined by

$$\varepsilon_{\alpha\beta} = a^{1/2} e_{\alpha\beta}. \qquad (3.23)$$

The tensorial character of $\varepsilon_{\alpha\beta}$ is easily established, using the transformation law for covariant tensors.

It is useful to note the following properties of $\varepsilon_{\alpha\beta}$, which follow immediately from the definition,

$$\varepsilon_{\alpha\beta} = -\varepsilon_{\beta\alpha} \qquad \text{(the tensor is skew-symmetrical)} \qquad (3.24)$$

$$\varepsilon^{\alpha\beta} = a^{-1/2} e_{\alpha\beta} = a^{-1} \varepsilon_{\alpha\beta} \quad \text{(both indices are raised when the tensor is divided by } a\text{)} \qquad (3.25)$$

$$\varepsilon_{\alpha\gamma} \varepsilon^{\beta\gamma} = \delta^\beta_\alpha \qquad \text{(Kronecker's delta)} \qquad (3.26)$$

$$D_\alpha \varepsilon_{\beta\gamma} = D_\alpha \varepsilon^{\beta\gamma} = 0 \qquad \text{(the tensor behaves like a constant under covariant differentiation).} \qquad (3.27)$$

Since $\varepsilon_{\alpha\beta}A^\alpha A^\beta = 0$ where A^α is any contravariant vector, it follows that the vector $\varepsilon_{\alpha\beta}A^\alpha$ is perpendicular to A^β. Thus *inner multiplication by $\varepsilon_{\alpha\beta}$ rotates a vector $\tfrac{1}{2}\pi$*. Equation (3.22) shows that r_α is the same vector as q_α, but rotated $\tfrac{1}{2}\pi$. This explains why we called q_α the rotation. In terms of displacements we have

$$r^\alpha = \varepsilon^{\alpha\beta}(d_{\beta\gamma}v^\gamma + w_{,\beta}). \tag{3.28}$$

The surface element is not rigid and hence its 'rotation' around the normal cannot be defined unambiguously, but we can, of course, calculate the rotation of a given and invariantly defined line-element on the surface around the normal. Let us therefore take the element PQ with end-points u^α and $u^\alpha + du^\alpha$. The direction of this element is $t^\alpha = du^\alpha/ds$ with the cartesian components $f^i_{,\alpha}t^\alpha$ before deformation and $f^{*i}_{,\alpha}t^\alpha$ after deformation. The angle of rotation ϕ around the normal axis X^i is equal to the cross-product of those two vectors, projected on X^i, i.e.,

$$\phi = e_{ijk}X^i f^j_{,\alpha} f^{*k}_{,\beta} t^\alpha t^\beta = e_{ijk}X^i f^j_{,\alpha} \bar{v}^k_{,\beta} t^\alpha t^\beta$$
$$= e_{ijk}X^i f^j_{,\alpha}(f^k_{,\gamma}p_\beta{}^\gamma + X^k q_\beta)t^\alpha t^\beta = \varepsilon_{\alpha\gamma}p_\beta{}^\gamma t^\alpha t^\beta$$

and hence

$$\phi = \varepsilon_{\alpha\gamma}(D_\beta v^\gamma - d^\gamma_\beta w)t^\alpha t^\beta. \tag{3.29}$$

If by Θ in (3.17) we mean the *average rotation* of two elements PQ and RS perpendicular to each other, then we get

$$\Theta = \tfrac{1}{2}\varepsilon^{\alpha\beta}D_\alpha v_\beta. \tag{3.30}$$

According to (3.9) the strain tensor $E_{\alpha\beta}$ is the symmetrical part of the displacement gradient $p_{\alpha\beta}$ (for small displacements). Multiplying (3.4) through by $\varepsilon^{\alpha\beta}$ and taking (3.26) and (3.25) into account, we get

$$\Theta = \tfrac{1}{2}\varepsilon^{\alpha\beta}p_{\alpha\beta}$$

so that Θ is the skew-symmetrical part of $p_{\alpha\beta}$. Thus

$$p_{\alpha\beta} = E_{\alpha\beta} + \varepsilon_{\alpha\beta}\Theta. \tag{3.31}$$

To sum up, the rotation of a small surface element can be resolved in an unambiguously defined rotation perpendicular to the normal X^i of the surface, given by the vector r^α or q^α and a rotation Θ around the normal. The possibility of resolving the rotation into components is, of course, limited to small rotations.

This closes our parenthesis and we return to a study of the deformation of the middle-surface under displacements of arbitrary size.

5. Bending of the middle-surface

A surface is characterized by its two fundamental tensors, which are $a^*_{\alpha\beta}$ and $d^*_{\alpha\beta}$ for the deformed middle-surface. The deformation, i.e., the change from $a_{\alpha\beta}, d_{\alpha\beta}$ to $a^*_{\alpha\beta}, d^*_{\alpha\beta}$, is given by the strain tensor $E_{\alpha\beta}$, which we have just determined, and the *bending tensor* $K_{\alpha\beta}$, given by

$$K_{\alpha\beta} = d^*_{\alpha\beta} - d_{\alpha\beta}. \tag{3.32}$$

Clearly, the bending tensor is symmetrical,

$$K_{\alpha\beta} = K_{\beta\alpha}.$$

We proceed now to calculate $d^*_{\alpha\beta}$ in terms of the displacements. According to the rule for covariant derivatives we have

$$f^{*i}_{,\alpha\beta} = D_\beta f^{*i}_{,\alpha} + \{^{\gamma}_{\alpha\,\beta}\} f^{*i}_{,\gamma},$$

which on multiplication with X^{*i} yields

$$X^{*i} f^{*i}_{,\alpha\beta} = X^{*i} D_\beta f^{*i}_{,\alpha} + \{^{\gamma}_{\alpha\,\beta}\} f^{*i}_{,\gamma} X^{*i}.$$

Since $f^{*i}_{,\gamma}$ is orthogonal to X^{*i} the last term vanishes. But according to definition (2.13), the left-hand side is the curvature tensor $d^*_{\alpha\beta}$ of the deformed middle-surface and hence

$$d^*_{\alpha\beta} = X^{*i} D_\beta f^{*i}_{,\alpha}. \tag{3.33}$$

We proceed to calculate the right-hand side in terms of the displacement gradient and the rotation. According to (2.7) and (3.6) we have

$$X^{*i} = a^{*-1/2} e_{ijk} (f^j_{,1} + f^j_{,\gamma} p_1^\gamma + X^j q_1)(f^k_{,2} + f^k_{,\gamma} p_2^\gamma + X^k q_2).$$

Since $e_{ijk} X^j f^k_{,\alpha}$ is the cross-product between X^j and the tangent vector $f^k_{,\alpha}$, it is another tangent vector, which we obtain by rotating $f^k_{,\alpha}$ the angle $\frac{1}{2}\pi$, i.e.,

$$e_{ijk} X^j f^k_{,\alpha} = \varepsilon_{\alpha\beta} a^{\beta\gamma} f^i_{,\gamma}, \tag{3.34}$$

a formula, which is easily verified with the help of (2.7) and (3.23). Expanding the product, we find

$$X^{*i} = \left(\frac{a}{a^*}\right)^{1/2} [X^i(1 + p_\alpha^\alpha + \mu/a) - f^i_{,\alpha}(q^\alpha + \varepsilon^{\alpha\beta}\varepsilon^{\gamma\delta} q_\gamma p_{\delta\beta})],$$

where

$$\mu = \det(p_{\alpha\beta}) = p_{11}p_{22} - p_{12}p_{21}. \tag{3.35}$$

At this stage we may check that the linear terms of X^{*i} give the same result as equation (3.20) for $\delta X^i = X^{*i} - X^i$.

We proceed to expand the second factor of (3.33),

$$D_\beta f^{*i}_{,\alpha} = D_\beta(f^i_{,\alpha} + f^i_{,\gamma} p_\alpha^\gamma + X^i q_\alpha)$$

$$= D_\beta D_\alpha f^i + p_\alpha^\gamma D_\beta D_\gamma f^i + f^i_{,\gamma} D_\beta p_\alpha^\gamma + X^i_{,\beta} q_\alpha + X^i D_\beta q_\alpha.$$

Using (2.25), (2.26), (2.28) and the orthogonality properties of $f^i_{,\alpha}$ and X^i, we get, after substitution in (3.33),

$$d^*_{\alpha\beta} = \left(\frac{a}{a^*}\right)^{1/2} [(1 + p_\varepsilon^\varepsilon + \mu/a)(d_{\alpha\beta} + D_\beta q_\alpha + d^\gamma_\beta p_{\alpha\gamma})$$

$$- (q^\rho + \varepsilon^{\rho\beta}\varepsilon^{\gamma\delta} q_\gamma p_{\delta\beta})(D_\beta p_{\alpha\rho} - d_{\beta\rho} q_\alpha)]$$

and hence

$$K_{\alpha\beta} = \left(\frac{a}{a^*}\right)^{1/2}[(1 + p_\varepsilon^\varepsilon + \rlap{/}{\ell}/a)(d_{\alpha\beta} + D_\beta q_\alpha + d_\beta^\gamma p_{\alpha\gamma})$$
$$- (q^\rho + \varepsilon^{\rho\beta}\varepsilon^{\gamma\delta}q_\gamma p_{\delta\beta})(D_\beta p_{\alpha\rho} - d_{\beta\rho}q_\alpha)] - d_{\alpha\beta}, \quad (3.36)$$

which is the exact nonlinear expression for the bending tensor. The symmetry in α and β of the term $D_\beta q_\alpha + d_\beta^\gamma p_{\alpha\gamma}$ can be verified with the help of Codazzi's equations (2.33) and of the term $D_\beta p_{\alpha\gamma} - d_{\gamma\beta} q_\alpha$ with the help of Gauss's equation in the form (2.16).

The linear part of the right-hand side is

$$D_\beta q_\alpha + d_\beta^\gamma p_{\alpha\gamma},$$

and hence the linear approximation for $K_{\alpha\beta}$ is given by

$$K_{\alpha\beta} \simeq D_\alpha D_\beta w + d_{\alpha\gamma} D_\beta v^\gamma + d_{\beta\gamma} D_\alpha v^\gamma + v^\gamma D_\beta d_{\gamma\alpha} - d_{\beta\gamma} d_\alpha^\gamma w. \quad (3.37)$$

By means of (3.32) we defined the bending of the shell in terms of the difference between the covariant components of the second fundamental tensors in the deformed and the initial state. Had we instead selected the contravariant or mixed components, the result would have been somewhat different. Thus, except for the term $D_\alpha D_\beta w$, the remaining linear terms of $K_{\alpha\beta}$ and, of course, the nonlinear terms would differ essentially. That there would be a difference is not surprising since these three ways of defining $K_{\alpha\beta}$ are not identical. For instance, let us take the mixed form

$$d_\beta^{*\alpha} - d_\beta^\alpha = a^{*\alpha\gamma} d_{\gamma\beta}^* - d_\beta^\alpha$$
$$= (a^{\alpha\gamma} - 2E^{\alpha\gamma} + \cdots)(d_{\gamma\beta} + K_{\gamma\beta}) - d_\beta^\alpha$$
$$= K_\beta^\alpha - 2E^{\alpha\gamma} d_{\gamma\beta} + \cdots,$$

where terms of higher order in the strain are indicated by the dots. We find that the covariant components $K_{\alpha\beta}$ in this case would get an additional linear term $-2E_\alpha^\gamma d_{\gamma\beta}$ plus terms of higher order. The additional linear term amounts to

$$-d_\beta^\gamma D_\alpha v_\gamma - d_\beta^\gamma D_\gamma v_\alpha + 2d_{\alpha\gamma} d_\beta^\gamma w,$$

which should be added to the right-hand side of (3.37).

There is, of course, nothing surprising in the fact that different definitions of $K_{\alpha\beta}$ would give different expressions in terms of the displacements. However, this simple fact is of great importance to a shell theory in which it is asserted that 'the bending moment is proportional to the change of curvature', since there are at least three different measures of this 'change of curvature' that might with equal right claim to be the one referred to.

The term $D_\alpha D_\beta w$ in (3.37) is not affected by the different measures of bending, which indicates that it has a very special standing in $K_{\alpha\beta}$. In fact, in some applications of shell theory it is the only one retained.

6. Equations of compatibility

The local deformation of a shell at any typical point of the middle-surface is exhaustively described by six numbers, three strain components $E_{\alpha\beta}$ and three bending components $K_{\alpha\beta}$. At any point these six components are independent and may be prescribed arbitrarily; however, as functions of the coordinates u^α they cannot be independent since $a^*_{\alpha\beta}$ and $d^*_{\alpha\beta}$, which they determine, are not. In this section we shall establish the relations that connect the fields $E_{\alpha\beta}$ and $K_{\alpha\beta}$. These relations are called the *equations of compatibility*.

We have already seen that necessary and sufficient conditions for $a^*_{\alpha\beta}$ and $d^*_{\alpha\beta}$ to be the fundamental tensors of a surface are expressed by three conditions; firstly, Gauss's equation

$$B^*_{\delta\gamma\beta\alpha} = d^*_{\alpha\gamma}d^*_{\beta\delta} - d^*_{\beta\gamma}d^*_{\alpha\delta}, \qquad (3.38)$$

secondly, the Codazzi equations

$$D^*_\alpha d^*_{\beta\gamma} - D^*_\beta d^*_{\alpha\gamma} = 0, \qquad (3.39)$$

and, thirdly, the condition

$$\det(a^*_{\alpha\beta}) > 0. \qquad (3.40)$$

It is important to notice that the covariant derivative D^*_α in the deformed coordinate system differs from the covariant derivative in the undeformed system because the Christoffel symbols differ. The difference

$$\hat{E}^{\gamma}_{\alpha\beta} = \{^{\gamma}_{\alpha\ \beta}\}^* - \{^{\gamma}_{\alpha\ \beta}\} \tag{3.41}$$

is called the *Christoffel deviator* and we shall see that it is a tensor as indicated by the notation.

Multiplying (3.41) with $a^*_{\gamma\delta}$ yields

$$\begin{aligned}
a^*_{\gamma\delta}\hat{E}^{\gamma}_{\alpha\beta} &= \tfrac{1}{2}(a^*_{\alpha\delta,\beta} + a^*_{\beta\delta,\alpha} - a^*_{\alpha\beta,\delta}) - a^*_{\gamma\delta}\{^{\gamma}_{\alpha\ \beta}\} \\
&= \tfrac{1}{2}(a_{\alpha\delta,\beta} + a_{\beta\delta,\alpha} - a_{\alpha\beta,\delta} - 2a_{\gamma\delta}\{^{\gamma}_{\alpha\ \beta}\}) \\
&\quad + E_{\alpha\delta,\beta} + E_{\beta\delta,\alpha} - E_{\alpha\beta,\delta} - 2E_{\gamma\delta}\{^{\gamma}_{\alpha\ \beta}\} \\
&= \tfrac{1}{2}(D_{\beta}a_{\alpha\delta} + D_{\alpha}a_{\beta\delta} - D_{\delta}a_{\alpha\beta}) + D_{\beta}E_{\alpha\delta} + D_{\alpha}E_{\beta\delta} - D_{\delta}E_{\alpha\beta}.
\end{aligned}$$

Since the metric tensor is constant under covariant differentiation we obtain

$$a^*_{\gamma\delta}\hat{E}^{\gamma}_{\alpha\beta} = E_{\alpha\beta\delta}, \tag{3.42}$$

where

$$E_{\alpha\beta\delta} = D_{\beta}E_{\alpha\delta} + D_{\alpha}E_{\beta\delta} - D_{\delta}E_{\alpha\beta} \tag{3.43}$$

so that

$$\hat{E}^{\gamma}_{\alpha\beta} = a^{*\gamma\delta}E_{\alpha\beta\delta}. \tag{3.44}$$

Since $E_{\alpha\beta\delta}$ and $a^{*\gamma\delta}$ are tensors, $\hat{E}^{\gamma}_{\alpha\beta}$ is a tensor. From its definition (3.41) it follows that

$$\hat{E}^{\gamma}_{\alpha\beta} = \hat{E}^{\gamma}_{\beta\alpha}. \tag{3.45}$$

The rule for covariant differentiation now follows immediately from (1.31),

$$D^*_{\varepsilon}T^{\alpha\beta\cdots}_{\gamma\delta\cdots} = D_{\varepsilon}T^{\alpha\beta\cdots}_{\gamma\delta\cdots} + \hat{E}^{\alpha}_{\varepsilon\sigma}T^{\sigma\beta\cdots}_{\gamma\delta\cdots} + \hat{E}^{\beta}_{\varepsilon\sigma}T^{\alpha\sigma\cdots}_{\gamma\delta\cdots} \\ + \cdots - \hat{E}^{\sigma}_{\varepsilon\gamma}T^{\alpha\beta\cdots}_{\sigma\delta\cdots} - \hat{E}^{\sigma}_{\varepsilon\delta}T^{\alpha\beta\cdots}_{\gamma\sigma\cdots} - \cdots. \tag{3.46}$$

Now, let us first try to express Gauss's equation (3.38) in terms of strain and bending. Using (1.37) we get

$$B^{*\sigma}{}_{\gamma\beta\alpha} = \{{}^\rho_{\gamma\alpha}\}^* \{{}^\sigma_{\rho\beta}\}^* - \{{}^\rho_{\gamma\beta}\}^* \{{}^\sigma_{\rho\alpha}\}^* + \{{}^\sigma_{\gamma\alpha}\}^*_{,\beta} - \{{}^\sigma_{\gamma\beta}\}^*_{,\alpha} .$$

Using (1.37), (3.41), (3.42), (3.43) and (1.41) we get, after some tedious calculations,

$$B^*_{\delta\gamma\beta\alpha} = B_{\delta\gamma\beta\alpha} + E_{\rho\delta}B^\rho{}_{\gamma\beta\alpha} - E_{\rho\gamma}B^\rho{}_{\delta\beta\alpha} + D_\beta D_\gamma E_{\alpha\delta} - D_\alpha D_\gamma E_{\beta\delta}$$
$$- D_\beta D_\delta E_{\alpha\gamma} + D_\alpha D_\delta E_{\beta\gamma} - \hat{E}^\rho_{\delta\beta} E_{\gamma\alpha\rho} + \hat{E}^\rho_{\delta\alpha} E_{\gamma\beta\rho} .$$

The right-hand side of (3.38) can be written with the help of (3.32) in the form

$$B^*_{\delta\gamma\beta\alpha} = B_{\delta\gamma\beta\alpha} + d_{\alpha\gamma}K_{\beta\delta} + K_{\alpha\gamma}d_{\beta\delta} - d_{\beta\gamma}K_{\alpha\delta} - K_{\beta\gamma}d_{\alpha\delta} + K_{\alpha\gamma}K_{\beta\delta} - K_{\beta\gamma}K_{\alpha\delta} .$$

Combining these results we get

$$D_\beta D_\gamma E_{\alpha\delta} - D_\alpha D_\gamma E_{\beta\delta} - D_\beta D_\delta E_{\alpha\gamma} + D_\alpha D_\delta E_{\beta\gamma} + E_{\sigma\delta}B^\sigma{}_{\gamma\beta\alpha}$$
$$- E_{\sigma\gamma}B^\sigma{}_{\delta\beta\alpha} - d_{\alpha\gamma}K_{\beta\delta} - K_{\alpha\gamma}d_{\beta\delta} + d_{\beta\gamma}K_{\alpha\delta} + K_{\beta\gamma}d_{\alpha\delta}$$
$$= K_{\alpha\gamma}K_{\beta\delta} - K_{\beta\gamma}K_{\alpha\delta} + \hat{E}^\rho_{\delta\beta} E_{\gamma\alpha\rho} - \hat{E}^\rho_{\delta\alpha} E_{\gamma\beta\rho} , \qquad (3.47)$$

where all nonlinear terms have been collected on the right-hand side. Of these sixteen equations only one is significant, all the others being either repetitions of it or else identities. We may write the significant equation by selecting the values 1212 for the indices $\delta\gamma\beta\alpha$, or what essentially amounts to the same thing, contract (3.47). Thus, multiplication of (3.47) by $\frac{1}{2}a^{\delta\beta}a^{\gamma\alpha}$ results in the following equation,

$$D_\alpha D^\alpha E^\beta_\beta - D_\alpha D^\beta E^\alpha_\beta + d^\alpha_\alpha K^\beta_\beta - d^\beta_\alpha K^\alpha_\beta - KE^\alpha_\alpha$$
$$= \tfrac{1}{2}[K^\alpha_\beta K^\beta_\alpha - K^\alpha_\alpha K^\beta_\beta] + \tfrac{1}{2}[\hat{E}^{\rho\beta}_\beta E^\alpha{}_{\alpha\rho} - \hat{E}^{\rho\beta}_\alpha E^\alpha{}_{\beta\rho}] , \qquad (3.48)$$

where all nonlinear terms are on the right-hand side. To obtain the last term on the left-hand side the identity $E^\alpha_\alpha K = E^\gamma_\beta(d^\alpha_\alpha d^\beta_\gamma - d^\alpha_\gamma d^\beta_\alpha)$ has been used. Here K is the Gaussian curvature of the middle-surface, given by (2.21). This is the equation of compatibility corresponding to Gauss's equation (3.38).

The equations of Codazzi (3.39) are transformed similarly. Thus

$$D_\alpha^* d_{\beta\gamma}^{*\delta} = D_\alpha d_{\beta\gamma}^{*\delta} - \hat{E}_{\alpha\beta}^\delta d_{\delta\gamma}^{*\delta} - \hat{E}_{\alpha\gamma}^\delta d_{\beta\delta}^{*\delta}$$
$$= D_\alpha[d_{\beta\gamma}^\delta + K_{\beta\gamma}^\delta] - E_{\alpha\beta\delta} d_\gamma^{*\delta} - E_{\alpha\gamma\delta} d_\beta^{*\delta},$$

and hence the equations (3.39) take the form

$$D_\alpha K_{\beta\gamma} - D_\beta K_{\alpha\gamma} = d_\beta^{*\delta}[D_\gamma E_{\alpha\delta} + D_\alpha E_{\gamma\delta} - D_\delta E_{\alpha\gamma}]$$
$$- d_\alpha^{*\delta}[D_\gamma E_{\beta\delta} + D_\beta E_{\gamma\delta} - D_\delta E_{\beta\gamma}]. \qquad (3.49)$$

Of these eight equations only two are independent. The remaining equations are either repetitions of these or else identities. The equations contain the mixed curvature tensor $d_\beta^{*\alpha}$ in the deformed state. We can calculate this in the following way,

$$d_\beta^{*\delta} = (a^{\delta\rho} - 2E^{\delta\rho} + 4E_\sigma^\delta E^{\sigma\rho} - \cdots)(d_{\rho\beta} + K_{\rho\beta})$$
$$= d_\beta^\delta + K_\beta^\delta - 2E^{\delta\rho} d_{\rho\beta} + \cdots,$$

and the linear terms of (3.49) are thus obtained if the starred quantities $d_\beta^{*\delta}$ and $d_\alpha^{*\delta}$ are replaced by the corresponding unstarred tensors.

The final condition (3.40) is easily established using (3.14) and will restrict the strains according to the condition

$$1 + 2E_\gamma^\gamma + 4\det(E_\beta^\alpha) > 0. \qquad (3.50)$$

This condition reflects the requirement that no finite portion of the undeformed middle-surface may shrink to a point during deformation.

Equation (3.14) shows that the relative increment of area due to stretching of the middle-surface $(a^*/a)^{1/2}$ is a scalar function. Its gradient $(a^*/a)^{1/2}_{,\alpha}$ will repeatedly appear in the subsequent analysis. We have

$$a_{,\alpha} = (a_{11}a_{22} - a_{12}a_{21})_{,\alpha}$$
$$= a_{11}a_{22,\alpha} + a_{22}a_{11,\alpha} - a_{12}a_{21,\alpha} - a_{21}a_{12,\alpha}$$

and, hence,

$$a_{,\alpha}/a = a^{\beta\gamma}a_{\beta\gamma,\alpha} = 2\{{}^{\beta}_{\beta\,\alpha}\}.$$

But,

$$(\log \sqrt{a})_{,\alpha} = \tfrac{1}{2}(\log a)_{,\alpha} = \tfrac{1}{2}a_{,\alpha}/a = \{{}^{\beta}_{\beta\,\alpha}\},$$

or

$$(\log \sqrt{a})_{,a} = \{{}^{\beta}_{\beta\,\alpha}\}. \tag{3.51}$$

Therefore, we have

$$[\log(a^*/a)^{1/2}]_{,a} = \hat{E}^{\beta}_{\beta\alpha}, \tag{3.52}$$

a formula that will prove useful in the following.

To sum up, the six functions $E_{\alpha\beta}$ and $K_{\alpha\beta}$ cannot be arbitrarily assigned. They must satisfy three equations of compatibility (3.48) and (3.49), otherwise the tensors $a^*_{\alpha\beta}$ and $d^*_{\alpha\beta}$ obtained by (3.8) and (3.32) would not be the fundamental tensors of a surface. In many applications only the linear terms of the equations are needed, but we shall see that in certain problems it is necessary to retain some of the nonlinear terms as well.

7. Summary

It is assumed that the state of strain everywhere in the shell is determined by the deformation of the middle-surface. This deformation is completely described by the displacement functions, a vector function v^α for the tangential displacement and a scalar function w for the normal displacement. From these three functions we can calculate a more sophisticated measure, the local deformation of the middle-surface, characterized by six numbers, three strain components $E_{\alpha\beta}$ and three bending components $K_{\alpha\beta}$. Eventually it is these measures of local deformation that we relate to the inner stresses of a thin elastic shell. We have seen that although these six numbers can be arbitrarily prescribed at any one point, they are not independent as functions of u^α, but have to satisfy three equations of compatibility.

CHAPTER 4

THE STATICS OF A SHELL

1. Introduction 75
2. An element of the shell 75
3. The divergence theorem 77
4. Statically equivalent forces and moments 79
5. External loads 82
6. Equations of equilibrium 83
7. Summary 87

CHAPTER 4

THE STATICS OF A SHELL

1. Introduction

Shell theory is based on a reduction of the equations of elasticity from three to two dimensions. In the previous chapter we have discussed one side of this problem, the kinematics of the middle-surface; in this chapter we shall deal with the other side, the statics. From the three-dimensional state of stress we shall derive statically equivalent forces and moments, acting at the middle-surface and deduce the two-dimensional equations of equilibrium, expressed in these integrated quantities.

Throughout this chapter and the next we shall assume that the shell is in equilibrium under the action of external and internal forces.

For simplicity, all asterisks are omitted in this chapter. This should not cause any confusion since we are dealing with only one coordinate system, even though it happens to be the deformed equilibrium system.

First, we shall deduce some formulas needed in the analysis.

2. An element of the shell

Let D be any connected, but not necessarily simply connected domain or part of the middle-surface and let C be the boundary of D. We shall assume that C is sufficiently smooth. The normals to the middle-surface on C and the two free surfaces of the shell define a proper element of the shell (Fig. 6).

It is convenient to define the boundary curve C by two functions $u^\alpha(s)$, where s is the arc-length of C of the middle-surface, measured from any invariantly defined point on the curve.

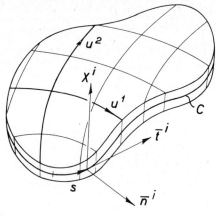

Fig. 6.

At any given point of the boundary curve C there are three mutually perpendicular unit vectors X^i, $\bar{n}^i$, and $\bar{t}^i$, where X^i is the normal to the middle-surface, $\bar{t}^i$ is a tangent vector to C, and $\bar{n}^i$ a normal vector to C, perpendicular to X^i. The positive direction of $\bar{n}^i$ is taken to be the direction out of the material of the element considered and for $\bar{t}^i$ we take the positive direction such that $\bar{n}^i, \bar{t}^i, X^i$ in that order will be right-handed.

Let the positive direction of s be that of the tangent vector, such that

$$t^\alpha = \frac{du^\alpha}{ds}. \tag{4.1}$$

The normal n^α is perpendicular to t^α and hence

$$n^\alpha = \varepsilon^{\alpha\beta} t_\beta \tag{4.2}$$

which fulfills the sign-convention just adopted. It is easily verified that the tangent and normal vectors so defined are unit vectors. To solve t_β from (4.2) multiply by $\varepsilon_{\alpha\gamma}$ and raise index γ. After changing indices we get

$$t^\alpha = \varepsilon^{\beta\alpha} n_\beta \tag{4.3}$$

The formulas (4.2) and (4.3) look alike. The reader should, however, notice the position of the dummy index in the alternating tensor.

3. The divergence theorem

Let V be a domain in three-dimensional Euclidean space and A the boundary surface of V. For any vector-field $B = (B_x, B_y, B_z)$ we have the equality

$$\iiint_V \left(\frac{\partial B_x}{\partial x} + \frac{\partial B_y}{\partial y} + \frac{\partial B_z}{\partial z}\right) dV = \iint_A (B_x n_x + B_y n_y + B_z n_z) dA$$

where (n_x, n_y, n_z) is the unit normal to A pointing in the direction out of the volume. The equality, proved in textbooks in analysis, states that the volume-integral of the divergence of a vector-field equals the flux out of the boundary. In tensor form the equality is written

$$\iiint_V D_j B^j \, dV = \iint_A B^j n_j \, dA$$

and in this form it holds good in any coordinate system in three-dimensional Euclidean space.[1]

In two dimensions the *divergence theorem* takes the form

$$\iint_D D_\alpha B^\alpha \, dA = \oint_C B^\alpha n_\alpha \, ds \qquad (4.4)$$

and holds good even if the surface is non-Euclidean. To verify this, let us take a three-dimensional domain bounded by two parallel surfaces $z = z_1$ and $z = z_2$ in normal coordinates to any given surface and by the normals to that surface along a closed curve C. Furthermore, let $B^i = (B^1, B^2, 0)$. Then

$$D_i B^i = D_\alpha B^\alpha + \{{}^{\ 3}_{3\ \beta}\} B^\beta = D_\alpha B^\alpha$$

since $\{{}^{\ 3}_{3\ \beta}\} = 0$, in normal coordinates. When the three-dimensional

---

[1] That it in fact also holds good in non-Euclidean space is of no interest for us in our analysis.

divergence theorem is applied, we get

$$\int_{z_1}^{z_2} \iint_D D_\alpha B^\alpha \, dA \, dz = \int_{z_1}^{z_2} \oint_C B^\alpha n_\alpha \, ds \, dz.$$

The contributions to the surface-integral on the right-hand side from the two parallel surfaces $z = z_1$ and $z = z_2$ did vanish because the normal to both surfaces is perpendicular to B^i. Since the equality holds independently of z_1 and z_2, the two integrands to dz must be equal and equation (4.4) follows. The reader should note that surface was not assumed to be Euclidean.

In the subsequent analysis we shall also make use of the following formula

$$\oint_C B^\alpha \phi_{,\alpha} \, ds = \oint_C B^\alpha n_\alpha \frac{\partial \phi}{\partial n} \, ds - \oint_C \phi \frac{\partial}{\partial s}(B^\alpha t_\alpha) \, ds \qquad (4.5)$$

where B^α is defined on C and ϕ is a single-valued function defined in D and on C. Let us verify this formula. The gradient $\phi_{,\alpha}$ may be resolved in the normal and tangential directions to the boundary

$$\phi_{,\alpha} = n_\alpha \frac{\partial \phi}{\partial n} + t_\alpha \frac{\partial \phi}{\partial s}.$$

Hence,

$$\oint_C B^\alpha \phi_{,\alpha} \, ds = \oint_C B^\alpha n_\alpha \frac{\partial \phi}{\partial n} \, ds + \oint_C B^\alpha t_\alpha \, d\phi$$

$$= \oint_C B^\alpha n_\alpha \frac{\partial \phi}{\partial n} \, ds + [B^\alpha t_\alpha \phi] - \oint_C \phi \frac{\partial}{\partial s}(B^\alpha t_\alpha) \, ds.$$

The term $[B^\alpha t_\alpha \phi]$ vanishes since ϕ is single-valued and the integral is taken the whole way around C. This proves equation (4.5).

We now turn to the main theme of this chapter.

4. Statically equivalent forces and moments

The surface-tractions acting on the boundary-surface between two adjacent normals to the middle-surface at u^α and $u^\alpha + du^\alpha$ of C respectively, are statically equivalent to a resultant force and a resultant moment at u^α. When $u^\alpha + du^\alpha$ approaches u^α, both force and moment become proportional to the distance between the points. It is, therefore, convenient to define them per unit length of the boundary curve C.

Let σ^{ij} be the components[2] of the three-dimensional stress tensor in normal coordinates at the point (u^1, u^2, z) and let n^i be the unit normal to the surface element $\mathbf{ds}\, dz$ (see Fig. 7). Furthermore, let A^i define a direction in space, i.e., let A^i be a unit parallel vector-field. Then $\sigma^{ij} n_i A_j\, \mathbf{ds}\, dz$ is the component of the force, acting on the element $\mathbf{ds}\, dz$, in the direction of A^i.

We obtain the component of the resultant in the direction A^i by summing up the contributions from $z = -\tfrac{1}{2}h$ to $z = +\tfrac{1}{2}h$. Letting R_A denote this component per unit length of C, we get

$$R_A\, ds = \int_{-\frac{1}{2}h}^{\frac{1}{2}h} \sigma^{ij} n_i A_j\, \mathbf{ds}\, dz, \qquad (4.6)$$

where ds is the distance $\mathbf{ds}$ at $z = 0$. We proceed to evaluate the

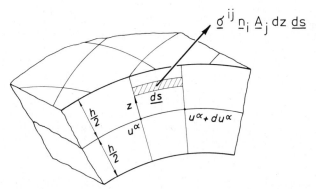

FIG. 7. (*Note*: Underlined quantities in Fig. 7 appear as boldface characters in text.)

---

[2] Boldface quantities are functions of z. Evaluated at $z = 0$, the quantities appear in normal type-font. The rule does not apply to the previously introduced quantities $g^{\alpha\beta}$, $g_{\alpha\beta}$ and g.

integral and observe that every factor of the integrand is a function of z.

Since A^i does not undergo any absolute change, we must have $D_j A^i = 0$. One of these equations states that

$$D_3 A^3 = A^3_{,3} + \{^3_{3\,k}\} A^k = 0.$$

However, since $\{^3_{3\,k}\}$ vanishes in normal coordinates, we find that A^3 must be independent of z. It is therefore convenient to consider two separate cases; the first characterized by the parallel vector-field $A^i = (A^1, A^2, 0)$ and the second by the field $A^i = (0, 0, A^3)$.

In the first case we have

$$D_3 A^\alpha = A^\alpha_{,3} + \{^\alpha_{3\,k}\} A^k = 0$$

or

$$A^\alpha_{,3} = -\tfrac{1}{2} g^{\alpha\beta} g_{\beta\gamma,3} A^\gamma. \tag{4.7}$$

We proceed to solve this system of differential equations with variable coefficients for the unknowns A^α. Multiplying (4.7) by $g_{\alpha\delta}$ and using (2.36), we get

$$(a_{\delta\varepsilon} - d_{\delta\varepsilon} z)(a^\varepsilon_\alpha - d^\varepsilon_\alpha z) A^\alpha_{,3} = d^\varepsilon_\gamma (a_{\delta\varepsilon} - d_{\delta\varepsilon} z) A^\gamma.$$

Since the determinant

$$\det(a_{\delta\varepsilon} - d_{\delta\varepsilon} z) = (ag)^{1/2} \neq 0,$$

we find that

$$(a^\varepsilon_\alpha - d^\varepsilon_\alpha z) A^\alpha_{,3} = d^\varepsilon_\gamma A^\gamma,$$

which may also be written as

$$[(a^\varepsilon_\alpha - d^\varepsilon_\alpha z) A^\alpha]_{,3} = 0.$$

The integral is given by

$$A_\beta = (a_\beta^\gamma - d_\beta^\gamma z)A_\gamma,$$

where $A_\gamma = A_\gamma(u^1, u^2, 0)$ and we see that A_β is a linear function of z. From (4.2) we have

$$n_\alpha = \varepsilon_{\alpha\delta} \frac{du^\delta}{ds} = g^{1/2} e_{\alpha\delta} \frac{du^\delta}{ds},$$

and substitution into (4.6) yields

$$R_A \, ds = \int_{-\frac{1}{2}h}^{\frac{1}{2}h} \sigma^{\alpha\beta} g^{1/2} e_{\alpha\delta} \, du^\delta (a_\beta^\gamma - d_\beta^\gamma z) A_\gamma \, dz.$$

Since

$$n_\alpha = n_\alpha(u^1, u^2, 0) = a^{1/2} e_{\alpha\delta} \frac{du^\delta}{ds}$$

and with

$$R_A = n_\alpha A_\gamma \mathcal{N}^{\alpha\gamma}$$

we get the *membrane stress tensor*

$$\mathcal{N}^{\alpha\gamma} = \int_{-\frac{1}{2}h}^{\frac{1}{2}h} \sigma^{\alpha\beta} (\delta_\beta^\gamma - d_\beta^\gamma z)(1 - 2Hz + Kz^2) \, dz, \tag{4.8}$$

where $(g/a)^{1/2}$ has been taken from (2.37). It is important to note that $\mathcal{N}^{\alpha\beta}$ is not symmetrical in general.

We now return to the second case in which the surface forces are projected on the normal direction, i.e., $A^i = (0, 0, A^3)$. We get

$$R_A \, ds = \int_{-\frac{1}{2}h}^{\frac{1}{2}h} \sigma^{\alpha 3} n_\alpha \, ds \, dz$$

or

$$R_A = \mathcal{Q}^\alpha n_\alpha,$$

where the *shear-force vector* $\mathcal{Q}^\alpha$ is given by

$$\mathcal{Q}^\alpha = \int_{-\frac{1}{2}h}^{\frac{1}{2}h} \sigma^{\alpha 3}(1 - 2Hz + Kz^2)\,dz. \tag{4.9}$$

The statically equivalent moment vector lies in the tangent plane to the middle-surface and may be decomposed[3] in a *twisting moment* M_V in the direction of the normal n^α and a *bending moment* M_B in the negative direction of the tangent t^α. By defining the *moment tensor* $\mathcal{M}^{\alpha\beta}$ by

$$\mathcal{M}^{\alpha\gamma} = -\int_{-\frac{1}{2}h}^{\frac{1}{2}h} \sigma^{\alpha\beta}(\delta^\gamma_\beta - d^\gamma_\beta z)(1 - 2Hz + Kz^2)z\,dz \tag{4.10}$$

we see that

$$M_B = \mathcal{M}^{\alpha\gamma} n_\alpha n_\gamma \tag{4.11}$$

and

$$M_V = \mathcal{M}^{\alpha\gamma} n_\alpha t_\gamma. \tag{4.12}$$

The *moment vector* M_γ is the vector sum of the twisting moment and the negative bending moment,

$$M_\gamma = M_V n_\gamma - M_B t_\gamma = \mathcal{M}^{\beta\delta} n_\beta t_\delta n_\gamma - \mathcal{M}^{\beta\delta} n_\beta n_\delta t_\gamma$$

or, since

$$t_\delta n_\gamma - n_\delta t_\gamma = \varepsilon_{\gamma\delta}$$

we have

$$M_\gamma = \varepsilon_{\gamma\beta}\mathcal{M}^{\alpha\beta} n_\alpha. \tag{4.13}$$

5. External loads

We assume that the shell is loaded by distributed forces over the free surfaces of the shell in addition to which there are mass- or inertia-forces. All the external loads are statically equivalent to a distributed

---

[3] See Fig. 8, Chapter 6.

resultant force and a resultant moment per unit area of the middle-surface, whatever proper part of the shell is considered.

The cartesian components $\bar{F}^i$ of the resultant force are

$$\bar{F}^i = f^i_{,\alpha}\hat{F}^\alpha + X^i\hat{p}, \qquad (4.14)$$

where

$$\hat{F}^\alpha = \int_{-\frac{1}{2}h}^{\frac{1}{2}h} p^\gamma(\delta^\alpha_\gamma - d^\alpha_\gamma z)(1 - 2Hz + Kz^2)\,dz$$

$$+ \sum (\hat{F}^\alpha)_{\pm\frac{1}{2}h}(1 \mp Hh + \tfrac{1}{4}Kh^2) \qquad (4.15)$$

and

$$\hat{p} = \int_{-\frac{1}{2}h}^{\frac{1}{2}h} p^3(1 - 2Hz + Kz^2)\,dz + \sum (\hat{F}^3)_{\pm\frac{1}{2}h}(1 \mp Hh + \tfrac{1}{4}Kh^2) \qquad (4.16)$$

are found following a procedure analogous to the one in previous section. Here p^i denotes the components of the volume-force in normal coordinates and Σ indicates a sum over the two free surfaces with the upper sign belonging to the outer surface ($z = \tfrac{1}{2}h$) and the lower sign to the inner surface ($z = -\tfrac{1}{2}h$).

In addition to the resultant force $\bar{F}^i$ there is the resultant moment

$$f^i_{,\alpha}\varepsilon^{\alpha\beta}m_\beta,$$

where

$$m^\beta = -\int_{-\frac{1}{2}h}^{\frac{1}{2}h} p^\gamma(\delta^\beta_\gamma - d^\beta_\gamma z)(1 - 2Hz + Kz^2)z\,dz$$

$$- \sum (\hat{F}^\beta)_{\pm\frac{1}{2}h}(1 \mp Hh + \tfrac{1}{4}Kh^2)(\pm\tfrac{1}{2}h). \qquad (4.17)$$

6. Equations of equilibrium

Consider an element D of the shell bounded by the closed curve C. One requirement for equilibrium of this element is that the resultant force vanishes,

$$\oint_C (f^i_{,\beta}\mathcal{N}^{\alpha\beta}n_\alpha + X^i\mathcal{Q}^\alpha n_\alpha)\,\mathrm{d}s + \iint_D (f^i_{,\alpha}\hat{F}^\alpha + X^i\hat{p})\,\mathrm{d}A = 0.$$

The first integral can be transformed into a surface integral by using the divergence theorem

$$\iint_D (D_\alpha f^i_{,\beta}[\mathcal{N}^{\alpha\beta} + X^i\mathcal{Q}^\alpha] + f^i_{,\alpha}\hat{F}^\alpha + X^i\hat{p})\,\mathrm{d}A = 0.$$

If this is to hold for any domain D, the integral must vanish,

$$D_\alpha(f^i_{,\beta}\mathcal{N}^{\alpha\beta} + X^i\mathcal{Q}^\alpha) + f^i_{,\alpha}\hat{F}^\alpha + X^i\hat{p} = 0 \tag{4.18}$$

or

$$\mathcal{N}^{\alpha\beta}D_\alpha f^i_{,\beta} + f^i_{,\beta}D_\alpha\mathcal{N}^{\alpha\beta} + X^i D_\alpha\mathcal{Q}^\alpha + \mathcal{Q}^\alpha X^i_{,\alpha} + f^i_{,\alpha}\hat{F}^\alpha + X^i\hat{p} = 0.$$

Multiplication by $f^i_{,\gamma}$ and raising index, yields

$$D_\alpha\mathcal{N}^{\alpha\beta} + d^\beta_\alpha\mathcal{Q}^\alpha + \hat{F}^\beta = 0 \tag{4.19}$$

(where (2.27) and (2.28) have been utilized). Multiplication by X^i instead yields

$$D_\alpha\mathcal{Q}^\alpha + d_{\alpha\beta}\mathcal{N}^{\alpha\beta} + \hat{p} = 0. \tag{4.20}$$

These are two equations of equilibrium in the tangent plane (4.19) and one in the direction of the normal (4.20).

Another and final requirement for the equilibrium of the element D under consideration is that the resultant moment vanishes.

Let us calculate the resultant moment of all forces at the origin of our cartesian coordinate system x^i. The moment of the external forces, acting on an element $\mathrm{d}A$ of the middle-surface is

$$e_{ijk}f^j_{,\alpha}(f^k_{,\alpha}\hat{F}^\alpha + X^k\hat{p})\,\mathrm{d}A + f^i_{,\alpha}\varepsilon^{\alpha\beta}m_\beta\,\mathrm{d}A.$$

From the membrane stress tensor $\mathcal{N}^{\alpha\beta}$ and the shear-force vector $\mathcal{Q}^\alpha$

on a segment ds of C the contribution is

$$e_{ijk}f^j(f^k_{,\beta}\mathcal{N}^{\alpha\beta}n_\alpha + X^k\mathcal{Q}^\alpha n_\alpha)\,\mathrm{d}s$$

and finally, from the moment vector M^γ, we have

$$f^i_{,\gamma}M^\gamma\,\mathrm{d}s = f^i_{,\gamma}a^{\gamma\delta}\varepsilon_{\delta\beta}\mathcal{M}^{\alpha\beta}n_\alpha\,\mathrm{d}s.$$

Adding up and using the identity (3.43),

$$f^i_{,\gamma}a^{\gamma\delta}\varepsilon_{\delta\beta} = e_{ijk}f^j_{,\beta}X^k,$$

we get

$$\oint_C e_{ijk}[f^jf^k_{,\beta}\mathcal{N}^{\alpha\beta} + f^jX^k\mathcal{Q}^\alpha + f^j_{,\beta}X^k\mathcal{M}^{\alpha\beta}]n_\alpha\,\mathrm{d}s$$

$$+ \iint_D e_{ijk}[f^jf^k_{,\alpha}\hat{F}^\alpha + f^jX^k\hat{p} + f^j_{,\beta}X^km^\beta]\,\mathrm{d}A = 0. \qquad (4.21)$$

Again we apply the divergence theorem to the first integral and transform it to a surface integral over D. Since D is arbitrary the integral must vanish.

$$e_{ijk}[D_\alpha(f^jf^k_{,\beta}\mathcal{N}^{\alpha\beta} + f^jX^k\mathcal{Q}^\alpha + f^j_{,\beta}X^k\mathcal{M}^{\alpha\beta})$$
$$+ f^jf^k_{,\alpha}\hat{F}^\alpha + f^jX^k\hat{p} + f^j_{,\beta}X^km^\beta] = 0.$$

But,

$$D_\alpha(f^jf^k_{,\beta}\mathcal{N}^{\alpha\beta} + f^jX^k\mathcal{Q}^\alpha) + f^jf^k_{,\alpha}\hat{F}^\alpha + f^jX^k\hat{p}$$
$$= f^j[D_\alpha(f^k_{,\beta}\mathcal{N}^{\alpha\beta} + X^k\mathcal{Q}^\alpha) + f^k_{,\alpha}\hat{F}^\alpha + X^k\hat{p}] + f^j_{,\alpha}f^k_{,\beta}\mathcal{N}^{\alpha\beta} + f^j_{,\alpha}X^k\mathcal{Q}^\alpha$$

and here the expression between brackets vanishes due to (4.18). Thus we have

$$e_{ijk}(f^j_{,\alpha}f^k_{,\beta}\mathcal{N}^{\alpha\beta} + f^j_{,\alpha}X^k\mathcal{Q}^\alpha + f^j_{,\beta}X^kD_\alpha\mathcal{M}^{\alpha\beta}$$
$$- d^\gamma_\alpha f^j_{,\beta}f^k_{,\gamma}\mathcal{M}^{\alpha\beta} + \mathcal{M}^{\alpha\beta}X^kd_{\alpha\beta}X^j + f^j_{,\alpha}X^km^\alpha) = 0.$$

Multiplication by X^i yields

$$\varepsilon_{\alpha\beta}N^{\alpha\beta} - d^\gamma_\alpha \varepsilon_{\beta\gamma} M^{\alpha\beta} = 0$$

or

$$\varepsilon_{\alpha\beta}(N^{\alpha\beta} - d^\beta_\gamma M^{\gamma\alpha}) = 0, \qquad (4.22)$$

which is the equation of moment equilibrium around the normal to the middle-surface. Multiplication by $f^i_{,\gamma}$ instead gives the projection of the resultant moment on the tangent plane,

$$\varepsilon_{\gamma\alpha} \mathcal{Q}^\alpha + \varepsilon_{\gamma\beta} D_\alpha M^{\alpha\beta} + \varepsilon_{\gamma\alpha} m^\alpha = 0$$

or

$$D_\alpha M^{\alpha\beta} + \mathcal{Q}^\beta + m^\beta = 0. \qquad (4.23)$$

The shear-force vector $\mathcal{Q}^\alpha$ can be eliminated from (4.19) and (4.20) using (4.23). We get

$$D_\alpha N^{\alpha\beta} + d^\beta_\alpha D_\gamma M^{\gamma\alpha} + F^\beta = 0, \qquad (4.24)$$

$$D_\alpha D_\beta M^{\beta\alpha} - d_{\alpha\beta} N^{\alpha\beta} - p = 0, \qquad (4.25)$$

where

$$F^\beta = \hat{F}^\beta + d^\beta_\alpha m^\alpha \qquad (4.26)$$

and

$$p = \hat{p} - D_\alpha m^\alpha. \qquad (4.27)$$

It is useful to note that the distributed moment m^α which had to be introduced since the external loads can be—and usually are—applied off the middle-surface is 'absorbed' in the *effective loads* F^β and p applied to the middle-surface.

7. Summary

From the internal stresses we have derived a set of statically equivalent forces and moments acting at the middle-surface and from considerations regarding the equilibrium of the shell we have established certain (differential) relations between these quantities.

We have noted that, in general, neither the membrane stress tensor $\mathcal{N}^{\alpha\beta}$ nor the moment tensor $\mathcal{M}^{\alpha\beta}$ is symmetrical.

CHAPTER 5

EFFECTIVE FORCES AND MOMENTS

1. Introduction 91
2. The principle of virtual work 91
3. Equations of equilibrium 99
4. Summary 99

CHAPTER 5

EFFECTIVE FORCES AND MOMENTS

1. Introduction

Due to lack of symmetry, the number of independent components of $\mathcal{N}^{\alpha\beta}$ and $\mathcal{M}^{\alpha\beta}$ is eight in all and it is therefore not possible to relate these components through a one-to-one relation with the components of the deformation tensors $E_{\alpha\beta}$ and $K_{\alpha\beta}$. This characteristic difficulty of shell theory is solved, or perhaps rather circumvented by the introduction of so-called 'effective' quantities. This is done with the help of the *principle of virtual work*. In this chapter, like in the previous one, we shall deal with the three-dimensional as well as the two-dimensional state. In the following we shall be forced to take certain precautions with regards to our notations, otherwise some confusion may arise.

2. The principle of virtual work

Let $\sigma^{ij}(u^1, u^2, z)$ represent the three-dimensional state of stress in equilibrium with the external volume forces $p^i(u^1, u^2, z)$. Let furthermore $\delta v_i(u^1, u^2, z)$ be a set of arbitrary virtual displacements. Then

$$[\mathcal{D}_j \sigma^{ij}(u^1, u^2, z) + p^i(u^1, u^2, z)] \delta v_i(u^1, u^2, z) \equiv 0, \qquad (5.1)$$

since the expression between brackets vanishes identically due to the equations of equilibrium in the three-dimensional state. Here, to avoid confusion with the notation D_α for the covariant derivative on the middle-surface, we use $\mathcal{D}_i$ in three-dimensional space, described by the system of normal coordinates.

Integrating (5.1) over the region in space occupied by a proper element of the shell, and applying the divergence theorem, we get

$$\iiint \sigma^{ij}\mathcal{D}_j\, \delta v_i g^{1/2}\, du^1 du^2 dz$$

$$= \iiint p^i\, \delta v_i g^{1/2}\, du^1 du^2 dz + \iint \sigma^{ij} n_i\, \delta v_j\, dz\, ds,$$

where the last term is a surface integral over the boundary surface.

The last integral can be transformed similarly as the right-hand side of (4.6),

$$\iint \sigma^{ij} n_i\, \delta v_j\, dz\, ds = \oint_C \left[\int_{-\frac{1}{2}h}^{\frac{1}{2}h} \sigma^{\alpha j}\, \delta v_j (1 - 2Hz + Kz^2)\, dz \right] n_\alpha\, ds.$$

Using (2.37) we therefore get the following equation for the principle of virtual work,

$$\iint_D \int_{-\frac{1}{2}h}^{\frac{1}{2}h} \sigma^{ij}\mathcal{D}_j\, \delta v_i (1 - 2Hz + Kz^2)\, dz\, dA$$

$$= \iint_D \int_{-\frac{1}{2}h}^{\frac{1}{2}h} p^i\, \delta v_i (1 - 2Hz + Kz^2)\, dz\, dA$$

$$+ \oint_C \left[\int_{-\frac{1}{2}h}^{\frac{1}{2}h} \sigma^{\alpha j}\, \delta v_j (1 - 2Hz + Kz^2)\, dz \right] n_\alpha\, ds. \qquad (5.2)$$

Let us now apply this formula to the following set of virtual displacements,

$$\delta v_\alpha(u^1, u^2, z) = (a_\alpha^\gamma - z d_\alpha^\gamma)(\delta v_\gamma - z\, \delta q_\gamma)$$

$$\delta v_3(u^1, u^2, z) = \delta w, \qquad (5.3)$$

where the rotation δq_γ is given by (compare with equation (3.5))

$$\delta q_\gamma = d_{\beta\gamma}\, \delta v^\beta + \delta w_{,\gamma}. \tag{5.4}$$

An examination of (5.3) and (5.4) will reveal that the virtual displacements $\delta v_i(u^1, u^2, z)$ represent the displacements according to the Kirchhoff assumptions throughout the shell when the displacments of the middle-surface are δv_α and δw. Thus, points on a normal to the undeformed middle-surface will remain on one and the same normal to the deformed middle-surface and also retain their original distance from it. But it should be stressed that by introducing these virtual displacements we make no assumption whatsoever regarding the true displacements. Our restricted choice of virtual displacements is only a restriction of our considerations!

The substitution of (5.3) into (5.2) is a rather complicated process, the details of which we shall give below.

According to the rule for covariant differentiation, we have

$$\mathscr{D}_\alpha\, \delta v_\beta = \delta v_{\beta,\alpha} - \{_{\alpha\ \beta}^{\ \gamma}\}\, \delta v_\gamma - \{_{\alpha\ \beta}^{\ 3}\}\, \delta v_3$$

or

$$\mathscr{D}_\alpha\, \delta v_\beta = (a_\beta^\gamma - zd_\beta^\gamma)(\delta v_{\gamma,\alpha} - z\, \delta q_{\gamma,\alpha})$$
$$+ (a_{\beta,\alpha}^\gamma - zd_{\beta,\alpha}^\gamma)(\delta v_\gamma - z\, \delta q_\gamma) - \{_{\alpha\ \beta}^{\ \gamma}\}\, \delta v_\gamma + \tfrac{1}{2} g_{\alpha\beta,3}\, \delta w.$$

But the fundamental tensor $g_{\alpha\beta}$ is a known function of z according to (2.36). Using this relation we get

$$\mathscr{D}_\alpha\, \delta v_\beta = (a_\beta^\gamma - zd_\beta^\gamma)(\delta v_{\gamma,\alpha} - d_{\gamma\alpha}\, \delta w - z\, \delta q_{\gamma,\alpha})$$
$$- zd_{\beta,\alpha}^\gamma(\delta v_\gamma - z\, \delta q_\gamma) - \{_{\alpha\ \beta}^{\ \gamma}\}\, \delta v_\gamma. \tag{5.5}$$

Using (1.26) we get

$$\{_{\alpha\ \beta}^{\ \gamma}\} = g^{\gamma k} f^i_{,k} f^i_{,\alpha\beta} = g^{\gamma\sigma} f^i_{,\sigma} f^i_{,\alpha\beta}$$

and, according to (2.8),

$$\{_{\alpha\ \beta}^{\ \gamma}\} = g^{\gamma\sigma}(f^i_{,\sigma} + zX^i_{,\sigma})(f^i_{,\alpha\beta} + zX^i_{,\alpha\beta}).$$

Substitution of (2.27) yields

$$\{{}^{\gamma}_{\alpha\beta}\} = g^{\gamma\sigma}(f^i_{,\sigma} - zd^\rho_\sigma f^i_{,\rho})(f^i_{,\alpha\beta} - zd^\rho_{\alpha,\beta} f^i_{,\rho} - zd^\rho_\alpha f^i_{,\rho\beta}).$$

Writing the second factor as $f^i_{,\varepsilon}(a^\varepsilon_\sigma - zd^\varepsilon_\sigma)$ and multiplying the last factor through by $f^i_{,\varepsilon}$ and rearranging the dummy index ε, we get

$$\{{}^{\gamma}_{\alpha\beta}\} = g^{\gamma\sigma}(a_{\varepsilon\sigma} - zd_{\varepsilon\sigma})(\{{}^{\varepsilon}_{\alpha\beta}\} - zd^\varepsilon_{\alpha,\beta} - zd^\rho_\alpha\{{}^{\varepsilon}_{\rho\beta}\}).$$

Hence,

$$\{{}^{\gamma}_{\alpha\beta}\}\delta v_\gamma = g^{\gamma\sigma}(a_{\varepsilon\sigma} - zd_{\varepsilon\sigma})(a^\nu_\gamma - zd^\nu_\gamma)(\delta v_\nu - z\,\delta q_\nu)$$
$$\cdot(\{{}^{\varepsilon}_{\alpha\beta}\} - zd^\varepsilon_{\alpha,\beta} - zd^\rho_\alpha\{{}^{\varepsilon}_{\rho\beta}\}).$$

With the help of (2.36) the following identity is easily established,

$$[g^{\gamma\sigma}(a_{\varepsilon\sigma} - zd_{\varepsilon\sigma})(a^\nu_\gamma - zd^\nu_\gamma) - \delta^\nu_\varepsilon](a_{\nu\beta} - zd_{\nu\beta}) = 0.$$

But this is a linear homogeneous system of equations for any fixed value of ε and for $\beta = 1, 2$. Since

$$\det(a_{\nu\beta} - zd_{\nu\beta}) \neq 0,$$

it follows that

$$g^{\gamma\sigma}(a_{\varepsilon\sigma} - zd_{\varepsilon\sigma})(a^\nu_\gamma - zd^\nu_\gamma) = \delta^\nu_\varepsilon$$

and hence

$$\{{}^{\gamma}_{\alpha\beta}\}\delta v_\gamma = (\delta v_\varepsilon - z\,\delta q_\varepsilon)(\{{}^{\varepsilon}_{\alpha\beta}\} - zd^\varepsilon_{\alpha,\beta} - zd^\rho_\alpha\{{}^{\varepsilon}_{\rho\beta}\}).$$

Substitution into (5.5) yields

$$\mathscr{D}_\alpha\delta v_\beta = (a^\gamma_\beta - zd^\gamma_\beta)[(\delta v_\gamma - z\,\delta q_\gamma)_{,\alpha} - d_{\gamma\alpha}\delta w]$$
$$- (\delta v_\varepsilon - z\,\delta q_\varepsilon)(a^\gamma_\beta - zd^\gamma_\beta)\{{}^{\varepsilon}_{\alpha\beta}\}$$

and finally

$$\mathcal{D}_\alpha \, \delta v_\beta = (a^\gamma_\beta - z d^\gamma_\beta)(D_\alpha \, \delta v_\gamma - z D_\alpha \, \delta q_\gamma - d_{\gamma\alpha} \, \delta w). \tag{5.6}$$

Furthermore, we find in a similar way, but with considerably less effort,

$$\mathcal{D}_3 \, \delta v_\alpha + \mathcal{D}_\alpha \, \delta v_3 = 0, \qquad \mathcal{D}_3 \, \delta v_3 = 0. \tag{5.7}$$

Let δV denote the work done by the internal stresses $\sigma^{ij}(u^1, u^2, z)$ on the virtual displacements $\delta v_i(u^1, u^2, z)$. This is the left-hand side of equation (5.2) and by (5.6) and (5.7) we get

$$\delta V = \iint_D [\mathcal{N}^{\alpha\gamma}(D_\alpha \, \delta v_\gamma - d_{\gamma\alpha} \, \delta w) + \mathcal{M}^{\alpha\gamma} D_\alpha \, \delta q_\gamma] \, dA, \tag{5.8}$$

where (4.8) and (4.10) have been used.

According to (5.4) we have

$$D_\alpha \, \delta q_\gamma = \delta v^\beta D_\alpha d_{\beta\gamma} + d_{\beta\gamma} D_\alpha \, \delta v^\beta + D_\alpha D_\gamma \, \delta w$$

and (3.37) yields

$$\delta K_{\alpha\gamma} = D_\alpha D_\gamma \, \delta w + d_{\alpha\beta} D_\gamma \, \delta v^\beta + d_{\gamma\beta} D_\alpha \, \delta v^\beta$$
$$+ \delta v^\beta D_\gamma d_{\beta\alpha} - d_{\beta\gamma} d^\beta_\alpha \, \delta w.$$

Eliminating $D_\alpha D_\gamma \, \delta w$ we get

$$D_\alpha \, \delta q_\gamma = \delta K_{\alpha\gamma} - d^\beta_\alpha (D_\gamma \, \delta v_\beta - d_{\beta\gamma} \, \delta w)$$

and hence

$$\delta V = \iint_D [(\mathcal{N}^{\alpha\gamma} - d^\gamma_\beta \mathcal{M}^{\beta\alpha})(D_\alpha \, \delta v_\gamma - d_{\alpha\gamma} \, \delta w) + \mathcal{M}^{\alpha\gamma} \, \delta K_{\alpha\gamma}] \, dA. \tag{5.9}$$

The equation of equilibrium (4.22) shows that the tensor

$$\mathcal{N}^{\alpha\gamma} - d^\gamma_\beta \mathcal{M}^{\beta\alpha}$$

is symmetrical and therefore only the symmetrical part

$$\tfrac{1}{2}(D_\alpha \, \delta v_\gamma + D_\gamma \, \delta v_\alpha) - d_{\alpha\gamma} \, \delta w$$

of the second factor contributes to the product. But according to (3.11) this is precisely $\delta E_{\alpha\gamma}$. Likewise, since $\delta K_{\alpha\gamma}$ is symmetrical, only the symmetrical part of $\mathscr{M}^{\alpha\gamma}$ namely

$$\tfrac{1}{2}(\mathscr{M}^{\alpha\gamma} + \mathscr{M}^{\gamma\alpha})$$

contributes to the second product. We shall therefore write (5.9) in the form

$$\delta V = \iint_D [(\mathscr{N}^{\alpha\gamma} - d^\gamma_\beta \mathscr{M}^{\beta\alpha}) \, \delta E_{\alpha\gamma} + \tfrac{1}{2}(\mathscr{M}^{\alpha\gamma} + \mathscr{M}^{\gamma\alpha}) \, \delta K_{\alpha\gamma}] \, dA$$

or

$$\delta V = \iint_D [N^{\alpha\gamma} \, \delta E_{\alpha\gamma} + M^{\alpha\gamma} \, \delta K_{\alpha\gamma}] \, dA , \qquad (5.10)$$

where

$$N^{\alpha\beta} = \mathscr{N}^{\alpha\beta} - d^\beta_\gamma \mathscr{M}^{\gamma\alpha} \qquad (5.11)$$

and

$$M^{\alpha\beta} = \tfrac{1}{2}(\mathscr{M}^{\alpha\beta} + \mathscr{M}^{\beta\alpha}) \qquad (5.12)$$

are symmetrical tensors. They will be called the *effective membrane stress tensor* and the *effective moment tensor* respectively.

To finish our reduction of the principle of virtual work we must substitute the virtual displacements (5.3) into the right-hand side of (5.2). With the help of (4.15)–(4.17) and (4.8)–(4.10) we get

$$\delta V = \iint_D (\hat{F}^\alpha \, \delta v_\alpha + m^\alpha \, \delta q_\alpha + \hat{p} \, \delta w) \, dA$$

$$+ \oint_C (\mathscr{N}^{\alpha\beta} \, \delta v_\beta + \mathscr{M}^{\alpha\beta} \, \delta q_\beta + \mathscr{Q}^\alpha \, \delta w) n_\alpha \, ds .$$

But

$$\iint_D m^\alpha \, \delta q_\alpha \, dA = \iint_D d^\beta_\alpha m^\alpha \, \delta v_\beta + \iint_D m^\alpha D_\alpha \, \delta w \, dA$$

and

$$\iint_D m^\alpha D_\alpha \, \delta w \, dA = \oint_C m^\alpha n_\alpha \, \delta w \, ds - \iint_D D_\alpha m^\alpha \, \delta w \, dA .$$

Hence, using (4.26) and (4.27) we get

$$\delta V = \iint_D (F^\alpha \, \delta v_\alpha + p \, \delta w) \, dA$$

$$+ \oint_C [\mathcal{N}^{\alpha\beta} \, \delta v_\beta + \mathcal{M}^{\alpha\beta} \, \delta q_\beta + (\mathcal{Q}^\alpha + m^\alpha) \, \delta w] n_\alpha \, ds . \quad (5.13)$$

Let T^α denote the *membrane force vector*, acting on the boundary and taken per unit length of the boundary curve C. Then

$$T^\alpha = \mathcal{N}^{\beta\alpha} n_\beta . \quad (5.14)$$

Also, from (3.22) we get

$$\delta q_\beta = \varepsilon_{\gamma\beta} \, \delta r^\gamma , \quad (5.15)$$

where δr^α is the rotation of the normal due to the virtual displacements. Then

$$\mathcal{M}^{\alpha\beta} n_\alpha \, \delta q_\beta = \mathcal{M}^{\alpha\beta} n_\alpha \varepsilon_{\gamma\beta} \, \delta r^\gamma = M_\gamma \, \delta r^\gamma ,$$

where

$$M_\gamma = \varepsilon_{\gamma\beta} \mathcal{M}^{\alpha\beta} n_\alpha \quad (5.16)$$

denotes the moment vector, defined in (4.13).

Finally, let $\hat{T}$ denote the *shear-force* per unit length of C, acting in the direction of the normal to the middle-surface,

$$\hat{T} = \mathcal{Q}^\alpha n_\alpha. \tag{5.17}$$

Then (5.13) reduces to

$$\delta V = \iint_D (F^\alpha \,\delta v_\alpha + p\,\delta w)\,\mathrm{d}A$$

$$+ \oint_C [T^\alpha \,\delta v_\alpha + M_\alpha \,\delta r^\alpha + (\hat{T} + m^\alpha n_\alpha)\,\delta w]\,\mathrm{d}s$$

and the principle of virtual work will appear in the following shape,

$$\iint_D [N^{\alpha\beta}\,\delta E_{\alpha\beta} + M^{\alpha\beta}\,\delta K_{\alpha\beta}]\,\mathrm{d}A$$

$$= \iint_D [F^\alpha \,\delta v_\alpha + p\,\delta w]\,\mathrm{d}A + \oint_C [T^\alpha \,\delta v_\alpha + M^\alpha \,\delta r_\alpha + T\,\delta w]\,\mathrm{d}s, \tag{5.18}$$

where T stands for the *supplemented shear-force*,

$$T = \hat{T} + m^\alpha n_\alpha. \tag{5.19}$$

Equation (5.18) holds exactly for any arbitrary set of virtual displacements (5.3) with $\delta v_\alpha, \delta w$ on the middle-surface. By regarding the virtual displacements arbitrary we shall derive from this equation the field equations expressed in the effective membrane stress tensor and effective moment tensor and a *proper set of boundary conditions*. At this point we recall that although we have omitted an indication for this, the analysis leading to equation (5.18) has been performed in the deformed coordinate system. In the next chapter we shall take care to reinstate proper notations for the deformed state.

3. Equations of equilibrium

The equations of equilibrium (4.24) and (4.25) can also be written in terms of the effective membrane stresses and moments. Since $M^{\alpha\beta}$ is the symmetrical part of $\mathcal{M}^{\alpha\beta}$, we have

$$\mathcal{M}^{\alpha\beta} = M^{\alpha\beta} + \varepsilon^{\alpha\beta}\Phi, \qquad (5.20)$$

where the skew-symmetrical part of $\mathcal{M}^{\alpha\beta}$ is written in terms of the alternating tensor and a scalar function Φ. Solving (5.11) for $\mathcal{N}^{\alpha\beta}$, we get

$$\mathcal{N}^{\alpha\beta} = N^{\alpha\beta} + d_\gamma^\beta(M^{\gamma\alpha} + \varepsilon^{\gamma\alpha}\Phi) \qquad (5.21)$$

and substitution into (4.24) and (4.25) yields, using (2.33),

$$D_\alpha N^{\alpha\beta} + 2d_\gamma^\beta D_\alpha M^{\gamma\alpha} + M^{\gamma\alpha}D_\alpha d_\gamma^\beta + F^\beta = 0, \qquad (5.22)$$

$$D_\alpha D_\beta M^{\alpha\beta} - d_{\alpha\gamma}d_\beta^\gamma M^{\alpha\beta} - d_{\alpha\beta}N^{\alpha\beta} - p = 0. \qquad (5.23)$$

We note that the function Φ does not appear in the equations of equilibrium.

4. Summary

We have applied the principle of virtual work to an arbitrary proper part of the shell. The virtual displacements of the middle-surface are left arbitrary but corresponding displacements off the middle-surface have been selected such that they satisfy the Kirchhoff assumptions. By integrating the work over the thickness of the shell we transformed the principle of virtual work into a form adapted for two dimensions. Again we stress that the restriction on the virtual displacements off the middle-surface does not imply any restriction on the true displacements—it is merely a restriction on our range of considerations.

In performing the *dimensional reduction* we found that the internal virtual work was produced by the virtual strain and bending on two symmetrical tensors, the effective membrane stress tensor $N^{\alpha\beta}$ and the effective moment tensor $M^{\alpha\beta}$. These tensors are easily determined from the tensors $\mathcal{N}^{\alpha\beta}$ and $\mathcal{M}^{\alpha\beta}$, but conversely $\mathcal{N}^{\alpha\beta}$ and $\mathcal{M}^{\alpha\beta}$ can only be determined from $N^{\alpha\beta}$ and $M^{\alpha\beta}$ up to an unknown scalar function Φ. The equations of equilibrium are readily found in terms of $N^{\alpha\beta}$ and $M^{\alpha\beta}$ and do not depend on the function Φ.

CHAPTER 6

BOUNDARY CONDITIONS

1. Introduction 103
2. External forces 103
3. Boundary conditions 105
4. Stress-distribution 109
5. Summary 110

CHAPTER 6

BOUNDARY CONDITIONS

1. Introduction

In this chapter we proceed to derive the boundary conditions appropriate to the equations of equilibrium of the previous chapter. Since we would like to make sure that the equations can be applied also to highly deformed shells, we shall reinstate the convention that all quantities referring to the deformed coordinate system are denoted by an asterisk.

2. External forces

The shell is assumed to be loaded with loads distributed over the middle-surface in the domain D and along the boundary curve C.

In some cases the loads are given per unit area of the deformed middle-surface like in the case of hydrostatic pressure but in other cases the loads are given per unit mass or, what is essentially the same, per unit area of the undeformed middle-surface. In our analysis we shall adhere to the convention that all *external loads are expressed per unit area of the deformed middle-surface or per unit length of the deformed boundary* as the case may be.

Let us assume that a load with the cartesian components $\bar{F}^i \, dA^*$ acts on an element of the middle-surface of area dA^* in the (deformed) equilibrium state. Furthermore, let us assume that the component p^* in the direction of the normal to the deformed middle-surface and $F^{*\alpha}$ perpendicular to it are given. Then,

$$\bar{F}^i = f^{*i}_{,\alpha} F^{*\alpha} + X^{*i} p^* \,. \tag{6.1}$$

Similarly, let us assume that a load with the cartesian components $\bar{T}^i \, ds^*$ acts on an element of length ds^* of the deformed boundary in the equilibrium state. Let T^* be the component in the direction of the normal to the deformed middle-surface and $T^{*\alpha}$ in the tangent plane. Then

$$\bar{T}^i = f^{*\,i}_{,\,\alpha} T^{*\alpha} + X^{*i} T^* . \tag{6.2}$$

In addition, the shell can support a distribution of moments along the boundary.

Let the components of the moment vector in the tangent plane be the twisting moment M^*_V perpendicular to the boundary and the bending moment M^*_B in the (negative) direction of the tangent to C (see Fig. 8).

The moment vector is then

$$M^{*\alpha} = M^*_V n^{*\alpha} - M^*_B t^{*\alpha} , \tag{6.3}$$

from which it follows that

$$M^*_V = M^{*\alpha} n^*_\alpha \quad \text{and} \quad M^*_B = -M^{*\alpha} t^*_\alpha . \tag{6.4}$$

The cartesian components of the boundary moment per unit length of the (deformed) boundary are then

$$\bar{M}^i = f^{*\,i}_{,\,\alpha} M^{*\alpha} . \tag{6.5}$$

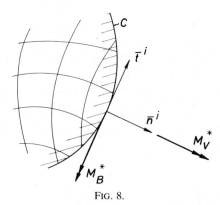

Fig. 8.

It is of course possible to apply a moment $X^{*i}M^*$ in the direction of the normal to the middle-surface along the boundary. But it can be verified that this has essentially the same effect as an additional force $\partial(M^*n^{*\alpha})/\partial s^*$ to be added to $T^{*\alpha}$ and we therefore omit it.

We are now prepared to calculate the work done by the external forces during virtual displacements.

3. Boundary conditions

Let us return to equation (5.18) and calculate both sides. On the left-hand side the increments $\delta E^*_{\alpha\beta}$ and $\delta K^*_{\alpha\beta}$ are the linear terms of the corresponding tensors in the deformed state,

$$\delta E^*_{\alpha\beta} = \tfrac{1}{2}(D^*_\alpha \delta v^*_\beta + D^*_\beta \delta v^*_\alpha) - d^*_{\alpha\beta} \delta w^*, \tag{6.6}$$

$$\delta K^*_{\alpha\beta} = D^*_\alpha D^*_\beta \delta w^* + d^*_{\alpha\gamma} D^*_\beta \delta v^{*\gamma} + d^*_{\beta\gamma} D^*_\alpha \delta v^{*\gamma}$$
$$+ \delta v^{*\gamma} D^*_\beta d^*_{\gamma\alpha} - d^*_{\beta\gamma} d^{*\gamma}_\alpha \delta w^*. \tag{6.7}$$

On the right-hand side of (5.18) we write

$$\delta V = \iint_D [F^{*\alpha} \delta v^*_\alpha + p^* \delta w^*] \, dA^*$$
$$+ \oint_C [T^{*\alpha} \delta v^*_\alpha + T^* \delta w^* + M^{*\alpha} \delta r^*_\alpha] \, ds^*, \tag{6.8}$$

where δr^*_α is the rotation vector due to the virtual displacements. From (3.28) we get

$$\delta r^{*\alpha} = \varepsilon^{*\alpha\beta}(d^*_{\beta\gamma} \delta v^{*\gamma} + \delta w^*_{,\beta}). \tag{6.9}$$

Let us first consider virtual displacements δw^* in the direction of the normal only. Then $\delta v^*_\alpha = 0$ and (5.18) becomes

$$\iint_D [M^{*\alpha\beta} D^*_\alpha D^*_\beta \delta w^* - (d^*_{\alpha\gamma} d^{*\gamma}_\beta M^{*\alpha\beta} + d^*_{\alpha\beta} N^{*\alpha\beta}) \delta w^*] \, dA^*$$
$$- \iint_D p^* \delta w^* \, dA^* - \oint_C (T^* \delta w^* + M^*_\alpha \varepsilon^{*\alpha\beta} \delta w^*_{,\beta}) \, ds^* = 0. \tag{6.10}$$

We recall that $N^{*\alpha\beta}$ and $M^{*\alpha\beta}$ are defined per unit length of the deformed section. They are sometimes called the *Cauchy stress tensors*. The corresponding quantities defined per unit length of the undeformed section are the *Piola–Kirchhoff stress tensors*,

$$\tilde{N}^{\alpha\beta} = (a^*/a)^{1/2} N^{*\alpha\beta}, \qquad (6.11)$$

$$\tilde{M}^{\alpha\beta} = (a^*/a)^{1/2} M^{*\alpha\beta}. \qquad (6.12)$$

The first term of (6.10) can be written as

$$\iint_D M^{*\alpha\beta} D^*_\alpha D^*_\beta \delta w^* \, dA^* = \iint_D D^*_\alpha (M^{*\alpha\beta} D^*_\beta \delta w^*) \, dA^*$$

$$- \iint_D D^*_\beta (\delta w^* D^*_\alpha M^{*\alpha\beta}) \, dA^* + \iint_D \delta w^* D^*_\alpha D^*_\beta M^{*\alpha\beta} \, dA^*$$

$$= \oint_C [M^{*\alpha\beta} n^*_\alpha \delta w^*_{,\beta} - (D^*_\alpha M^{*\alpha\beta}) n^*_\beta \delta w^*] \, ds^*$$

$$+ \iint_D \delta w^* D^*_\alpha D^*_\beta M^{*\alpha\beta} \, dA^*,$$

where the first two surface-integrals on the right-hand side have been transformed to boundary-integrals by applying the divergence theorem. Substitution into (6.10) yields

$$\iint_D [D^*_\alpha D^*_\beta M^{*\alpha\beta} - d^*_{\alpha\gamma} d^{*\gamma}_\beta M^{*\alpha\beta} - d^*_{\alpha\beta} N^{*\alpha\beta} - p^*] \delta w^* \, dA^*$$

$$+ \oint_C (M^{*\alpha\beta} n^*_\alpha - M^{*}_\alpha \varepsilon^{*\alpha\beta}) \delta w^*_{,\beta} \, ds^*$$

$$- \oint_C [T^* + (D^*_\alpha M^{*\alpha\beta}) n^*_\beta] \delta w^* \, ds^* = 0. \qquad (6.13)$$

This must vanish for all δw^*. First, let us take δw^* such that δw^* and its normal derivative $\partial(\delta w^*)/\partial n^*$ vanish everywhere on C. Then the two line-integrals of (6.13) vanish and hence the surface-integral must be zero for all δw^* that satisfy the boundary conditions $\delta w^* = \partial(\delta w^*)/\partial n^* = 0$ on C. However, this can only be true if the integrand of the surface-integral vanishes everywhere in the open domain D. Thus

$$D_\alpha^* D_\beta^* M^{*\alpha\beta} - d_{\alpha\gamma}^* d_\beta^{*\gamma} M^{*\alpha\beta} - d_{\alpha\beta}^* N^{*\alpha\beta} - p^* = 0 \quad \text{in } D, \quad (6.14)$$

which is the equation of equilibrium in the direction of δw^*, i.e., in the direction of the normal to the deformed middle-surface. It is actually the same equation as (5.23).

When (6.14) is substituted back into (6.13), the surface-integral vanishes and hence the sum of the two boundary-integrals must be zero. Thus

$$\oint_C (M^{*\alpha\beta} n_\alpha^* - M_\alpha^* \varepsilon^{*\alpha\beta}) n_\beta^* \frac{\partial}{\partial n^*} (\delta w^*) \, ds^*$$

$$- \oint_C [T^* + (D_\alpha^* M^{*\alpha\beta}) n_\beta^* + \frac{\partial}{\partial s^*} (M^{*\alpha\beta} n_\alpha^* t_\beta^* - M_\alpha^* \varepsilon^{*\alpha\beta} t_\beta^*)] \delta w^* \, ds^* = 0,$$

where equation (4.5) was used to transform the first boundary-integral of (6.13).

Since δw^* and $\partial(\delta w^*)/\partial n^*$ may be given independently of each other and since they are arbitrary, both integrands must vanish. Hence,

$$M^{*\alpha\beta} n_\alpha^* n_\beta^* = M_B^* \quad \text{on } C \quad (6.15)$$

and

$$-(D_\alpha^* M^{*\alpha\beta}) n_\beta^* - \partial(M^{*\alpha\beta} n_\alpha^* t_\beta^*)/\partial s^* = Q^* \quad \text{on } C, \quad (6.16)$$

where

$$Q^* = T^* - \partial(M_V^*)/\partial s^* \quad (6.17)$$

is the *effective shear-force* per unit length of the (deformed) boundary.

This concludes for the time being our investigation of transverse equilibrium.

Two more equations of equilibrium are obtained when we consider virtual displacements δv_α^* in the tangent plane only, keeping $\delta w^* = 0$ everywhere.

For a variation of δv_α^* in the tangent plane, equation (5.18) takes the form

$$\iint_D [(N^{*\beta\gamma} + 2M^{*\alpha\beta} d_\alpha^{*\gamma}) D_\beta^* \delta v_\gamma^* + (M^{*\alpha\beta} D_\beta^* d_\alpha^{*\gamma} - F^{*\gamma}) \delta v_\gamma^*] \, dA^*$$

$$- \oint_C (T^{*\gamma} + M_\alpha^* \varepsilon^{*\alpha\beta} d_\beta^{*\gamma}) \delta v_\gamma^* \, ds^* = 0 .$$

Using the divergence theorem we get

$$\iint_D [-D_\beta^*(N^{*\beta\gamma} + 2M^{*\alpha\beta} d_\alpha^{*\gamma}) + M^{*\alpha\beta} D_\beta^* d_\alpha^{*\gamma} - F^{*\gamma}] \delta v_\gamma^* \, dA^*$$

$$+ \oint_C [(N^{*\beta\gamma} + 2M^{*\alpha\beta} d_\alpha^{*\gamma}) n_\beta^* - T^{*\gamma} - M_\alpha^* \varepsilon^{*\alpha\beta} d_\beta^{*\gamma}] \delta v_\gamma^* \, ds^* = 0 .$$

Using an argument like the one before, we find that both integrals must vanish. This leads us to the following two equations of equilibrium in the tangent plane,

$$D_\beta^* N^{*\beta\alpha} + 2d_\gamma^{*\alpha} D_\beta^* M^{*\gamma\beta} + M^{*\gamma\beta} D_\beta^* d_\gamma^{*\alpha} + F^{*\alpha} = 0 \quad \text{in } D \quad (6.18)$$

(which should be compared with equation (5.22)) and the boundary conditions

$$(N^{*\beta\alpha} + 2M^{*\gamma\beta} d_\gamma^{*\alpha}) n_\beta^* = T^{*\alpha} + d_\beta^{*\alpha} (t^{*\beta} M_\nu^* + n^{*\beta} M_B^*) .$$

With the help of (6.15) the bending moment can be eliminated and we get

$$[N^{*\beta\alpha} + (2d_\gamma^{*\alpha} - d_\rho^{*\alpha}n^{*\rho}n_\gamma^*)M^{*\beta\gamma}]n_\beta^* = \hat{T}^{*\alpha} \quad \text{on } C, \quad (6.19)$$

where

$$\hat{T}^{*\alpha} = T^{*\alpha} + d_\beta^{*\alpha}t^{*\beta}M_V^* \quad (6.20)$$

is the *effective membrane force* per unit length of the (deformed) boundary.

The virtual work of the stresses on the virtual displacements at the boundary, which is the second term on the right-hand side of (6.8) can now be written in terms of the effective boundary quantities $\hat{T}^{*\alpha}$, Q^* and M_B^* using (6.20), (6.17) and (6.3). We get

$$\oint_C [T^{*\alpha}\delta v_\alpha^* + T^*\delta w^* + M^{*\alpha}\delta r_\alpha^*]\,ds^*$$

$$= \oint_C [\hat{T}^{*\alpha}\delta v_\alpha^* + Q^*\delta w^* + M_B^*\delta\psi^*]\,ds^*, \quad (6.21)$$

where $\delta\psi^*$ is the rotation of the boundary-edge

$$\delta\psi^* = \delta q_\beta^* n^\beta = (d_\beta^{*\alpha}\delta v_\alpha^* + \delta w_{,\beta}^*)n^{*\beta}. \quad (6.22)$$

In retrospect it appears that our precaution to use asterisks has been unnecessary, since almost all quantities in this section are affixed with an asterisk. Leaving them all out however, might lead to a misinterpretation of (6.1), (6.2), (6.5), (6.11) and (6.12). In addition, the asterisks may serve a useful purpose as a warning and a remainder of the correct interpretation of the final equations.

4. Stress-distribution

The relation between the three-dimensional Cauchy stresses $\sigma^{\alpha\beta}$ and the two-dimensional tensors $N^{\alpha\beta}$ and $M^{\alpha\beta}$ are found when (4.8) and (4.10) are substituted into (5.11) and (5.12). We get, when all asterisks are omitted,

$$N^{\alpha\beta} = \int_{-\frac{1}{2}h}^{\frac{1}{2}h} (\sigma^{\alpha\beta} - \sigma^{\gamma\delta} d^\alpha_\gamma d^\beta_\delta z^2)(1 - 2Hz + Kz^2)\,dz \qquad (6.23)$$

and

$$M^{\alpha\beta} = -\int_{-\frac{1}{2}h}^{\frac{1}{2}h} [\sigma^{\alpha\beta} - \tfrac{1}{2}(\sigma^{\alpha\gamma} d^\beta_\gamma + \sigma^{\beta\gamma} d^\alpha_\gamma)z](1 - 2Hz + Kz^2)z\,dz . \qquad (6.24)$$

The scalar field Φ is found according to (5.20) to be

$$\Phi = \frac{1}{2}\int_{-\frac{1}{2}h}^{\frac{1}{2}h} \varepsilon_{\alpha\beta}\sigma^{\alpha\gamma} d^\beta_\gamma (1 - 2Hz + Kz^2)z^2\,dz . \qquad (6.25)$$

Knowledge of the six components $N^{\alpha\beta}$ and $M^{\alpha\beta}$ does not provide us with enough information to determine $\sigma^{\alpha\beta}$ as function of z. However, (6.23) and (6.24) are necessary and sufficient to determine a stress-distribution, *that is assumed to be linear in z*.

An approximate linear distribution is therefore given by

$$\sigma^{\alpha\beta} \simeq N^{\alpha\beta}/h - 12zM^{\alpha\beta}/h^3 , \qquad (6.26)$$

where terms of relative order Hh and Kh^2 have been omitted.

The shear-stresses $\sigma^{\alpha 3}$ must vanish at the stress-free surfaces and hence we would get from (4.9) the following approximate distribution,

$$\sigma^{\alpha 3} \simeq \frac{3}{2h}\mathcal{Q}^\alpha \left[1 - \left(\frac{2z}{h}\right)^2\right] . \qquad (6.27)$$

These estimates of the three-dimensional stress-distribution are consistent with a *first order theory*, in which errors of order h/R and $(h/L)^2$ are accepted (see remark after equation (7.8)).

5. Summary

It has been shown that the six functions $N^{*\alpha\beta}$ and $M^{*\alpha\beta}$ are not independent but satisfy three equations of equilibrium (6.14) and (6.18), which constitute a system of partial differential equations. In addition, they satisfy exactly four boundary conditions, given by (6.15), (6.16) and (6.19). The boundary conditions are given in terms of the bending moment M_B^*, the effective shear force Q^* and the two components of the effective membrane force $\hat{T}^{*\alpha}$. The reader should notice that four of the given boundary loads, namely M_V^*, T^* and $T^{*\alpha}$ have been absorbed into three 'effective' loads and that neither M_V^* nor T^* can be recovered from the field quantities $N^{*\alpha\beta}$ and $M^{*\alpha\beta}$ while $T^{*\alpha}$ can only be recovered on boundaries where $d_\beta^{*\alpha} t^{*\beta} = 0$ (like in plates, for example).

The four boundary conditions (6.15), (6.16) and (6.19) derived by means of the principle of virtual work in this chapter could also have been derived directly from (5.20) and (5.21). Thus, it is possible to express five boundary conditions (three components of the force and two of the moment) in terms of $\mathcal{N}^{\alpha\beta}$ and $\mathcal{M}^{\alpha\beta}$. Elimination of Φ by means of the condition for the torsional moment leads to the four conditions (6.15), (6.16) and (6.19).[1]

---

[1] See KOITER, W.T. (1964), "On the dynamic boundary conditions in the theory of thin shells", *Proc. Kon. Ned. Akad. Wetenschappen* **B67**, pp. 117–126.

CHAPTER 7

STRESS–STRAIN RELATIONS

1. Introduction 115
2. The elastic energy 115
3. The mathematical problem 122
4. Summary 123

CHAPTER 7

STRESS–STRAIN RELATIONS

1. Introduction

The kinematics and the statics of surfaces have been the two main themes of our analysis of shells up to now. In this chapter we shall establish the link between these sections which is needed to complete the theory. So far we have made no assumptions whatsoever regarding the material or materials of which the shell is made. All what has been said holds good for a shell of any conceivable material. But now we shall restrict our considerations to shells made of a homogeneous, linearly elastic material and our analysis to small strain.

2. The elastic energy

In an elastic shell the work done by the external forces is conserved and stored in the shell as an internal energy, usually called the *elastic energy*.

Let σ^{ij} be the components of stress[1] in normal coordinates and η^{ij} corresponding components of strain. For a material that obeys *Hooke's law*, we have

$$\sigma^{ij} = \frac{E}{1+\nu}\left[\eta^{ij} + \frac{\nu}{1-2\nu}g^{ij}\eta^{\kappa}_{\kappa}\right], \tag{7.1}$$

where E is *Young's modulus* and ν *Poisson's ratio*.

---

[1] Since the analysis is restricted to small strain we need not distinguish between the undeformed and the deformed coordinate systems.

The *strain-energy density* W is given by

$$W = \frac{1}{2}\int_{-\frac{1}{2}h}^{\frac{1}{2}h} \sigma^{ij}\eta_{ij}(g/a)^{1/2}\,dz, \tag{7.2}$$

where due to (7.1) the integrand is a quadratic function of the components of strain η^{ij}.

At any point of the middle-surface the deformation of this surface is described by $E_{\alpha\beta}$, $K_{\alpha\beta}$ and their covariant derivatives of all orders.

For a linearly elastic shell, W must therefore be a quadratic function of these tensors. The expression for the strain energy density W can therefore be written in the following dimensionless form,

$$W/Eh = C^{\alpha\beta\gamma\delta}E_{\alpha\beta}E_{\gamma\delta} + D^{\alpha\beta\gamma\delta}E_{\alpha\beta}K_{\gamma\delta} + F^{\alpha\beta\gamma\delta}K_{\alpha\beta}K_{\gamma\delta} + r, \tag{7.3}$$

where the fourth-order tensors $C^{\alpha\beta\gamma\delta}$, $D^{\alpha\beta\gamma\delta}$ and $F^{\alpha\beta\gamma\delta}$ are functions of the geometry of the shell ($a_{\alpha\beta}$, $d_{\alpha\beta}$ and h) and of Poisson's ratio ν. The last term r is a quadratic function of the covariant derivatives of $E_{\alpha\beta}$ and $K_{\alpha\beta}$ of all orders.

The invariance of (7.3) with respect to coordinate transformations must hold good for any given state of deformation. But this can only be so if each term on the right-hand side of (7.3) is invariant. We may thus proceed to examine the terms one by one.

The properties of symmetry of $E_{\alpha\beta}E_{\gamma\delta}$ imply that only the corresponding symmetrical part of $C^{\alpha\beta\gamma\delta}$ contributes to the inner product. Taking advantage of this fact we can—without loss of generality—express the fourth-order tensor $C^{\alpha\beta\gamma\delta}$ in the following form,

$$C^{\alpha\beta\gamma\delta} = C_1 a^{\alpha\beta}a^{\gamma\delta} + C_2 a^{\alpha\gamma}a^{\beta\delta} + C_3 h a^{\alpha\beta}d^{\gamma\delta} \\ + C_4 h a^{\alpha\gamma}d^{\beta\delta} + C_5 h^2 d^{\alpha\beta}d^{\gamma\delta} + C_6 h^2 d^{\alpha\gamma}d^{\beta\delta}, \tag{7.4}$$

where the coefficients $C_1, \ldots, C_6$ are dimensionless analytic functions of the dimensionless variables ν, Hh and Kh^2. Note that inner products of $d_{\alpha\beta}$ can be reduced with the help of Hamilton–Cayley's theorem (2.42). Thus, for instance,

$$d^{\alpha}_{\xi}d^{\xi}_{\eta}d^{\eta\beta} = (4H^2 - K)d^{\alpha\beta} - 2HKa^{\alpha\beta}.$$

When the direction of the normal (which depends on the choice of coordinates u^1, u^2) is reversed, $d^{\alpha\beta}$ changes sign and C_3 and C_4 must therefore also change signs, while their absolute value remains unchanged. The coefficients C_1, C_2, C_5 and C_6 do not undergo any change. Clearly, this restricts the dependence of the C's on the variable Hh and implies that

$$C_1(\nu, Hh, Kh^2) = \bar{C}_1(\nu, (Hh)^2, Kh^2)$$

and similarly for C_2, C_5 and C_6, whereas

$$C_3(\nu, Hh, Kh^2) = Hh\,\bar{C}_3(\nu, (Hh)^2, Kh^2)$$

and similarly for C_4.

It is seen now that the first term $C^{\alpha\beta\gamma\delta}E_{\alpha\beta}E_{\gamma\delta}$ of the strain energy expression may be considerably simplified if terms of relative order $(h/R)^2$ are omitted.[2] As the functions $\bar{C}_1, \ldots, \bar{C}_6$ are regular at $(Hh)^2 = Kh^2 = 0$, we are justified in evaluating them at $(Hh)^2 = Kh^2 = 0$ within this approximation. In the following we shall denote the coefficients $\bar{C}_1, \ldots, \bar{C}_6$ evaluated at $(Hh)^2 = Kh^2 = 0$ by $C_{10}, \ldots, C_{60}$.

Now we shall show that the contributions from all terms except the first two terms of (7.4) are, at most, of relative order $(h/R)^2$.

Let us multiply (7.4) through by $E_{\alpha\beta}E_{\gamma\delta}$. The first two terms yield

$$C_{10}(e_1 + e_2)^2 + C_{20}(e_1^2 + e_2^2), \tag{7.5}$$

where e_1 and e_2 are the principal strains[3] ($|e_1| \geq |e_2|$), and where C_{20} is positive and C_{10} nonnegative since the strain energy is positive definite.

The contribution from the remaining terms, one by one, will be compared with (7.5). As

$$|d^{\gamma\delta}E_{\gamma\delta}| \leq 2|e_1|/R,$$

we have

$$|C_3 h a^{\alpha\beta} d^{\gamma\delta} E_{\alpha\beta} E_{\gamma\delta}| \leq |C_{30} Hh^2 (2e_1)^2/R|$$

---

[2] Here R denotes the smallest principal radius of curvature.
[3] See definition on p. 59, Chapter 3.

and furthermore, as

$$|a^{\alpha\gamma}d^{\beta\delta}E_{\alpha\beta}E_{\gamma\delta}| \leq 4e_1^2/R$$

and

$$|d^{\alpha\gamma}d^{\beta\delta}E_{\alpha\beta}E_{\gamma\delta}| \leq 4e_1^2/R^2,$$

we get the inequalities

$$|C_4 h a^{\alpha\gamma}d^{\beta\delta}E_{\alpha\beta}E_{\gamma\delta}| \leq |C_{40}Hh^2 4e_1^2/R|$$

and

$$|C_5 h^2 d^{\alpha\beta}d^{\gamma\delta}E_{\alpha\beta}E_{\gamma\delta}| \leq |C_{50}h^2(2e_1/R)^2|.$$

This proves that the relative error in omitting all terms except (7.5) is, at most, of relative order $(h/R)^2$.

Proceeding in this fashion we can show that the contribution to the strain energy from the third term of (7.3) is given by the expression

$$h^2[F_{10}(k_1 + k_2)^2 + F_{20}(k_1^2 + k_2^2)], \tag{7.6}$$

the terms omitted being at most of relative order $(h/R)^2$. Here k_1 and k_2 are the principal changes of curvature. Expression (7.6) is analogous to (7.5), however, the factor h^2 appears here, compensating the principal dimensions of k_1 and k_2. Like C_{10} and C_{20}, the coefficients F_{10} and F_{20} are dimensionless functions of ν.

The second (mixed) term of (7.3) can now be shown in a very analogous way to contribute to the strain energy at most with terms of relative order h/R.

Finally, let us examine the last term r on the right-hand side of (7.3). This term represents a quadratic function of the covariant derivatives of $E_{\alpha\beta}$ and $K_{\alpha\beta}$ of all orders. There are a great number of possibilities for constructing quadratic invariants from the fundamental tensors $a^{\alpha\beta}$ and $d^{\alpha\beta}$ and the covariant derivatives of $E_{\alpha\beta}$ and $K_{\alpha\beta}$ of any given order. However, it is easily seen that the two lowest order quadratic terms of the covariant derivatives of $E_{\alpha\beta}$ contain only products of the type $(E..,.E..,.)$ and $(E..E..,..)$, since the resulting tensor has to be of even order.

Corresponding sixth-order tensor coefficients multiplying these quadratic functions have the physical dimension (length)² and will therefore contain the factor h^2. From this it follows that the relative order of magnitude of these terms is on the average $(h/L)^2$ in comparison with the term $C^{\alpha\beta\gamma\delta}E_{\alpha\beta}E_{\gamma\delta}$, where L is a characteristic wavelength of the deformation pattern. Terms containing higher order derivatives of $E_{\alpha\beta}$ are of course smaller still.

The remaining terms of r contain derivatives of $K_{\alpha\beta}$ and are compared with either $D^{\alpha\beta\gamma\delta}E_{\alpha\beta}K_{\gamma\delta}$ or $F^{\alpha\beta\gamma\delta}K_{\alpha\beta}K_{\gamma\delta}$, depending on whether the strain tensor $E_{\alpha\beta}$ (or its derivatives) is involved or not. It should be immediately clear that, on the average, r is at most of relative order $(h/L)^2$.

We conclude that by writing

$$W/Eh = (C_1 a^{\alpha\beta}a^{\gamma\delta} + C_2 a^{\alpha\gamma}a^{\beta\delta})E_{\alpha\beta}E_{\gamma\delta} \\ + h^2(F_1 a^{\alpha\beta}a^{\gamma\delta} + F_2 a^{\alpha\gamma}a^{\beta\delta})K_{\alpha\beta}K_{\gamma\delta} \qquad (7.7)$$

we have introduced errors of relative order at most $|h/R| + (h/L)^2$.

In (7.7) the coefficients are functions of ν only and can be determined from suitably selected solutions to the three-dimensional equations of elasticity. From a simple three-dimensional analysis of a rectangular body we get

(i) *pure shear* $(v^1 = \gamma u^2, v^2 = \gamma u^1, v^3 = 0)$:

$$\frac{W}{Eh} \underset{\text{(3-D)}}{=} \frac{1}{1+\nu}\gamma^2 \underset{\text{(2-D)}}{=} 2C_2\gamma^2 ;$$

(ii) *pure tension* $(v^1 = \varepsilon u^1, v^2 = -\nu\varepsilon u^2, v^3 = -\nu\varepsilon z)$:

$$\frac{W}{Eh} \underset{\text{(3-D)}}{=} \frac{1}{2}\varepsilon^2 \underset{\text{(2-D)}}{=} [C_1(1-\nu)^2 + C_2(1+\nu^2)]\varepsilon^2 ;$$

(iii) *pure twist* $(v^1 = \phi u^2 z, v^2 = \phi u^1 z, v^3 = -\phi u^1 u^2)$:

$$\frac{W}{Eh} \underset{\text{(3-D)}}{=} \frac{1}{12}\frac{h^2}{1+\nu}\phi^2 \underset{\text{(2-D)}}{=} 2h^2 F_2 \phi^2 ;$$

(iv) *pure bending* ($v^1 = -2\kappa u^1 z$, $v^2 = 0$, $v^3 = \kappa(u^1)^2$):

$$\frac{W}{Eh} \underset{\text{(3-D)}}{=} \frac{1}{6}\frac{h^2}{1-\nu^2}\kappa^2 \underset{\text{(2-D)}}{=} 4(F_1+F_2)h^2\kappa^2,$$

where γ, ε, ϕ and κ are the shear-strain, the normal strain, the angle of twist per unit length, and the curvature respectively. Solving for C_1, C_2, F_1 and F_2 we get

$$C_1 = \frac{1}{2}\frac{\nu}{1-\nu^2}, \qquad C_2 = \frac{1}{2}\frac{1}{1+\nu},$$

$$F_1 = \frac{1}{24}\frac{\nu}{1-\nu^2}, \qquad F_2 = \frac{1}{24}\frac{1}{1+\nu},$$

which substituted into (7.7) yields

$$W = \frac{Eh}{2(1-\nu^2)}[(1-\nu)E^\beta_\alpha E^\alpha_\beta + \nu E^\alpha_\alpha E^\beta_\beta]$$

$$+ \frac{Eh^3}{24(1-\nu^2)}[(1-\nu)K^\beta_\alpha K^\alpha_\beta + \nu K^\alpha_\alpha K^\beta_\beta]. \qquad (7.8)$$

This expression gives the strain energy density with a relative error of order $|h/R| + (h/L)^2$. In our analysis of thin elastic shells we shall accept errors of this magnitude. *This is the point at which we introduce an approximation in our theory of shells.* The way to improve the accuracy of the theory is apparent: by including terms of order h/R and $(h/L)^2$ in the energy expression.

If we now apply the principle of virtual work we find

$$\delta W = \frac{\partial W}{\partial E_{\alpha\beta}}\delta E_{\alpha\beta} + \frac{\partial W}{\partial K_{\alpha\beta}}\delta K_{\alpha\beta}. \qquad (7.9)$$

But this could only hold good (compare with equation (5.10)) if

$$N^{\alpha\beta} = \frac{\partial W}{\partial E_{\alpha\beta}} \quad \text{and} \quad M^{\alpha\beta} = \frac{\partial W}{\partial K_{\alpha\beta}}, \qquad (7.10)$$

from which we derive the *constitutive equations* for thin elastic shells,

$$N^{\alpha\beta} = \frac{Eh}{1-\nu^2}[(1-\nu)E^{\alpha\beta} + \nu a^{\alpha\beta}E^\gamma_\gamma],$$

$$M^{\alpha\beta} = D[(1-\nu)K^{\alpha\beta} + \nu a^{\alpha\beta}K^\gamma_\gamma], \qquad (7.11)$$

$$D = \frac{Eh^3}{12(1-\nu^2)}.$$

Solving for $E^{\alpha\beta}$ and $K^{\alpha\beta}$ we get

$$E^{\alpha\beta} = \frac{1}{Eh}[(1+\nu)N^{\alpha\beta} - \nu a^{\alpha\beta}N^\gamma_\gamma],$$

$$K^{\alpha\beta} = \frac{1}{D(1-\nu^2)}[(1+\nu)M^{\alpha\beta} - \nu a^{\alpha\beta}M^\gamma_\gamma]. \qquad (7.12)$$

It should be noted that in these one-to-one relations the effective membrane stress tensor $N^{\alpha\beta}$ determines the strain tensor $E^{\alpha\beta}$ and, correspondingly, $M^{\alpha\beta}$ determines $K^{\alpha\beta}$. The constitutive equations are *uncoupled* in this sense. This property considerably contributes to the relative simplicity of problems in the classical theory of thin elastic shells.

If terms of order h/R are retained in the energy expression, the term $D^{\alpha\beta\gamma\delta}E_{\alpha\beta}K_{\gamma\delta}$ in (7.3) cannot be neglected and clearly (7.10) would not give uncoupled constitutive equations. However, W would still be a function of $E_{\alpha\beta}$ and $K_{\alpha\beta}$ only and (7.9) could still represent (5.10), i.e., our choice of virtual displacements off the middle-surface is sufficiently general to encompass any strain energy expression of the general form

$$W = W(E_{\alpha\beta}, K_{\alpha\beta}). \qquad (7.13)$$

Should a still higher degree of accuracy be necessary, then terms of order $(h/L)^2$ must be included in the strain energy expression, and the form (7.13) is not adequate for that. We would have to include the dependence of W on $D_\alpha K^\gamma_\gamma$ and $D_\alpha D_\beta K^\gamma_\gamma$ but (5.10) does not allow for that. We would have to select the virtual displacements off the middle-surface from a wider class. But this falls outside the scope of this book.

3. The mathematical problem

We now have the complete theory to be able to translate a shell problem into a well-posed mathematical problem. In principle we can follow two different lines.

According to the first one we select the displacements v^α and w as unknowns. By (3.9) and (3.36) we can express $E_{\alpha\beta}$ and $K_{\alpha\beta}$ in terms of v^α and w. Next, the constitutive equations (7.11) give $N^{\alpha\beta}$ and $M^{\alpha\beta}$ in terms of v^α and w and finally the equations of equilibrium (6.14) and (6.18) yield three partial differential equations for the three unknowns. When linearized the equations are of fourth order in w and of second order in v^α. There are precisely four boundary conditions of either kinematic or static type. From (6.21) we see that we can prescribe

$$\begin{array}{ll} \text{either} & \hat{T}^\alpha \text{ or } v^\alpha \\ \text{and either} & Q \text{ or } w \\ \text{and either} & M_B \text{ or } \psi \end{array}$$

at the boundary.[4]

The second way to formulate a well-posed mathematical problem consists in taking the six functions $E_{\alpha\beta}$ and $K_{\alpha\beta}$ as unknown. Using the constitutive equations (7.11) and the equations of equilibrium (6.14) and (6.18) we get three partial differential equations for the unknowns. The equations of compatibility (3.48) and (3.49) are the remaining three differential equations.

The choice of procedure usually depends on the boundary conditions. Mostly we prefer a formulation in which the boundary conditions are as simple as possible—usually at the cost of more complicated differential equations. Hence, if the boundary conditions are kinematical, the first line above is probably to prefer, and the second one if the boundary conditions are given in terms of stresses.

Whichever way we choose, the derivation, although straightforward, is extremely tedious, except for the simplest cases. Fortunately, there are nowadays very good tools available to perform all this routine work on formulas, namely *symbolic and algebraic manipulation* performed with the help of computers (see, for instance, JENSEN and NIORDSON (1977)).[5]

---

[4] We could of course prescribe any linear combination of $\hat{T}^\alpha$ and v^α instead of either of them and similarly for the other.

[5] See bibliography on p. 133.

4. Summary

Using dimensional analysis and invariance properties we have found that the strain energy density for thin elastic shells can be written as a simple quadratic function of $E_{\alpha\beta}$ and $K_{\alpha\beta}$. This has lead us to simple linear uncoupled equations, one group relating the components of strain $E_{\alpha\beta}$ with the components of the effective stress tensor $N^{\alpha\beta}$ and one group relating the bending tensor $K_{\alpha\beta}$ with the effective moment tensor $M^{\alpha\beta}$. With the help of these equations we can now formulate well-posed mathematical problems for the solution of shell problems.

CHAPTER 8

THE STATIC-GEOMETRIC ANALOGY

1. Introduction 127
2. Alternative measures of bending 127
3. Static-geometric analogy 129
4. Stress-functions 131
5. Summary and discussion 133
   Bibliography 133

CHAPTER 8

THE STATIC-GEOMETRIC ANALOGY

1. Introduction

The three equations of equilibrium have very little formal resemblance to the three equations of compatibility as they appear in our analysis. However, in the linear case we can actually put them in identical shapes, as we shall see in this chapter. This fact is of some importance because it leads in a natural way to the introduction of *stress-functions*.

2. Alternative measures of bending

We have already discussed[1] the arbitrariness in the definition of the bending tensor $K_{\alpha\beta}$. From a purely formal point of view we may take any measure of bending $\tilde{K}_{\alpha\beta}$ and strain $\tilde{E}_{\alpha\beta}$ from which our original measures $K_{\alpha\beta}$ and $E_{\alpha\beta}$ can be recovered and the new measures will serve our purpose just as well. In particular, it seems clear that there is no physical ground for selecting the covariant components of the second fundamental tensor for the definition of $K_{\alpha\beta}$. Thus, had we for instance taken the difference between the mixed tensors as our definition for bending, it would differ from $K_{\alpha\beta}$ with an amount of about $2E_{\alpha\gamma}d_\beta^\gamma$ (linearized difference). But how would that affect the constitutive equations? Would these still remain uncoupled? We shall show that they indeed would be uncoupled provided that we still neglect terms of relative order h/R and $(h/L)^2$ in the expression for the strain energy density.

---

[1] Chapter 3, §5.

There seems to be no reason for using other measures than $E_{\alpha\beta}$ for the strain but things are entirely different for the measure of bending and in the literature many different measures are used. These are all of the following general class,

$$\tilde{K}_{\alpha\beta} = K_{\alpha\beta} + C^{\gamma\delta}_{\alpha\beta} E_{\gamma\delta}, \qquad (8.1)$$

where $C^{\gamma\delta}_{\alpha\beta} E_{\gamma\delta}$ is a linear combination of the terms

$$d^{\gamma}_{\gamma} E_{\alpha\beta}, \quad d_{\alpha\beta} E^{\gamma}_{\gamma}, \quad d^{\gamma}_{\alpha} E_{\beta\gamma}, \quad a_{\alpha\beta} d^{\delta}_{\gamma} E^{\gamma}_{\delta}, \quad a_{\alpha\beta} d^{\gamma}_{\gamma} E^{\delta}_{\delta}.$$

If (8.1) is solved for $K_{\alpha\beta}$ and substituted into (7.3), we get

$$\frac{W}{Eh} = \tilde{C}^{\alpha\beta\gamma\delta} E_{\alpha\beta} E_{\gamma\delta} + \tilde{D}^{\alpha\beta\gamma\delta} E_{\alpha\beta} \tilde{K}_{\gamma\delta} + F^{\alpha\beta\gamma\delta} \tilde{K}_{\alpha\beta} \tilde{K}_{\gamma\delta} + r, \qquad (8.2)$$

where neither $F^{\alpha\beta\gamma\delta}$ nor r have been affected. Following the same arguments as in Chapter 7 we see that if terms of order h/R and $(h/L)^2$ are neglected, the strain energy density can be written as (7.8) but with $\tilde{K}_{\alpha\beta}$ instead of $K_{\alpha\beta}$.

This shows—what we in fact have claimed—that the uncoupled constitutive equations are not a consequence of our particular choice $K_{\alpha\beta}$ for the measure of bending but that there are many other measures that would permit us to use the same constitutive equations with the same accuracy.

Particular interest is attached to the measure

$$\hat{K}_{\alpha\beta} = K_{\alpha\beta} - \tfrac{1}{2}(d^{\gamma}_{\alpha} E_{\gamma\beta} + d^{\gamma}_{\beta} E_{\alpha\gamma}), \qquad (8.3)$$

which is a special case of (8.1). This measure has the property that it is unaffected by a constant displacement w. For that case we have namely, according to (3.37) and (3.11),

$$K_{\alpha\beta} = -d_{\alpha\gamma} d^{\gamma}_{\beta} w, \qquad E_{\alpha\beta} = -d_{\alpha\beta} w$$

and hence

$$\hat{K}_{\alpha\beta} = 0.$$

It is difficult to find a sound physical reason for preferring a measure having this particular property, but we shall find another useful property of $\hat{K}_{\alpha\beta}$, which is not shared by $K_{\alpha\beta}$.

3. Static-geometric analogy

When $K_{\alpha\beta}$ from (8.3) is substituted into (5.10) we get

$$\delta V = \iint_D [\hat{N}^{\alpha\beta} E_{\alpha\beta} + M^{\alpha\beta} \hat{K}_{\alpha\beta}] \, dA , \tag{8.4}$$

where $\hat{N}^{\alpha\beta}$ denotes a new membrane stress tensor defined by

$$\hat{N}^{\alpha\beta} = N^{\alpha\beta} + \tfrac{1}{2}(d^\alpha_\gamma M^{\gamma\beta} + d^\beta_\gamma M^{\gamma\alpha}) . \tag{8.5}$$

Solving for $N^{\alpha\beta}$ and substituting into the equations of equilibrium (5.22) and (5.23), we get

$$D_\alpha \hat{N}^{\alpha\beta} + d^\beta_\gamma D_\alpha M^{\alpha\gamma} + \tfrac{1}{2} D_\alpha(M^{\alpha\gamma} d^\beta_\gamma - M^{\beta\gamma} d^\alpha_\gamma) + F^\beta = 0 \tag{8.6}$$

and

$$D_\alpha D_\beta M^{\alpha\beta} - d_{\alpha\beta} \hat{N}^{\alpha\beta} - p = 0 , \tag{8.7}$$

where the second one has gained in simplicity.

Substitution of $K_{\alpha\beta}$ from (8.3) into the linearized equation of compatibility (3.48), in which the quadratic terms on the right-hand side have been thrown out, yields

$$D_\alpha D^\alpha E^\beta_\beta - D_\alpha D^\beta E^\alpha_\beta = d^\alpha_\beta \hat{K}^\beta_\alpha - d^\alpha_\alpha \hat{K}^\beta_\beta$$

or

$$D_\alpha D_\beta (a^{\alpha\beta} E^\gamma_\gamma - E^{\alpha\beta}) - d_{\alpha\beta}(\hat{K}^{\alpha\beta} - a^{\alpha\beta} \hat{K}^\gamma_\gamma) = 0 .$$

By defining new *reversed* quantities

$$\bar{K}^{\alpha\beta} = -\varepsilon^{\alpha\xi} \varepsilon^{\beta\eta} \hat{K}_{\xi\eta} = -a^{\alpha\beta} \hat{K}^\gamma_\gamma + \hat{K}^{\alpha\beta} \tag{8.8}$$

and

$$\bar{E}^{\alpha\beta} = \varepsilon^{\alpha\xi}\varepsilon^{\beta\eta}E_{\xi\eta} = a^{\alpha\beta}E^{\gamma}_{\gamma} - E^{\alpha\beta} \tag{8.9}$$

and inserting them into the equation of compatibility above, we get

$$D_{\alpha}D_{\beta}\bar{E}^{\alpha\beta} - d_{\alpha\beta}\bar{K}^{\alpha\beta} = 0. \tag{8.10}$$

This equation has precisely the same shape as the third equation of equilibrium (8.7) in the homogeneous case $p = 0$. We shall now show that the remaining linearized equations of compatibility (3.49) may be put in precisely the same shape as the two first equations of equilibrium (8.6), when the measures $\bar{K}^{\alpha\beta}$ and $\bar{E}^{\alpha\beta}$ are used.

Substitution of $K_{\alpha\beta}$ from (8.3) into (3.49) yields

$$D_{\alpha}\hat{K}_{\beta\gamma} - D_{\beta}\hat{K}_{\alpha\gamma} + \tfrac{1}{2}D_{\alpha}(d^{\delta}_{\beta}E_{\delta\gamma} + d^{\delta}_{\gamma}E_{\delta\beta})$$
$$- \tfrac{1}{2}D_{\beta}(d^{\delta}_{\alpha}E_{\delta\gamma} + d^{\delta}_{\gamma}E_{\delta\alpha}) - d^{\delta}_{\beta}(D_{\gamma}E_{\alpha\delta} + D_{\alpha}E_{\gamma\delta} - D_{\delta}E_{\alpha\gamma})$$
$$+ d^{\delta}_{\alpha}(D_{\gamma}E_{\beta\delta} + D_{\beta}E_{\gamma\delta} - D_{\delta}E_{\beta\gamma}) = 0.$$

Of these eight equations only two are independent. We obtain these by multiplying through by $a^{\alpha\gamma}$ which leaves β as the only free index. With a further factor $a^{\beta\eta}$ the free index is raised. We get

1st and 2nd term:

$$a^{\beta\eta}a^{\alpha\gamma}(D_{\alpha}\hat{K}_{\beta\gamma} - D_{\beta}\hat{K}_{\alpha\gamma}) = a^{\beta\eta}(D_{\alpha}\hat{K}^{\alpha}_{\beta} - D_{\beta}\hat{K}^{\gamma}_{\gamma}) = D_{\alpha}\bar{K}^{\alpha\eta} \ ;$$

3rd and 4th term:

$$\tfrac{1}{2}D_{\alpha}(d^{\delta\eta}E^{\alpha}_{\delta} + d^{\delta\alpha}E^{\eta}_{\delta}) - D_{\beta}(d^{\delta}_{\alpha}E^{\alpha}_{\delta}a^{\beta\eta})$$
$$= \tfrac{1}{2}D_{\alpha}(d^{\eta}_{\delta}E^{\alpha\delta} + d^{\alpha}_{\delta}E^{\eta\delta} - 2d^{\delta}_{\gamma}E^{\gamma}_{\delta}a^{\alpha\eta})$$
$$= \tfrac{1}{2}D_{\alpha}(d^{\alpha}_{\delta}E^{\eta\delta} - d^{\eta}_{\delta}E^{\alpha\delta}) + D_{\alpha}(d^{\eta}_{\delta}E^{\alpha\delta} - d^{\gamma}_{\delta}E^{\delta}_{\gamma}a^{\alpha\eta})$$
$$= \tfrac{1}{2}D_{\alpha}[d^{\eta}_{\delta}(a^{\alpha\delta}E^{\gamma}_{\gamma} - E^{\alpha\delta}) - d^{\alpha}_{\delta}(a^{\eta\delta}E^{\gamma}_{\gamma} - E^{\eta\delta})]$$
$$+ d^{\eta}_{\delta}D_{\alpha}E^{\alpha\delta} - d^{\gamma}_{\delta}a^{\alpha\eta}D_{\alpha}E^{\delta}_{\gamma} + E^{\alpha\delta}a^{\eta\gamma}(D_{\alpha}d_{\gamma\delta} - D_{\gamma}d_{\alpha\delta})$$
$$= \tfrac{1}{2}D_{\alpha}(d^{\eta}_{\delta}\bar{E}^{\alpha\delta} - d^{\alpha}_{\delta}\bar{E}^{\eta\delta}) + d^{\eta}_{\delta}D_{\alpha}E^{\alpha\delta} - d^{\gamma}_{\delta}a^{\alpha\eta}D_{\alpha}E^{\delta}_{\gamma} \ ;$$

5th and 6th term:

$$-d^\delta_\beta a^{\beta\eta}(D_\gamma E^\gamma_\delta + D_\alpha E^\alpha_\delta - D_\delta E^\alpha_\alpha) + d^\delta_\alpha a^{\alpha\gamma}(D_\gamma E^\eta_\delta + a^{\beta\eta} D_\beta E_{\gamma\delta} - D_\delta E^\eta_\gamma)$$
$$= -2d^\eta_\delta D_\gamma E^{\gamma\delta} + d^\delta_\gamma D_\delta(a^{\gamma\eta} E^\alpha_\alpha) + d^\alpha_\delta a^{\beta\eta} D_\beta E^\delta_\alpha.$$

Summing up, we have

$$D_\alpha \bar{K}^{\alpha\eta} + \tfrac{1}{2} D_\alpha (d^\eta_\delta \bar{E}^{\alpha\delta} - d^\alpha_\delta \bar{E}^{\eta\delta}) + d^\eta_\delta D_\beta (a^{\delta\beta} E^\gamma_\gamma - E^{\delta\beta}) = 0$$

and, finally,

$$D_\alpha \bar{K}^{\alpha\beta} + d^\beta_\gamma D_\alpha \bar{E}^{\alpha\gamma} + \tfrac{1}{2} D_\alpha (\bar{E}^{\alpha\gamma} d^\beta_\gamma - \bar{E}^{\beta\gamma} d^\alpha_\gamma) = 0. \qquad (8.11)$$

Again, these equations are precisely of the same form as the first two equations of equilibrium (8.6) in the homogeneous case $F^\beta = 0$.

All six equations can now be condensed into the complex form

$$D_\alpha W^{\alpha\beta} + d^\beta_\gamma D_\alpha Z^{\alpha\gamma} + \tfrac{1}{2} D_\alpha (Z^{\alpha\gamma} d^\beta_\gamma - Z^{\beta\gamma} d^\alpha_\gamma) = 0 \qquad (8.12)$$

and

$$D_\alpha D_\beta Z^{\alpha\beta} - d_{\alpha\beta} W^{\alpha\beta} = 0, \qquad (8.13)$$

where

$$\begin{aligned} W^{\alpha\beta} &= \hat{N}^{\alpha\beta} + i\bar{K}^{\alpha\beta}, \\ Z^{\alpha\beta} &= M^{\alpha\beta} + i\bar{E}^{\alpha\beta}, \end{aligned} \quad i = \sqrt{-1}. \qquad (8.14)$$

To this set of equations we can add uncoupled constitutive relations and still argue that the theory is adequate for thin elastic shells, since the relative errors are of order h/R and $(h/L)^2$.

4. Stress-functions

Apart from the somewhat curious fact that the equations of compatibility can be put in a formally identical shape with the equations of equilibrium, the static-geometric analogy has the following useful consequence.

Since the equations of compatibility are identically satisfied whenever the strain tensor $E_{\alpha\beta}$ and the bending tensor $K_{\alpha\beta}$ are derived from a set of arbitrarily selected displacements v^α and w, the membrane stress tensor and the moment tensor must also be derivable from three functions, a vector and a scalar function, and so that the equations of equilibrium become identically satisfied.

To write $\hat{K}_{\alpha\beta}$ in terms of the displacements, we make use of (3.11) and (3.37), which substituted in (8.3) yield

$$\hat{K}_{\alpha\beta} = D_\alpha D_\beta w + \tfrac{1}{4} d_\alpha^\gamma (3 D_\beta v_\gamma - D_\gamma v_\beta)$$
$$+ \tfrac{1}{4} d_\beta^\gamma (3 D_\alpha v_\gamma - D_\gamma v_\alpha) + v^\gamma D_\gamma d_{\alpha\beta} \tag{8.15}$$

and hence, due to (8.8),

$$\bar{K}^{\alpha\beta} = -\varepsilon^{\alpha\xi} \varepsilon^{\beta\eta} [D_\xi D_\eta w + \tfrac{1}{4} d_\xi^\gamma (3 D_\eta v_\gamma - D_\gamma v_\eta)$$
$$+ \tfrac{1}{4} d_\eta^\gamma (3 D_\xi v_\gamma - D_\gamma v_\xi) + v^\gamma D_\gamma d_{\xi\eta}]. \tag{8.16}$$

Similarly, using (8.9) and (3.11) we get

$$\bar{E}^{\alpha\beta} = \varepsilon^{\alpha\xi} \varepsilon^{\beta\eta} [\tfrac{1}{2}(D_\xi v_\eta + D_\eta v_\xi) - d_{\xi\eta} w]. \tag{8.17}$$

Now, let $\psi_\alpha(u^1, u^2)$ be an arbitrary vector function and $\Phi(u^1, u^2)$ an arbitrary scalar function. Furthermore, define the membrane stress tensor and moment tensor by the relations

$$\hat{N}^{\alpha\beta} = -\varepsilon^{\alpha\xi} \varepsilon^{\beta\eta} [D_\xi D_\eta \Phi + \tfrac{1}{4} d_\xi^\gamma (3 D_\eta \psi_\gamma - D_\gamma \psi_\eta)$$
$$+ \tfrac{1}{4} d_\eta^\gamma (3 D_\xi \psi_\gamma - D_\gamma \psi_\xi) + \psi^\gamma D_\gamma d_{\xi\eta}] \tag{8.18}$$

and

$$M^{\alpha\beta} = \varepsilon^{\alpha\xi} \varepsilon^{\beta\eta} [\tfrac{1}{2}(D_\xi \psi_\eta + D_\eta \psi_\xi) - d_{\xi\eta} \Phi]. \tag{8.19}$$

Then, since (8.16) and (8.17) satisfy the equations of compatibility identically, and since the equations of equilibrium have exactly the same shape as the equations of compatibility, the static measures $\hat{N}^{\alpha\beta}$ and $M^{\alpha\beta}$, defined by (8.18) and (8.19), *will satisfy the equations of equilibrium identically irrespective of the choice of* ψ_α *and* Φ.

To formulate a problem in terms of stress-functions we use (8.18) and (8.19) to find the membrane stress tensor and the moment tensor. Using the constitutive equations (7.12) and the equations of compatibility (3.48) and (3.49) we get a system of three partial differential equations for the three unknown functions ψ_α and Φ. If the boundary conditions are given in terms of stresses, this formulation would normally be preferable to one in terms of displacements.

5. Summary and discussion

The somewhat arbitrary way in which the bending tensor may be defined leads to a number of possibilities. Even if we give a strong preference to a theory of thin elastic shells which has uncoupled constitutive equations, there is still a considerable freedom in the way the measure for the bending tensor can be defined. This latitude can apparently be used to impose some extra conditions on our theory, which we might find desirable.

Such a condition may for instance be that our theory possesses a static-geometric analogy but even that restriction would not make our choice unique. This has been shown by BUDIANSKY and SANDERS (1962) who argue that an additional advantage would be if the theory when applied to the symmetrical bending of shells of revolution, would have stress and strain measures that agree with those of the classical theory of LOVE. That is actually the case of (8.3) but of no other bending measure of the class considered.

The static-geometric analogy has deep roots in the theory of plates and appeared early in Russian literature on shells (see, for example, LURIE (1961)).

We should, however, be aware of the fact that as soon as we require a more accurate theory, the freedom to impose extra conditions evaporates quickly.

Bibliography

BUDIANSKY, B. and J.L. SANDERS (1962), "On the 'best' first-order linear shell theory", Office of Naval Research, Technical Report No. 14.

JENSEN, J. and F.I. NIORDSON (1977), "Symbolic and algebraic manipulation languages and their applications in mechanics", *Structural Mechanics, Software Series, Vol. I*, N. Perrone and B. Pilkey, Eds., Univ. Press of Virginia, Charlottesville, pp. 541–576.

KOITER, W.T. (1959), "A consistent first approximation in the general theory of thin elastic shells", *Proc. Symp. on the Theory of Thin Elastic Shells*, W.T. Koiter, Ed., North-Holland, Amsterdam.

KOITER, W.T. and J.G. SIMMONDS (1973), "Foundations of shell theory", *Proc. 13th Internat. Congr. of Theoretical and Applied Mechanics*, E. Becker and G.K. Mikhailov, Eds., Springer, Berlin.

LURIE, A.I. (1961), "On the static-geometric analogue of shell theory", *Problems of Continuum Mechanics*, SIAM, Philadelphia.

NIORDSON, F.I. (1971), "A note on the strain energy of elastic shells", *Internat. J. Solids Structures* **7**, pp. 1573–1579.

CHAPTER 9

PLATES

1. Introduction 137
2. Separation of the plate problem 137
3. In-plane loaded plates 138
4. Bending theory of plates 144
5. Bending of plates in rectangular coordinates 144
6. Corners 145
7. Bending of a simply supported rectangular plate 148
8. The clamped and uniformly loaded elliptical plate 150
9. Vibrating plates 151
10. Energy methods 152
11. Bending of a clamped rectangular plate 160
12. Lowest natural frequency of a skew plate 161
13. Bending of plates in polar coordinates 165
14. Axisymmetrical bending of a circular plate 166
15. Free vibrations of a circular plate 168
16. Behaviour at a corner of a simply supported plate 172
    Bibliography 175

CHAPTER 9

PLATES

1. Introduction

The mathematical problem to which the theory of thin elastic shells leads is far too complicated for a general solution. We shall therefore proceed to develop the theory in certain directions that eventually lead to the solution of a number of problems of engineering interest.

A case that involves considerable simplifications is the special case of *flat shells* or *plates*, i.e., the case when the middle-surface of the undeformed shell is plane. As we shall see, this leads under rather general assumptions to the linear theories of in-plane loaded plates and transversely loaded plates.

2. Separation of the plate problem

If the middle-surface of the undeformed shell is plane, the curvature tensor $d_{\alpha\beta}$ vanishes everywhere and the linearized equations of equilibrium (5.22) and (5.23) reduce to

$$D_\alpha N^{\alpha\beta} + F^\beta = 0, \qquad (9.1)$$

$$D_\alpha D_\beta M^{\alpha\beta} - p = 0. \qquad (9.2)$$

When expressed in the displacements, the strain tensor (3.11) and the bending tensor (3.37) take the form

$$E_{\alpha\beta} = \tfrac{1}{2}(D_\alpha v_\beta + D_\beta v_\alpha), \qquad (9.3)$$

$$K_{\alpha\beta} = D_\alpha D_\beta w. \qquad (9.4)$$

If the constitutive equations (7.11) are employed, equations (9.1) and (9.2) can be written as three partial differential equations in terms of the three unknowns v^α and w. It is easily verified that unless there is a coupling through the external forces the complete system will consist of a single differential equation in w only, and two equations (in general coupled, to be sure) in v^α. If, in addition, the boundary conditions are uncoupled between w and v^α, the problem is completely separated into one for v^α and one for w.

In the first case, the problem of in-plane loading, we take $w \equiv 0$ and consider the displacements v^α only, while in the second case, the problem of transversely loaded plates, we consider the displacements w only, taking $v^\alpha \equiv 0$.

3. In-plane loaded plates

In the case of in-plane loading it is generally preferable to formulate the problem in terms of the stress-functions and in this case we can in fact express all dependent variables in terms of one single scalar stress function Φ. Thus, according to (8.5) and (8.18) we have, when $d_{\alpha\beta} \equiv 0$,

$$N^{\alpha\beta} = -\varepsilon^{\alpha\xi}\varepsilon^{\beta\eta}D_\xi D_\eta \Phi. \tag{9.5}$$

Clearly, this satisfies the homogeneous equations of equilibrium

$$D_\alpha N^{\alpha\beta} = 0, \tag{9.6}$$

but we must also satisfy the equation of compatibility (3.48). Substituting (9.5) into (7.12) and the strain tensor $E^{\alpha\beta}$ into (3.48), taking only the linear terms into account, yields

$$\Delta^2 \Phi = 0. \tag{9.7}$$

A function that is four times continuously differentiable and satisfies (9.7) is called *biharmonic*.[1]

---

[1] This function is often called Airy's stress function after the British astronomer G.B. AIRY, who first observed that the stress components in plane stress can be expressed as the second partial differential quotients of a single function (1863). AIRY was not aware of the requirement given by equation (9.7).

The complete solution of (9.1) is obtained now as the sum of a particular integral of (9.1) and the membrane tensor (9.5), where Φ is a biharmonic function.

This opens the way to the solution of a number of technically important problems. For rectangular coordinates, one way is through the application of complex function theory. With $z = x + iy$ and $\bar{z} = x - iy$ we get

$$\frac{\partial}{\partial x} = \frac{\partial}{\partial z} + \frac{\partial}{\partial \bar{z}}, \qquad \frac{\partial^2}{\partial x^2} = \frac{\partial^2}{\partial z^2} + 2\frac{\partial^2}{\partial z \partial \bar{z}} + \frac{\partial^2}{\partial \bar{z}^2}$$

$$\frac{\partial}{\partial y} = i\frac{\partial}{\partial z} - i\frac{\partial}{\partial \bar{z}}, \qquad \frac{\partial^2}{\partial y^2} = -\frac{\partial^2}{\partial z^2} + 2\frac{\partial^2}{\partial z \partial \bar{z}} - \frac{\partial^2}{\partial \bar{z}^2}$$

and hence

$$\frac{\partial^2}{\partial x^2} + \frac{\partial^2}{\partial y^2} = 4\frac{\partial^2}{\partial z \partial \bar{z}}.$$

From this it follows that

$$\Delta^2 \Phi = 16 \frac{\partial^4 \Phi}{\partial z^2 \partial \bar{z}^2}$$

and the equation

$$\frac{\partial^4 \Phi}{\partial z^2 \partial \bar{z}^2} = 0$$

can be integrated directly, giving

$$\Phi = \chi_1(z) + \chi_2(\bar{z}) + \bar{z}\phi_1(z) + z\phi_2(\bar{z}),$$

where $\phi_1, \phi_2, \chi_1, \chi_2$ are arbitrary analytic functions. Since Φ is real, we must take

$$\phi_2(\bar{z}) = \overline{\phi_1(z)}, \qquad \chi_2(\bar{z}) = \overline{\chi_1(z)}$$

and can thus write

$$2\Phi = \bar{z}\phi(z) + z\overline{\phi(z)} + \chi(z) + \overline{\chi(z)}$$

or

$$\Phi = \mathrm{Re}\{\bar{z}\phi(z) + \chi(z)\}. \tag{9.8}$$

The expression was first obtained by E. GOURSAT in 1898. It is now possible to express all relevant quantities in terms of the two complex functions ϕ and χ. Thus,

$$N^{11} + N^{22} = -\Delta\Phi = -4\frac{\partial^2 \Phi}{\partial z \partial \bar{z}}$$

and

$$N^{11} - N^{22} - 2\mathrm{i}N^{12} = \frac{\partial^2 \Phi}{\partial x^2} - \frac{\partial^2 \Phi}{\partial y^2} - 2\mathrm{i}\frac{\partial^2 \Phi}{\partial x \partial y} = 4\frac{\partial^2 \Phi}{\partial z^2}.$$

Therefore, we get

$$N^{11} + N^{22} = -2[\phi'(z) + \bar{\phi}'(\bar{z})] \tag{9.9}$$

and

$$N^{11} - N^{22} - 2\mathrm{i}N^{12} = 2[\bar{z}\phi''(z) + \chi''(z)]. \tag{9.10}$$

From these two expressions all three components of the membrane stress tensor $N^{\alpha\beta}$ are easily determined. Similarly, the displacements, represented by the complex displacement vector $V = u + iv$, can be expressed in terms of ϕ and χ. We find

$$2\frac{\partial V}{\partial \bar{z}} = \frac{\partial V}{\partial x} + \mathrm{i}\frac{\partial V}{\partial y} = \frac{\partial u}{\partial x} - \frac{\partial v}{\partial y} + \mathrm{i}\left(\frac{\partial u}{\partial y} + \frac{\partial v}{\partial x}\right)$$
$$= E_{11} - E_{22} + 2\mathrm{i}E_{12},$$

where (9.3) has been used. With the help of Hooke's law (7.11) the right-hand side can be expressed in terms of the complex conjugate of

(9.10), and we therefore get

$$\frac{\partial V}{\partial \bar{z}} = \frac{1+\nu}{Eh}[z\bar{\phi}''(\bar{z}) + \bar{\chi}''(\bar{z})].$$

Integrating, we find that

$$V = \frac{1+\nu}{Eh}[z\bar{\phi}'(\bar{z}) + \bar{\chi}'(\bar{z}) + \psi(z)],$$

where ψ is a so far undetermined function of z. To determine this function, we take

$$2\frac{\partial V}{\partial z} = \frac{\partial V}{\partial x} - i\frac{\partial V}{\partial y} = \frac{\partial u}{\partial x} + \frac{\partial v}{\partial y} + i\left(\frac{\partial v}{\partial x} - \frac{\partial u}{\partial y}\right)$$

$$= E_{11} + E_{22} + 2i\theta$$

where θ is the rotation around the normal to the middle-plane according to (3.30). Applying Hooke's law (7.11) again, we can write, using (9.9),

$$2\frac{\partial V}{\partial z} = \frac{1-\nu}{Eh}(N^{11} + N^{22}) + 2i\theta$$

$$= -2\frac{1-\nu}{Eh}[\phi'(z) + \bar{\phi}'(\bar{z})] + 2i\theta.$$

Hence

$$\frac{1+\nu}{Eh}[\bar{\phi}'(\bar{z}) + \psi'(z)] = -\frac{1-\nu}{Eh}[\phi'(z) + \bar{\phi}'(\bar{z})] + 2i\theta.$$

Equating the real parts of both sides, we get

$$\frac{1}{2}\frac{1+\nu}{Eh}[\bar{\phi}'(\bar{z}) + \psi'(z) + \phi'(z) + \bar{\psi}'(\bar{z})] = -\frac{1-\nu}{Eh}[\phi'(z) + \bar{\phi}'(\bar{z})]$$

and hence

$$\psi'(z) + \phi'(z) + 2\frac{1-\nu}{1+\nu}\phi'(z) = -\bar{\psi}'(\bar{z}) - \bar{\phi}'(\bar{z}) - 2\frac{1-\nu}{1+\nu}\bar{\phi}'(\bar{z}).$$

Since the left-hand side is a function of z only and the right-hand side is a function of $\bar{z}$ only, each side must equal a (complex) constant A_1. But clearly, $A_1 = -\bar{A}_1$ and A_1 is therefore purely imaginary. Integration yields

$$\psi(z) = -\frac{3-\nu}{1+\nu}\phi(z) + A_1 z + B_1.$$

This leads to the following equation for the displacement vector,

$$u + iv = \frac{1+\nu}{Eh}\left[z\bar{\phi}'(\bar{z}) + \bar{\chi}'(\bar{z}) - \frac{3-\nu}{1+\nu}\phi(z)\right] + Az + B. \quad (9.11)$$

The displacements are determined up to a rigid body motion, represented by the rotation A (an imaginary number) and the translation B.

The problem of determining the in-plane stresses and displacements of a plate is reduced to determination of two complex analytic functions ϕ and χ, that fulfil certain conditions on the boundary.

Turning the problem around, we might take any two analytic functions ϕ and χ and determine the boundary stresses and the displacements from (9.9), (9.10) and (9.11), hoping that the result would be of interest. Actually, some interesting results can be obtained immediately if we take some low order polynomials for ϕ and χ. Thus, taking for example

$$\phi(z) = -2icz^3 \quad \text{and} \quad \chi(z) = ic(z^4 + 3a^2 z^2)$$

we get from (9.9) and (9.10)

$$N^{11} = -48cxy, \quad N^{12} = 24cy^2 - 6ca^2, \quad N^{22} = 0.$$

When this result is applied to the rectangular region

$$-l \leq x \leq 0, \quad -\tfrac{1}{2}a \leq y \leq \tfrac{1}{2}a$$

the sides $y = \pm\tfrac{1}{2}a$ turn out to be stress-free. Furthermore, on the sides $x = 0$ and $x = -l$ there will be a parabolic shear stress distribution of equal magnitude. Finally, on the side $x = -l$ there will also be a linearly varying normal stress distribution.

The result can be interpreted as the 'ideal' stress distribution in a beam of rectangular cross-section (of width h and height a), which transmits a constant shear force

$$P = \int_{-\frac{1}{2}a}^{\frac{1}{2}a} N^{12}\,dy = -4ca^3.$$

This shear force is balanced by a bending moment

$$M = \int_{-\frac{1}{2}a}^{\frac{1}{2}a} N^{11} y\,dy = -4ca^3 x$$

that increases linearly with x. The displacements are found from (9.11) to be

$$u = \frac{2c}{Eh}[4y^3(2+\nu) - 3a^2 y(1+\nu) - 12x^2 y]$$

and

$$v = \frac{2c}{Eh}[4x^3 - 3a^2 x(1+\nu) + 12\nu xy^2]$$

where we have taken $A = B = 0$.

This result serves to illustrate how the true state of stress and deformation deviates from the assumed state in elementary beam theory. From this result some confidence is gained with respect to the approximate stress distribution assumed in bending of plates and shells (see equations (6.26) and (6.27)).

We shall not pursue this subject further here, except for emphasizing that the fact that any biharmonic function Φ can be written in terms of functions of a complex variable is of greatest importance, since the properties of such functions are generally well known. An extensive amount of work has been done in this field, and the student who wishes to follow this subject further is advised to consult a treatise on plane elasticity.[2]

4. Bending theory of plates

Substitution of (9.4) into (7.11) and $M^{\alpha\beta}$ from (7.11) into (9.2) yields the following fourth order partial differential equation for w,

$$\Delta^2 w = p/D. \qquad (9.12)$$

Again the complete solution to the problem can be obtained as a sum of a particular integral and a solution to the homogeneous equation

$$\Delta^2 w = 0$$

so that in fact the problem of bending leads to much the same problem as the one of in-plane loading.

It should be noted that in the case of plates the scalar field Φ in equation (5.20) vanishes identically and that therefore we have

$$\mathcal{M}^{\alpha\beta} = M^{\alpha\beta}.$$

The twisting moment is therefore given by the formula

$$M_V = M^{\alpha\beta} n_\alpha t_\beta$$

for any plate.

5. Bending of plates in rectangular coordinates

In rectangular coordinates the metric tensor is constant and all covariant derivatives reduce to partial derivatives.

Let $a_{\alpha\beta} = \delta^\alpha_\beta$. Then $a^{\alpha\beta} = a_{\alpha\beta}$ and the plate equation (9.12) takes the form

---

[2] A standard treatise on this subject is MUSKHELISHVILI, N.I. (1953), *Some Basic Problems of the Mathematical Theory of Elasticity*, Noordhoff, Groningen.

$$w_{,\alpha\alpha\beta\beta} = p/D. \tag{9.13}$$

Consider a straight portion of the boundary $u^1 = $ const with the normal $n_\alpha = (1, 0)$ and tangent $t_\alpha = (0, 1)$. Since $K_{\alpha\beta} = w_{,\alpha\beta}$, we get according to (6.15)[3]

$$M_B = M^{11} = D(w_{,11} + \nu w_{,22}) = D\left(\frac{\partial^2 w}{\partial n^2} + \nu \frac{\partial^2 w}{\partial t^2}\right) \tag{9.14}$$

at the boundary. According to (6.16) the effective shear force Q is

$$Q = -D(\Delta w)_{,1} - D(1-\nu)w_{,122} = -D[w_{,111} + (2-\nu)w_{,122}]$$

and hence

$$Q = -D\left[\frac{\partial^3 w}{\partial n^3} + (2-\nu)\frac{\partial^3 w}{\partial n \partial t^2}\right]. \tag{9.15}$$

We can summarize some often applied boundary conditions as follows:

Clamped edge:
$$w = 0 \quad \text{and} \quad \frac{\partial w}{\partial n} = 0;$$

Simply supported edge:
$$w = 0 \quad \text{and} \quad \frac{\partial^2 w}{\partial n^2} = 0;$$

Free edge:
$$\frac{\partial^2 w}{\partial n^2} + \nu \frac{\partial^2 w}{\partial t^2} = 0 \quad \text{and} \quad \frac{\partial^3 w}{\partial n^3} + (2-\nu)\frac{\partial^3 w}{\partial n \partial t^2} = 0.$$

6. Corners

Condition (6.16) was derived under the assumption that the boundary was smooth. At a corner there is a sudden change in the directions of the normal and tangent vectors and the second term of (6.16),

$$\partial(M^{\alpha\beta}n_\alpha t_\beta)/\partial s,$$

---

[3] In the linearized expressions the asterisks disappear.

becomes undefined. Since a rapid change of $M^{\alpha\beta}n_\alpha t_\beta$ along the boundary represents a large contribution to Q, one would expect a sudden change to correspond to a concentrated force. This is in fact so.

Let s_0 be the value of s at a corner. The shear-force P_C at the corner is then

$$P_C = \lim_{\varepsilon \to 0} \int_{s_0-\varepsilon}^{s_0+\varepsilon} Q \, ds$$

and according to (6.16) this yields

$$P_C = -M^{\alpha\beta}n_\alpha t_\beta \Big|_{s_0-0}^{s_0+0} \qquad (9.16)$$

since the first term of (6.16) is bounded.

We can write this force in terms of the twisting moment M_V. Using (4.12) and (5.20), equation (9.16) yields

$$P_C = M_V^{(-)} - M_V^{(+)}, \qquad (9.17)$$

where $^{(-)}$ indicates the value just before the corner (in the direction of s), and $^{(+)}$ the value just after the corner.

An example of this is the skew deformation of a square plate $(0 \leq u^1 \leq a\,;\, 0 \leq u^2 \leq a)$ defined by

$$w = cu^1u^2,$$

which clearly satisfies the plate equation (9.13) with $p = 0$ everywhere. It follows from (9.4) and (7.11) that

$$M^{11} = M^{22} = 0 \quad \text{and} \quad M^{12} = D(1-\nu)c$$

everywhere. Therefore, on all four boundaries we must have $Q = M_B = 0$. This, however, may not be interpreted to mean that the boundaries are unloaded. On the contrary, since M^{12} does not vanish, it will be seen that the boundary is loaded everywhere by a constant twisting moment $M_V = \pm D(1-\nu)c$ (see Fig. 9). But according to (6.16)

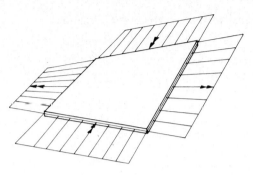

Fig. 9.

the contribution of this loading to Q is zero everywhere. However, due to (9.17) the distributed twisting moments are equivalent to four concentrated loads of magnitude $P_C = \pm 2D(1-\nu)c$, one at each corner (see Fig. 10).

We may therefore claim that whether we load the plate with distributed twisting moments according to Fig. 9 or with concentrated loads at the corners according to Fig. 10, the effect with regard to the deformation, the strains and the stresses will be essentially the same everywhere in the plate. This is clearly not true *at* the boundary and therefore certainly not true very close to the boundary, but except for a narrow boundary layer (comparable in width to the thickness of the plate) the equivalence of the two loading systems is a fact.

Since it is comparatively easy to test a square plate with the four corner loads in a testing machine, this case has been found suitable for testing the basic assumptions of plate theory.[4]

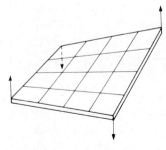

Fig. 10.

[4] See, for example. NADAI, A. (1925), *Die Elastischen Platten*, Springer, Berlin.

7. Bending of a simply supported rectangular plate

Let us consider a simply supported plate in the domain $0 \leq x \leq a$ and $0 \leq y \leq b$ loaded by the transverse normal load $p(x, y)$.[5] Following a procedure devised by C. NAVIER as early as 1820, we expanded the function $p(x, y)$ in a double Fourier series,

$$p(x, y) = \sum_m \sum_n A_{mn} \sin \frac{m\pi x}{a} \sin \frac{n\pi y}{b}, \qquad (9.18)$$

where the coefficients A_{mn} are given by the integrals

$$A_{mn} = \frac{4}{ab} \int_0^a \int_0^b p(x, y) \sin \frac{m\pi x}{a} \sin \frac{n\pi y}{b} \, dx \, dy. \qquad (9.19)$$

Assuming a solution $w(x, y)$ to the plate equation (9.13) also in the shape of a double Fourier series

$$w(x, y) = \sum_m \sum_n B_{mn} \sin \frac{m\pi x}{a} \sin \frac{n\pi y}{b} \qquad (9.20)$$

we determine the coefficients B_{mn} by substituting (9.19) and (9.20) into (9.13). Since the coefficient of term $\sin(m\pi x/a)\sin(n\pi y/b)$ must vanish for any m and n, we get

$$B_{mn} = \frac{A_{mn}}{\pi^4 D (m^2/a^2 + n^2/b^2)^2}. \qquad (9.21)$$

Since each term of (9.20) satisfies the conditions at a simply supported boundary ($w = \partial^2 w/\partial n^2 = 0$) we may hope that the infinite sum (9.20) does so too. The question of whether this is true or not must be answered by investigating the convergence of this series in each particular case.

For the case of a uniformly distributed load $p(x, y) = p_0$ we get

$$A_{mn} = \frac{16 p_0}{\pi^2 mn} \quad (m, n \text{ odd})$$

[5] We shall use the conventional notation (x, y) freely for rectangular coordinates instead of (u^1, u^2).

and

$$A_{mn} = 0 \quad (m \text{ or } n \text{ even}).$$

Thus we get

$$w(x, y) = \frac{16p_0}{\pi^6 D} \sum_m{}' \sum_n{}' \frac{\sin(m\pi x/a)\sin(n\pi y/b)}{mn(m^2/a^2 + n^2/b^2)^2}, \quad (9.22)$$

where Σ' indicates that only odd integers are considered. This is in fact a rapidly converging series and it is easily established that it satisfies the boundary conditions.

If the plate were loaded by a concentrated force P located at (x_0, y_0), we could write

$$p(x, y) = P\delta(x - x_0, y - y_0),$$

where $\delta(\xi, \eta)$ is the *Dirac delta function*.[6] This 'function' (which is not a function in the normal sense) can loosely be described as a function that is zero everywhere except at $\xi = \eta = 0$, where it is infinite in such a manner that its total volume $\iint \delta \, d\xi \, d\eta$ is unity. Hence, after integration of (9.19) we get

$$A_{mn} = \frac{4P}{ab} \sin\frac{n\pi x_0}{a} \sin\frac{m\pi y_0}{b} \quad (n, m \text{ odd}). \quad (9.23)$$

In particular, we find the deflection under a central force at $x_0 = x = \frac{1}{2}a$ and $y_0 = y = \frac{1}{2}b$ to be

$$w_0 = \frac{4P}{\pi^4 Dab} \sum_m{}' \sum_n{}' \frac{1}{(m^2/a^2 + n^2/b^2)^2}. \quad (9.24)$$

For a square plate $a = b$, this gives

$$w_0 = 0.01160 Pa^2/D.$$

Generally speaking, the Navier method, using a double Fourier

---

[6] See, for example, CARRIER, G., M. KROOK and C. PEARSON (1966), *Functions of a Complex Variable*, McGraw-Hill, New York, p. 318.

series expansion yields series for the deflection w that are rapidly converging. In fact, the series (9.22) may also be used for computing the components of the bending tensor $M^{\alpha\beta}$ and shear force vector Q^α, simply by taking the derivatives of w, term by term on the right-hand side.

8. The clamped and uniformly loaded elliptical plate

To solve the plate equation (9.12) with given boundary conditions, one is usually well advised to apply a coordinate system that fits the boundary of the plate. There is a great advantage in having simple expressions for the conditions at the boundary, and in general this advantage fully compensates for the disadvantage of a more complicated differential equation.

An exception to this rule is found in the case of an elliptic plate, clamped along the entire boundary and loaded by a uniformly distributed load p_0. For this special case, rectangular coordinates are very adequate.

We shall look for a solution in terms of a fourth degree polynomial in x and y, since such a polynomial would provide a right-hand side of equation (9.12) that is constant. In addition, the polynomial must vanish at the boundary

$$(x/a)^2 + (y/b)^2 - 1 = 0,$$

where a and b are the semi-axes of the elliptic boundary. Also, the normal derivative of the function must vanish on this ellipse.

All the requirements are clearly satisfied by the function

$$w = w_0[(x/a)^2 + (y/b)^2 - 1]^2, \qquad (9.25)$$

where w_0 is a constant. Substitution into (9.12) yields

$$w_0\left(\frac{24}{a^4} + \frac{24}{b^4} + \frac{16}{a^2 b^2}\right) = \frac{p}{D}, \qquad (9.26)$$

which determines the constant w_0. From (9.25) all other relevant quantities can be determined. For instance, the maximum bending moment at the boundary appears at the intersection of the boundary with the minor semi-axis, at $x = 0$ and $y = \pm b$ and is

$$(M_B)_{\max} = \frac{8w_0 D}{b^2}. \tag{9.27}$$

It must be emphasized that this example is rather an exception to the rule that the coordinate system should fit the boundary. For instance, a simply supported elliptic plate cannot be solved in this simple fashion and, generally speaking, elliptic plates are best studied in elliptic coordinate systems.

9. Vibrating plates

The dynamic load acting on a vibrating shell or plate is derived from the inertia forces acting on each element of the body. Assuming harmonic motion of angular frequency ω and a displacement amplitude[7] $v^i(u^1, u^2, z)$, the d'Alembert forces per unit volume are

$$p^i = -\gamma \frac{\partial^2}{\partial t^2}[v^i(u^1, u^2, z) \sin \omega t] = \omega^2 \gamma v^i,$$

where γ is the mass density of the body.

To proceed it is necessary to express the three-dimensional displacements v^i in terms of the displacements of the middle-surface. This we can only do approximately, assuming that the displacements are of the Kirchhoff type (see equation (5.3)). For a plate we therefore assume that

$$v^\alpha = v^\alpha - z D^\alpha w, \qquad v^3 = w.$$

From the volume forces p^i we now derive the following resultant force and moment (4.16)–(4.17),

$$\hat{p} = \omega^2 \gamma h w, \qquad m_\alpha = \omega^2 \gamma \frac{h^3}{12} w_{,\alpha},$$

---

[7] Boldface quantities are dependent on z.

corresponding to the effective load (4.27)

$$p = \hat{p} - D_\alpha m^\alpha = \omega^2 \gamma h w - \omega^2 \gamma \frac{h^3}{12} \Delta w.$$

The second term $-\omega^2\gamma(h^3/12)\Delta w$ represents the influence of rotational inertia. Since it is proportional to $h^3\Delta w$ and the first term to hw its relative importance is of order $(h/L)^2$, where L is a characteristic wave-length of the deformation pattern. For thin plates it is negligible. For thicker plates it may become of some importance, however, since terms of precisely this order of magnitude have been neglected in the stress–strain relations, there is no point in keeping it here.

We conclude that the expression

$$p = \omega^2 \gamma h w \tag{9.28}$$

for the effective load is consistent with the approximations introduced in our stress–strain relations.

With this, the plate equation (9.12) can be written as[8]

$$\Delta^2 w = \lambda w, \tag{9.29}$$

where

$$\lambda = \omega^2 \gamma h / D.$$

10. Energy methods

Solutions to the plate problem have been derived and are known in a number of primitive cases. Other cases can be solved by a variety of methods, and in particular energy methods are widely used. Here we shall confine ourselves to the *method of minimum potential energy* and the *Rayleigh–Ritz method*, which are approximative methods with special appeal to engineers.

The strain energy of a plate can be expressed in terms of the lateral

---

[8] For thick plates the corresponding equation was derived by the author (see NIORDSON, F.I. (1979), "An asymptotic theory for vibrating plates", *Internat. J. Solids & Structures* **15**, pp. 167–181.

deflection w in the form

$$\tfrac{1}{2}D \iint_{\mathcal{D}} [(1-\nu)(D^\alpha D_\beta w)(D_\alpha D^\beta w) + \nu(\Delta w)^2]\,dA\,,$$

when (9.4) is substituted into (7.8). Throughout the following we shall assume that the boundary conditions of our problem are such that either w or Q and either $\partial w/\partial n$ or M_B vanish at the boundary. This will include a free boundary ($Q = M_B = 0$), a simply supported boundary ($w = M_B = 0$), and a clamped boundary ($w = \partial w/\partial n = 0$). The case $Q = \partial w/\partial n = 0$ is also included, but has few practical applications. Conditions on w and $\partial w/\partial n$ will be called *kinematic* boundary conditions.

A logical consequence of our assumption with regards to the boundary conditions is that no work is produced by either Q or M_B at the boundary. The potential energy of the system is therefore

$$U[w] = \tfrac{1}{2}D \iint_{\mathcal{D}} [(1-\nu)(D^\alpha D_\beta w)(D_\alpha D^\beta w) + \nu(\Delta w)^2]\,dA - \iint_{\mathcal{D}} pw\,dA\,,$$

(9.30)

where the second term on the right-hand side is the potential of the external load.

We shall conceive $U[w]$ as a functional that maps the set of all twice continuously differentiable functions on the real numbers, $p(u^1, u^2)$ being a given function.

Let w_s denote the solution to the problem and $\bar{w}$ any twice continuously differentiable function that satisfies the kinematic boundary conditions of the problem. (Such a function will be called *admissible*.) Then

$$U[w_s + \bar{w}] = U[w_s] + U[\bar{w}] + \iint_{\mathcal{D}} M^{\alpha\beta} D_\alpha D_\beta \bar{w}\,dA\,,$$

where

$$M^{\alpha\beta} = D[(1-\nu)D^\alpha D^\beta w_s + \nu a^{\alpha\beta} \Delta w_s]$$

is the moment tensor of the solution. Now by repeated use of the divergence theorem, we find

$$\iint_\mathcal{D} M^{\alpha\beta} D_\alpha D_\beta \bar{w} \, dA = \oint_\mathcal{C} M_B \frac{\partial \bar{w}}{\partial n} \, ds + \oint_\mathcal{C} Q\bar{w} \, ds + \iint_\mathcal{D} p\bar{w} \, dA \,.$$

Due to the conditions at the boundary the products $M_B \partial \bar{w}/\partial n$ and $Q\bar{w}$ vanish everywhere and we get therefore

$$U[w_s + \bar{w}] = U[w_s] + \tfrac{1}{2} D \iint_\mathcal{D} [(1-\nu)(D^\alpha D_\beta \bar{w})(D_\alpha D^\beta \bar{w}) + \nu(\Delta \bar{w})^2] \, dA \,.$$

Hence

$$U[w_s + \bar{w}] \geq U[w_s], \tag{9.31}$$

where the equality sign holds if and only if $\bar{w} = 0$ everywhere. U has therefore a proper minimum at the solution w_s of the problem among the set of all admissible functions. This minimum is equal to

$$-\frac{1}{2} \iint_\mathcal{D} pw \, dA \,,$$

which is the negative of the *compliance* Φ of the plate for the load p. The inequality (9.31) therefore provides a lower bound for the compliance

$$\Phi \geq -U[w_s + \bar{w}] \,.$$

If the boundary is piecewise straight and the conditions are such that $w = 0$ everywhere on $\mathcal{C}$, the functional U can be considerably simplified. This will be the case when no portion of the polygonal boundary is free. Using the divergence theorem we find

$$\iint_\mathcal{D} M^{\alpha\beta} D_\alpha D_\beta w \, dA = \oint_\mathcal{C} M_B \frac{\partial w}{\partial n} \, ds + \oint_\mathcal{C} Qw \, ds + D \iint_\mathcal{D} w \Delta^2 w \, dA$$

and also

$$\iint_{\mathcal{D}} w\Delta^2 w \, dA = \oint_{\mathcal{C}} w \frac{\partial \Delta w}{\partial n} \, ds - \oint_{\mathcal{C}} \Delta w \frac{\partial w}{\partial n} + \iint_{\mathcal{D}} (\Delta w)^2 \, dA.$$

Hence

$$\iint_{\mathcal{D}} M^{\alpha\beta} D_\alpha D_\beta w \, dA = \oint_{\mathcal{C}} (M_B - D\Delta w) \frac{\partial w}{\partial n} \, ds$$

$$+ \oint_{\mathcal{C}} w \left(D \frac{\partial \Delta w}{\partial n} + Q \right) ds + D \iint_{\mathcal{D}} (\Delta w)^2 \, dA.$$

Since $w = 0$ on $\mathcal{C}$, we have $\Delta w = \partial^2 w/\partial n^2$ and, according to (9.14), $M_B = D\partial^2 w/\partial n^2$ on $\mathcal{C}$. Therefore, both line integrals vanish and we get

$$U[w] = \tfrac{1}{2} D \iint_{\mathcal{D}} (\Delta w)^2 dA - \iint_{\mathcal{D}} pw \, dA, \qquad (9.32)$$

which is considerably simpler than (9.30). It is, by the way, easy to see that if a portion of the boundary is free we cannot assert that the product $w(D\partial \Delta w/\partial n + Q)$ vanishes, since, according to (9.15),

$$D \frac{\partial \Delta w}{\partial n} + Q = -D(1 - \nu) \frac{\partial^3 w}{\partial n \partial t^2}.$$

In the energy method the minimum properties of the functional U are utilized to find an approximate solution. We take

$$w = \sum_{i=1}^{n} C_i \phi_i(u^1, u^2), \qquad (9.33)$$

where the functions ϕ_i are n given linearly independent admissible functions, and the C_i are coefficients to be determined. In selecting the functions ϕ_i the engineer is guided by his[9] experience, and the result

---

[9] "His" stands for "his or her" throughout this volume.

naturally depends on his choice. He will usually take ϕ_1 to represent his mental image of the solution and add such functions as he has reason to believe will improve the result taking, for instance, properties of symmetry and anti-symmetry into account.

When (9.33) is substituted into (9.30) or (9.32) as the case may be, we get a polynomial of second degree in the unknowns C_i on the right-hand side. The best possible approximation that we can obtain with the given functions ϕ_i is determined as the minimum value with respect to all coefficients C_i.

This leads to the following system of linear equations,

$$\sum_{j=1}^{n} A_{ij}C_j - B_i = 0, \quad i = 1, \ldots, n, \tag{9.34}$$

where the matrix A_{ij} is found either from (9.30),

$$A_{ij} = D \iint_{\mathcal{D}} [(1-\nu)(D^\alpha D_\beta \phi_i)(D_\alpha D^\beta \phi_j) + \nu \Delta\phi_i \Delta\phi_j] \, dA, \tag{9.35}$$

or from (9.32) (if $w = 0$ on $\mathcal{C}$),

$$A_{ij} = D \iint_{\mathcal{D}} \Delta\phi_i \Delta\phi_j \, dA, \tag{9.36}$$

and where

$$B_i = \iint_{\mathcal{D}} p\phi_i \, dA. \tag{9.37}$$

The coefficients C_i are found as the solution of (9.34) and (9.33) will provide the (approximate) solution.

For the eigenvalue problem (9.29) we use the Rayleigh–Ritz method, which is similar, except that the result is obtained in a somewhat different manner.[10]

---

[10] For a discussion of Rayleigh's principle and the Rayleigh-Ritz method, see, for instance, TEMPLE, G. and W.G. BICKLEY (1956), *Rayleigh's Principle*, Dover Publications.

Multiplying both sides of (9.29) by w and integrating, we get

$$\iint_{\mathcal{D}} w\Delta^2 w \, dA = \lambda \iint_{\mathcal{D}} w^2 \, dA. \tag{9.38}$$

But $D\Delta^2 w = D_\alpha D_\beta M^{\alpha\beta}$ and hence the left-hand side can be written as

$$\iint_{\mathcal{D}} wD_\alpha D_\beta M^{\alpha\beta} \, dA = \oint_{\mathcal{C}} wD_\beta M^{\alpha\beta} n_\alpha \, ds$$

$$- \oint_{\mathcal{C}} w_{,\alpha} M^{\alpha\beta} n_\beta \, ds + \iint_{\mathcal{D}} K_{\alpha\beta} M^{\alpha\beta} \, dA.$$

Due to the boundary conditions, the line integrals vanish and we get, with the help of (7.11),

$$\iint_{\mathcal{D}} w\Delta^2 w \, dA = \frac{1}{D} \iint_{\mathcal{D}} K_{\alpha\beta} M^{\alpha\beta} \, dA$$

$$= \iint_{\mathcal{D}} [(1-\nu) K^{\alpha\beta} K_{\alpha\beta} + \nu K^\alpha_\alpha K^\beta_\beta] \, dA$$

$$= \iint_{\mathcal{D}} [(1-\nu)(D^\alpha D^\beta w)(D_\alpha D_\beta w) + \nu(\Delta w)^2] \, dA,$$

which is the strain energy of the plate. Hence from (9.38) we have $\lambda = R[w]$ where

$$R[w] = \iint_{\mathcal{D}} [(1-\nu)(D^\alpha D^\beta w)(D_\alpha D_\beta w) + \nu(\Delta w)^2] \, dA \Big/ \iint_{\mathcal{D}} w^2 \, dA. \tag{9.39}$$

This is the so-called *Rayleigh quotient* of the problem. If the boundary is piecewise straight and the conditions are such that $w = 0$ on $\mathscr{C}$, we can write the quotient in the much simpler form

$$R[w] = \iint_{\mathscr{D}} (\Delta w)^2 \, dA \Big/ \iint_{\mathscr{D}} w^2 \, dA. \tag{9.40}$$

Defining $R[w]$ as a functional on all admissible functions one can show in a manner similar to that shown for U that

$$\lambda_1 \leq R[w], \tag{9.41}$$

where λ_1 is the smallest eigenvalue of the problem and where the equality sign holds good if and only if w is the solution of (9.29) corresponding to λ_1.

Let us write

$$R = T/N$$

where

$$T = \iint_{\mathscr{D}} [(1 - \nu)(D^\alpha D_\beta w)(D_\alpha D^\beta w) + \nu(\Delta w)^2] \, dA$$

or

$$T = \iint_{\mathscr{D}} (\Delta w)^2 \, dA$$

as the case may be, and

$$N = \iint_{\mathscr{D}} w^2 \, dA.$$

Then

$$\frac{\partial \lambda}{\partial C_i} = \frac{1}{N^2} \left(N \frac{\partial T}{\partial C_i} - T \frac{\partial N}{\partial C_i} \right) = \frac{1}{N} \left(\frac{\partial T}{\partial C_i} - R \frac{\partial N}{\partial C_i} \right)$$

and therefore

$$\frac{\partial T}{\partial C_i} - R \frac{\partial N}{\partial C_i} = 0. \tag{9.42}$$

T and N are quadratic functions of the C_i's and equations (9.42) are therefore linear and homogeneous and can be written in the form

$$\sum_{i=1}^{n} (A_{ij} - RB_{ij})C_i = 0, \quad j = 1, \ldots, n, \tag{9.43}$$

where A_{ij} is the coefficient of C_i in $\partial T/\partial C_j$ and B_{ij} the coefficient of C_i in $\partial N/\partial C_j$,

$$A_{ij} = 2 \iint_{\mathcal{D}} [(1-\nu)(D^\alpha D_\beta \phi_i)(D_\alpha D^\beta \phi_j) + \nu \Delta \phi_i \Delta \phi_j] \, \mathrm{d}A \tag{9.44}$$

when (9.39) is used and

$$A_{ij} = 2 \iint_{\mathcal{D}} \Delta \phi_i \Delta \phi_j \, \mathrm{d}A \tag{9.45}$$

when (9.40) is used. Furthermore, we have

$$B_{ij} = 2 \iint_{\mathcal{D}} \phi_i \phi_j \, \mathrm{d}A . \tag{9.46}$$

The system (9.43) will have nontrivial solutions only when its determinant vanishes,

$$|A_{ij} - RB_{ij}| = 0, \tag{9.47}$$

which is an algebraic equation of degree n in R. It is well known that the n roots of this equation are all real and constitute upper bounds for the n lowest eigenvalues of equation (9.29).

In the following paragraphs we shall give an example concerning the static deflection of a plate under a given load, and another one for a vibrating plate, using the Rayleigh–Ritz procedure.

11. Bending of a clamped rectangular plate

As an example of the method of minimum potential energy we shall determine (approximately) the deflection of a clamped rectangular plate in the region $-a \leq x \leq a$ and $-b \leq y \leq b$, subject to a uniformly distributed transverse load p.

Let us take the following set of functions,

$$\phi_1 = (x^2 - a^2)^2(y^2 - b^2)^2, \qquad \phi_3 = (x^2 - a^2)^2(y^2 - b^2)^3,$$
$$\phi_2 = (x^2 - a^2)^3(y^2 - b^2)^2, \qquad \phi_4 = (x^2 - a^2)^3(y^2 - b^2)^3,$$

which are clearly admissible, fulfilling the kinematic boundary conditions $\phi = \partial\phi/\partial n = 0$ on all sides of the rectangular region. Substitution into (9.36) yields

$$A_{11} = 20.805(a^4 + b^4)a^5b^5D + 11.889a^7b^7D,$$
$$A_{12} = -18.914a^{11}b^5D - 11.889a^9b^7D - 17.833a^7b^9D,$$
$$A_{13} = -17.833a^9b^7D - 11.889a^7b^9D - 18.914a^5b^{11}D,$$
$$A_{14} = 16.212a^{11}b^7D + 11.889a^9b^9D + 16.212a^{11}b^7D,$$
$$A_{22} = 17.459a^{13}b^5D + 12.969a^{11}b^7D + 23.777a^9b^9D,$$
$$A_{23} = A_{14},$$
$$A_{24} = -14.965a^{13}b^7D - 12.969a^{11}b^9D - 21.616a^9b^{11}D,$$
$$A_{33} = 23.777a^9b^9D + 12.969a^7b^{11}D + 17.459a^5b^{13}D,$$
$$A_{34} = -21.616a^{11}b^9D - 12.969a^9b^{11}D - 14.965a^7b^{13}D,$$
$$A_{44} = 19.953(a^4 + b^4)a^9b^9D + 14.148a^{11}b^{11}D.$$

The remaining elements of the matrix A_{ij} are given by the symmetry relation

$$A_{ij} = A_{ji}.$$

In additon, we get the following elements of the vector B_i,

$$B_1 = 1.1778a^5b^5p, \qquad B_3 = 0.9752a^5b^7p,$$
$$B_2 = 0.9752a^7b^5p, \qquad B_4 = 0.8359a^7b^7p.$$

Solving the system for different ratios b/a we find the values presented in Table 1 for the deflection w_0 at the centre $x = 0$, $y = 0$.

In comparison with a more accurate solution [11] the errors are small. The largest error affects the value of w_0 for $b/a = 2$, being about 1.4 per cent. For engineering purposes such approximations are generally acceptable.

TABLE 1. Deflection w_0 at the centre of a clamped rectangular uniformly loaded plate.

b/a	w_0
1	$0.02023\, pa^4/D$
1.2	$0.02750\, pa^4/D$
1.4	$0.03302\, pa^4/D$
1.6	$0.03664\, pa^4/D$
1.8	$0.03884\, pa^4/D$
2.0	$0.04003\, pa^4/D$

12. Lowest natural frequency of a skew plate

As an example of the Rayleigh–Ritz method we shall determine an approximate value of the lowest natural frequency of a simply supported skew plate (see Fig. 11), having the side lengths a and b with the angle α between them.

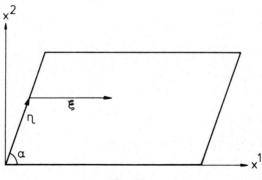

FIG. 11.

---

[11] See EVANS, T.H. (1939), *J. Appl. Mech.* **6**, p. A-7.

To fit the boundaries, we shall use an oblique coordinate system defined by

$$x^1 = \xi + \eta \cos \alpha, \qquad x^2 = \eta \sin \alpha, \qquad x^3 = 0, \tag{9.48}$$

where we have used (ξ, η) instead of (u^1, u^2). From these equations we derive the components of the metric tensor,

$$a_{\alpha\beta} = \begin{pmatrix} 1 & \cos \alpha \\ \cos \alpha & 1 \end{pmatrix}, \qquad a^{\alpha\beta} = \frac{1}{\sin^2 \alpha} \begin{pmatrix} 1 & -\cos \alpha \\ -\cos \alpha & 1 \end{pmatrix}. \tag{9.49}$$

Since the components of $a_{\alpha\beta}$ are constant, all Christoffel symbols vanish and the covariant derivatives reduce to partial ones. The Laplace operator is found to be

$$\Delta = \frac{1}{\sin^2 \alpha} \left(\frac{\partial^2}{\partial \xi^2} - 2 \cos \alpha \frac{\partial^2}{\partial \xi \partial \eta} + \frac{\partial^2}{\partial \eta^2} \right) \tag{9.50}$$

so that the plate equations can be written as

$$\frac{\partial^4 w}{\partial \xi^4} - 4 \cos \alpha \frac{\partial^4 w}{\partial \xi^3 \partial \eta} + 2(1 + 2 \cos^2 \alpha) \frac{\partial^4 w}{\partial \xi^2 \partial \eta^2}$$

$$- 4 \cos \alpha \frac{\partial^4 w}{\partial \xi \partial \eta^3} + \frac{\partial^4 w}{\partial \eta^4} = \frac{p \sin^4 \alpha}{D}. \tag{9.51}$$

Taking $p = \omega^2 \rho h w$ from (9.28) the plate equation can be written in the form (9.29) with

$$\omega^2 = \lambda \frac{E h^2}{12(1 - \nu^2)\rho}. \tag{9.52}$$

With the help of the oblique coordinates we have mapped the skew region of the plate into the rectangular domain $0 \leq \xi \leq a$, $0 \leq \eta \leq b$. The eigenfunction $w(\xi, \eta)$ can now be expanded in a double Fourier series (9.20) since any function of the type

$$\sin \frac{m \pi x}{a} \sin \frac{n \pi y}{b} \qquad (n, m \text{ integers})$$

is an admissible function. In applying the Rayleigh–Ritz method we only take the few first terms of the expansion (9.20). Clearly, a reasonably good first approximation of the fundamental mode is

$$\phi_1 = \sin\frac{\pi x}{a}\sin\frac{\pi y}{b}$$

and of the next terms of (9.20) we take only such terms where $n + m$ is even, since that is required by the conditions of symmetry. Taking the function

$$\phi_1 = \sin\frac{\pi x}{a}\sin\frac{\pi y}{b}$$

we get, from (9.45) and (9.46),

$$A_{11} = \frac{\pi^4}{2\sin^4\alpha}\left[\frac{b}{a^3} + \frac{2}{ab}(1+2\cos^2\alpha) + \frac{a}{b^3}\right], \quad B_{11} = \frac{\pi^4}{2ab}.$$

From the determinant (9.47) with one single element we determine R, an approximate value and an upper bound for λ_1, i.e.,

$$\lambda_1 \approx \frac{ab}{\sin^4\alpha}\left[\frac{b}{a^3} + \frac{2}{ab}(1+2\cos^2\alpha) + \frac{a}{b^3}\right]. \quad (9.53)$$

Taking $a = 1$ and $b = 2$ as an example we get the numerical values for λ_1 as given in the second column ($n = 1$) of Table 2 for different values of α.

TABLE 2. Approximate values of λ_1 for a skew plate with sides $a = 1$ and $b = 2$, using a 1, 2 and 5 term expansion.

α	$n = 1$	$n = 2$	$n = 5$
90°	25.00	25.00	25.00
85°	25.51	25.40	25.40
80°	27.09	26.66	26.64
75°	29.95	28.93	28.87
70°	34.46	32.54	32.33
65°	41.29	38.02	37.48
60°	51.56	46.29	45.10

To improve the result we take in additon to ϕ_1 a second function

$$\phi_2 = \sin\frac{2\pi x}{a} \sin\frac{2\pi y}{b}.$$

We find

$$A_{12} = A_{21} = \frac{9\pi^2 \cos\alpha}{80 \sin^4\alpha}\left(\frac{1}{a^2}+\frac{1}{b^2}\right),$$

$$A_{22} = \frac{8\pi^4}{\sin^4\alpha}\left[\frac{b}{a^3}+\frac{2}{ab}(1+2\cos^2\alpha)+\frac{a}{b^3}\right],$$

$$B_{22} = \frac{\pi^4}{2ab}.$$

The determinant now renders an equation for R of second degree, the smallest root of which is given in Table 2 in the third column ($n = 2$). There is a noticeable improvement and the values are rather close to the values of the last column, which was obtained using a five-term expansion by adding the functions

$$\phi_3 = \sin\frac{\pi x}{a}\sin\frac{3\pi y}{b},$$

$$\phi_4 = \sin\frac{3\pi x}{a}\sin\frac{\pi y}{b},$$

$$\phi_5 = \sin\frac{3\pi x}{a}\sin\frac{3\pi y}{b}$$

to the set.

Since our choice of functions ($n + m$ even) utilized the symmetry of the fundamental mode, the second root of the algebraic equation could not provide an acceptable approximation for λ_2. To obtain a good approximation for λ_2 we should instead take terms with $n + m$ odd, thereby utilizing the anti-symmetry of the second fundamental mode.

13. Bending of plates in polar coordinates

Let the polar coordinate system be defined by

$$x^1 = r\cos\phi, \qquad x^2 = r\sin\phi, \qquad x^3 = 0, \tag{9.54}$$

where the conventional notation (r, ϕ) has been used for (u^1, u^2). In this coordinate system we have

$$a_{\alpha\beta} = \begin{pmatrix} 1 & 0 \\ 0 & r^2 \end{pmatrix}, \qquad a^{\alpha\beta} = \begin{pmatrix} 1 & 0 \\ 0 & 1/r^2 \end{pmatrix} \tag{9.55}$$

and

$$\{{}^1_{22}\} = -r, \qquad \{{}^2_{12}\} = \{{}^2_{21}\} = 1/r, \tag{9.56}$$

while all the other Christoffel symbols vanish.

The Laplace operator takes the form

$$\Delta = \frac{\partial^2}{\partial r^2} + \frac{1}{r}\frac{\partial}{\partial r} + \frac{1}{r^2}\frac{\partial^2}{\partial \phi^2} \tag{9.57}$$

and the plate equation

$$\left(\frac{\partial^2}{\partial r^2} + \frac{1}{r}\frac{\partial}{\partial r} + \frac{1}{r^2}\frac{\partial^2}{\partial \phi^2}\right)^2 w = \frac{p}{D}. \tag{9.58}$$

On a boundary circle with $r = $ const we get the bending moment

$$M_B = D\left[\frac{\partial^2 w}{\partial r^2} + \frac{\nu}{r}\frac{\partial w}{\partial r} + \frac{\nu}{r^2}\frac{\partial^2 w}{\partial \phi^2}\right] \tag{9.59}$$

and the effective shear-force

$$Q = -D\left[\frac{\partial^3 w}{\partial r^3} + \frac{1}{r}\frac{\partial^2 w}{\partial r^2} - \frac{1}{r^2}\frac{\partial w}{\partial r} + \frac{2-\nu}{r^2}\frac{\partial^3 w}{\partial r \partial \phi^2} - \frac{3-\nu}{r^3}\frac{\partial^2 w}{\partial \phi^2}\right]. \tag{9.60}$$

On a radial section $\phi = $ const we get instead

$$M_B = D\left[\frac{1}{r^2}\frac{\partial^2 w}{\partial \phi^2} + \frac{1}{r}\frac{\partial w}{\partial r} + \nu\frac{\partial^2 w}{\partial r^2}\right] \quad (9.61)$$

and

$$Q = -D\left[\frac{1}{r^3}\frac{\partial^3 w}{\partial \phi^3} - \frac{1-\nu}{r^2}\frac{\partial^2 w}{\partial r\partial \phi} + \frac{2-\nu}{r^3}\frac{\partial w}{\partial \phi} + \frac{2-\nu}{r}\frac{\partial^3 w}{\partial r^2 \partial \phi}\right] \quad (9.62)$$

for the bending moment and the effective shear-force.

14. Axisymmetrical bending of a circular plate

In the case of axial symmetry, w is a function of r only and the differential equation (9.58) reduces to

$$\left(\frac{d^2}{dr^2} + \frac{1}{r}\frac{d}{dr}\right)^2 w = \frac{p}{D}.$$

But

$$\frac{d^2}{dr^2} + \frac{1}{r}\frac{d}{dr} = \frac{1}{r}\frac{d}{dr}r\frac{d}{dr}$$

and hence

$$\frac{1}{r}\frac{d}{dr}r\frac{d}{dr}\frac{1}{r}\frac{d}{dr}r\frac{dw}{dr} = \frac{p}{D}, \quad (9.63)$$

which can be solved by quadratures.

If the load is uniformly distributed ($p = $ const), we get upon integration

$$w = \frac{pr^4}{64D} + C_1 r^2 \log r + C_2 r^2 + C_3 \log r + C_4, \quad (9.64)$$

where C_1, C_2, C_3 and C_4 are arbitrary constants, determined by the boundary conditions.

As an example, let us determine the deflection $w_{\max}$ of an annular plate, clamped at its inner boundary $r = a$ and free at its outer rim $r = b$ (see Fig. 12). We find from the boundary conditions $dw/dr = 0$ at $r = a$,

$$\frac{pa^2}{16D} + 2C_1 a \log a + C_1 a + 2C_2 a + \frac{C_3}{a} = 0.$$

From the condition $M_B = 0$ at $r = b$ we get

$$\frac{(3+\nu)pb^2}{16D} + C_1[2(1+\nu)\log b + 3 + \nu] + 2(1+\nu)C_2 - \frac{1-\nu}{b^2}C_3 = 0,$$

and from $Q = 0$ at $r = b$ we get

$$\frac{pb}{2D} + \frac{4}{b}C_1 = 0.$$

From these three equations, the constants C_1, C_2 and C_3 are determined and substitution into

$$w_{\max} = \frac{p(b^4 - a^4)}{64D} + C_1(b^2 \log b - a^2 \log a) + C_2(b^2 - a^2) + C_3 \log \frac{b}{a}$$

yields

$$w_{\max} = \frac{pa^4}{D} f(\beta),$$

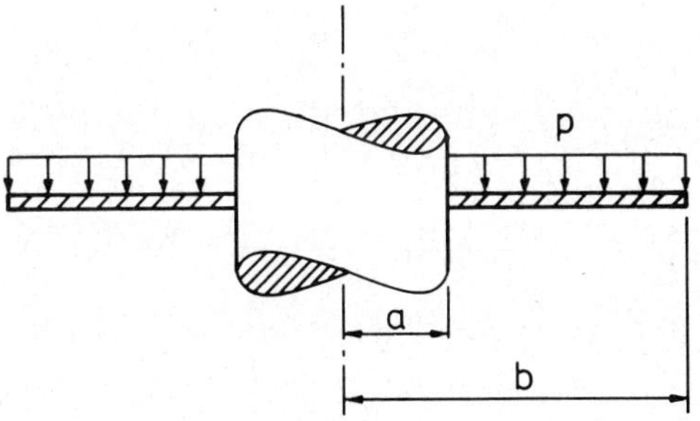

FIG. 12.

where

$$\beta = b/a$$

and

$$f(\beta) = \{1 - \nu - \beta^2\{7 - 5\nu + 4[1 - \nu + \beta^2(5 - \nu)]\log \beta\}$$
$$- \beta^4[1 + 7\nu + 16(1 + \nu)\log^2 \beta] + \beta^6(7 + 3\nu)\}$$
$$/\{64[1 - \nu + \beta^2(1 + \nu)]\}\,.$$

Other cases with clamped, simply supported, and free boundaries are solved in a similar way.

15. Free vibrations of a circular plate

Let us write the plate equation in the form

$$\Delta^2 w = k^4 w, \qquad (9.65)$$

where

$$k^4 = \omega^2 \gamma h/D. \qquad (9.66)$$

Factorizing the equation we get

$$(\Delta - k^2)(\Delta + k^2)w = 0,$$

where Δ is given by (9.57). Now we expand w in the Fourier series,

$$w = \sum_{n=-\infty}^{\infty} y_n(r)\, e^{in\phi}. \qquad (9.67)$$

Then every term alone must satisfy the differential equation and hence y_n must be a solution of

$$\left(\frac{d^2}{dr^2} + \frac{1}{r}\frac{d}{dr} - \frac{n^2}{r^2} - k^2\right)\left(\frac{d^2}{dr^2} + \frac{1}{r}\frac{d}{dr} - \frac{n^2}{r^2} + k^2\right)y = 0.$$

We find immediately two independent solutions of this equation, both regular for $r = 0$, namely $J_n(kr)$ and $J_n(ikr)$ with $i = \sqrt{-1}$, the Bessel functions of first kind and order n. Therefore, the function

$$w = A_1 J_n(kr) \cos n\phi + A_2 J_n(ikr) \cos n\phi \qquad (9.68)$$

is a solution of (9.65), i.e., an eigenfunction. The eigenvalue k is determined from the boundary conditions.

For a plate, which is *clamped* at its boundary $r = a$, we get

$$A_1 J_n(\mu) + A_2 J_n(i\mu) = 0$$

and

$$A_1 J'_n(\mu) + A_2 i J'_n(i\mu) = 0,$$

where $\mu = ka$.

The condition for a nontrivial solution is that the determinant vanishes, i.e.,

$$J'_n(\mu) J_n(i\mu) - i J'_n(i\mu) J_n(\mu) = 0.$$

But we can write

$$J_n(ix) = i^n \mathcal{B}_n(x),$$

where $\mathcal{B}_n(x)$ is a real function of a real argument and

$$J'_n(x) = \frac{n}{x} J_n(x) - J_{n+1}(x).$$

Using these formulas, the determinant can be rewritten in the form

$$J_n(\mu) \mathcal{B}_{n+1}(\mu) + J_{n+1}(\mu) \mathcal{B}_n(\mu) = 0. \qquad (9.69)$$

The roots of this transcendental equation determine the eigenfrequencies. They can be found by using a power series expansion of the Bessel functions. The few lowest roots are given in Table 3, and the corresponding natural frequencies are found from the formula

$$\omega = \frac{\mu^2}{a^2} \sqrt{\frac{D}{\gamma h}}. \qquad (9.70)$$

TABLE 3. Clamped plate. Roots μ of determinant equation.

m \ n	0	1	2	3
1	3.1962	4.6109	5.9057	7.1435
2	6.3064	7.7993	9.1969	10.5367
3	9.4395	10.9581	12.4022	13.7951
4	12.5771	14.1086	15.5795	17.0053
5	15.7164	17.2557	18.7440	
6	18.8565			

m \ n	4	5	6	7
1	8.3466	9.5257	10.6870	11.8345
2	11.8367	13.1074	14.3552	15.5846
3	15.1499	16.4751	17.7764	19.0581
4	18.3960	19.7583		

The number n is equal to the number of nodal diameters of the amplitude function and m in Table 3 indicates the number of nodal circles.

For a circular plate with a *free* boundary we have $M_B = Q = 0$ at $r = a$. Before substituting (9.68) into the boundary conditions, we rewrite (9.59) and (9.60) using (9.57) in the form

$$M_B = D\left[\Delta w - \frac{1-\nu}{r}\frac{\partial w}{\partial r} - \frac{1-\nu}{r^2}\frac{\partial^2 w}{\partial \phi^2}\right] \qquad (9.71)$$

and

$$Q = -D\left[\frac{\partial \Delta w}{\partial r} + \frac{2-\nu}{r^2}\frac{\partial^3 w}{\partial r \partial \phi^2} - \frac{1-\nu}{r^3}\frac{\partial^2 w}{\partial \phi^2}\right]. \qquad (9.72)$$

This is more convenient in this case since we already know that

$$\Delta J_n(kr) \cos n\phi = -k^2 J_n(kr) \cos n\phi$$

and
$$\Delta J_n(ikr)\cos n\phi = k^2 J_n(ikr)\cos n\phi.$$

Utilizing this fact we get the determinant equation

$$\{[(1-\nu)(n^2-n)-\mu^2]J_n(\mu)+(1-\nu)\mu J_{n+1}(\mu)\}$$
$$\times\{n[n^2+(1-\nu)(n^2-n)-\mu^2]\mathscr{B}_n(\mu)-\mu[\mu^2-(2-\nu)n^2]\mathscr{B}_{n+1}(\mu)\}$$
$$-\{[(1-\nu)(n^2-n)+\mu^2]\mathscr{B}_n(\mu)-(1-\nu)\mu\mathscr{B}_{n+1}(\mu)\}$$
$$\times\{n[n^2+(1-\nu)(n^2-n)+\mu^2]J_n(\mu)-\mu[\mu^2+(2-\nu)n^2]J_{n+1}(\mu)\}=0.$$
(9.73)

In contrast to the case of the clamped plate, the roots of the determinant equation are dependent on Poisson's ratio ν. For $\nu = 0.3$ the few lowest roots are given in Table 4. Again, n is the number of nodal diameters and m the number of nodal circles. The double root $\mu = 0$ corresponds to the zero frequencies of the rigid body motions possible for a free plate.

TABLE 4. Free plate. Roots μ of determinant equation for $\nu = 0.3$.

m \ n	0	1	2	3
1	0	0	2.3148	3.5269
2	3.0005	4.5249	5.9380	7.2806
3	6.2003	7.7338	9.1851	10.5804
4	9.3675	10.9068	12.3817	13.8091
5	12.5227	14.0667	15.5575	17.0070
6	15.6727	17.2203	18.7226	
7	18.8200			

m \ n	4	5	6	7
1	4.6728	5.7875	6.8832	7.9659
2	8.5757	9.8364	11.0711	12.2853
3	11.9344	13.2565	14.5530	15.8283
4	15.1997	16.5606	17.8969	19.2125
5	18.4232	19.8117		

16. Behaviour at a corner of a simply supported plate

Let us consider a plate in the shape of a sector OAB (see Fig. 13), simply supported along the two radial edges OA and OB and subject to given loads on the circumferential edge AB, corresponding to a bending moment M_B and an effective shear force Q.

Expanding the given moment and force in Fourier series, we get

$$M_B = \sum_{n=1}^{\infty} A_n \sin \frac{n\pi\phi}{\alpha}, \qquad Q = \sum_{n=1}^{\infty} B_n \sin \frac{n\pi\phi}{\alpha}, \qquad (9.74)$$

and we shall consider the numbers A_n and B_n to be given.

We seek a solution to the homogeneous plate equation

$$\Delta^2 w = 0, \qquad (9.75)$$

satisfying the conditions of simple support at the radial edges, and producing the given loads on the circumferential edge. Taking

$$w = \sum_{n=1}^{\infty} R_n(r) \sin \frac{n\pi\phi}{\alpha}, \qquad (9.76)$$

where the R_n are functions of r only, the boundary conditions on the radial edges will be satisfied by each term of this series.

To determine the functions R_n we take

$$y = r^m \sin \lambda\phi \qquad (9.77)$$

and calculate $\Delta^2 y$. We find

$$\Delta^2 y = [\lambda^4 - \lambda^2(2m^2 - 4m + 4) + m^2(m^2 - 4m + 4)]r^{m-4} \sin \lambda\phi$$

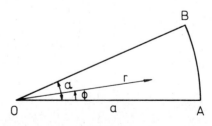

Fig. 13.

BEHAVIOUR OF A SIMPLY SUPPORTED PLATE

and conclude that y is a solution of (9.75) provided that m satisfies the characteristic equation

$$\lambda^4 - \lambda^2(2m^2 - 4m + 4) + m^2(m^2 - 4m + 4) = 0 .$$

The equation has the four roots

$$m = \pm\lambda, \qquad m = 2\pm\lambda, \tag{9.78}$$

and the function R_n is therefore

$$R_n = C_{1n}r^{\lambda_n} + C_{2n}r^{-\lambda_n} + C_{3n}r^{2+\lambda_n} + C_{4n}r^{2-\lambda_n}, \tag{9.79}$$

where

$$\lambda_n = n\pi/\alpha . \tag{9.80}$$

Since we require that w and dw/dr remain bounded as we approach the corner $r = 0$, we must take $C_{2n} = C_{4n} = 0$. Of the remaining terms the lowest power of r is obtained for $n = 1$ and we write this term as

$$w_1 = (C_1 r^\lambda + C_3 r^{\lambda+2}) \sin \lambda\phi , \tag{9.81}$$

where $\lambda = \pi/\alpha$.

Substitution of (9.81) into (9.59) and (9.60) yields the bending moment

$$M_B = D\{C_1(1-\nu)\lambda(\lambda-1)r^{\lambda-2} \\ + C_3[2(1+\nu) + \lambda(3+\nu) + \lambda^2(1-\nu)]r^\lambda\} \sin \lambda\phi \tag{9.82}$$

and the effective shear-force

$$Q = D\{C_1(1-\nu)\lambda^2(\lambda-1)r^{\lambda-3} \\ + C_3\lambda[(1-\nu)(\lambda-1) - 3(\lambda+1)]r^{\lambda-1}\} \sin \lambda\phi . \tag{9.83}$$

The conditions at $r = a$ provide two linear equations for C_1 and C_3. The determinant of the system is

$$2(1-\nu)(3+\nu)\lambda^2(\lambda-1)a^{2\lambda-3}$$

and cannot vanish except in the degenerate case $\lambda = 1$, corresponding to $\alpha = \pi$, i.e., no corner. Excluding this trivial case we see that the constants C_1 and C_3 can be expressed as linear homogeneous functions of A_1 and B_1 and normally neither C_1 nor C_3 vanish.

From the expressions (9.82) and (9.83) we find that the lowest power of r is $\lambda - 3$, appearing in the equation for Q. For $\lambda < 3$ the power is negative and therefore the effective shear force becomes unbounded as r approaches zero if $\alpha > \frac{1}{3}\pi$.

This, however, does not necessarily mean unbounded shear stresses, as we shall see. According to (6.17) we have

$$Q = T - \partial M_V / \partial s,$$

where the twisting moment on a boundary $r = $ const is

$$M_V = D(1-\nu)\frac{1}{r}\left(\frac{\partial^2 w}{\partial r \partial \phi} - \frac{1}{r}\frac{\partial w}{\partial \phi}\right). \tag{9.84}$$

In our case, with w given by (9.81), this equation yields

$$M_V = D(1-\nu)[C_1\lambda(\lambda-1)r^{\lambda-2} + C_3\lambda(\lambda+1)r^{\lambda}]\cos\lambda\phi. \tag{9.85}$$

Hence

$$\begin{aligned}T &= Q + \partial M_V/\partial s \\ &= C_3\lambda[(1-\nu)(\lambda-1) - 4(\lambda+1)]r^{\lambda-1}\sin\lambda\phi\end{aligned} \tag{9.86}$$

since the coefficient of $C_1 r^{\lambda-3}$ vanishes identically.

Therefore, the shear stresses due to M_V are of the order $r^{\lambda-2}$, precisely, as the bending stresses according to (9.82).

It is now clear that the stresses at the corner of a simply supported plate are bounded if the angle of the corner is right or acute ($\lambda \geq 2$, $\alpha \leq \frac{1}{2}\pi$), and unbounded if the angle is obtuse ($\lambda < 2$, $\alpha > \frac{1}{2}\pi$).

A question now arises of the significance of the singular behaviour of the solution at an obtuse corner. To what extent is it the result of an inadequate theory? Naturally, our theory is not a tool for analyzing the details of stress distribution, especially not close to a boundary. From our theory we cannot possibly extract much information about the stresses and strains in a boundary-layer of a width comparable to the thickness of the plate. But we may safely assume that the solution (9.82)–(9.83) is adequate for values of $r \gg h$ and at an obtuse corner of angle α we can therefore risk the stress-level raised at least by the factor

$$\left(\frac{r}{h}\right)^{2-\pi/\alpha}.$$

in comparison with the stresses at radius r, provided, of course that the assumptions of elastic behaviour still hold good.

It may be added that similar results are obtained for other boundary conditions. The stress concentrations at obtuse corners are therefore of great interest to engineers, and must be given serious attention especially in structures, where repeated loading may cause fatigue.

Bibliography

TIMOSHENKO, S. (1959), *Theory of Plates and Shells*, 2nd ed., McGraw-Hill, New York, Chapters 1–12.
NADAI, A. (1925), *Die Elastischen Platten*, Springer, Berlin (reprinted 1968).
MORLEY, L.S.D. (1963), *Skew Plates and Structures*, Pergamon Press, Oxford.
DONNELL, L.H. (1976), *Beams, Plates, and Shells*, McGraw-Hill, New York, Chapter 4.

CHAPTER 10

THE MEMBRANE STATE

1. Introduction 179
2. The membrane state in orthogonal coordinates 180
3. Axisymmetrical shells 181
4. Drop container 187
5. Domes 190
6. Conditions at the apex 192
7. Conditions at the base 193
8. Spherical dome of uniform thickness 193
9. Dome of uniform strength 194
10. Domes with given circumferential stresses 196
11. Deformation of axisymmetrical shells in the membrane state 197
12. General solution of the membrane state in axisymmetrical shells 199
13. Axisymmetrical loading 201
14. Wind load 201
15. The membrane state of a spherical shell 202
16. Membrane state of the toroidal shell 213
17. The membrane state of shells having zero Gaussian curvature 216
18. Discussion 221
    Bibliography 221

CHAPTER 10

THE MEMBRANE STATE

1. Introduction

A 'string' is a one-dimensional structure that resists stretching but not bending. We find this concept very useful, not only in the analysis of the thin wires in certain musical instruments, but also in other cases like for instance the analysis of thick cables in suspension bridges, power lines, etc.

The two-dimensional analogy to a string is a 'membrane', a surface that resists stretching, but not bending. For shells the 'membrane state' is an idealization that has been found very useful, since, in many cases of practical interest, the state of stress in a shell is such that the influence of the transverse shear forces can be neglected in the equations of equilibrium. Putting $M^{\alpha\beta} \equiv 0$ in the linear equations (5.22) and (5.23) we get the following system,

$$D_\alpha N^{\alpha\beta} + F^\beta = 0, \qquad (10.1)$$

$$d_{\alpha\beta} N^{\alpha\beta} + p = 0, \qquad (10.2)$$

which consists of three equations for the three unknown components of $N^{\alpha\beta}$. If they can be solved, the membrane stresses $N^{\alpha\beta}$ become *statically determined*, characterizing a 'membrane state' of the shell.

We shall look upon the membrane state as a first approximation of the true state. In actual cases of practical interest its justification can—strictly speaking—only be established a posteriori by the following procedure:

(1) $N^{\alpha\beta}$ is determined from (10.1) and (10.2).
(2) The strain tensor $E^{\alpha\beta}$ is found from Hooke's law (7.12).
(3) The displacements v_α and w are found by integrating (3.11).
(4) $K_{\alpha\beta}$ is determined from (3.37) and $M^{\alpha\beta}$ from (7.11).
(5) The terms, that were neglected in the equations of equilibrium (5.22) and (5.23), are calculated.

It is now possible to check if the approximation introduced in the equations of equilibrium were justified.

2. The membrane state in orthogonal coordinates

Lowering index β and performing the covariant differentiation in (10.1), we get

$$N^\alpha_{\beta,\alpha} + \{^{\ \alpha}_{\gamma\ \alpha}\}N^\gamma_\beta - \{^{\ \gamma}_{\beta\ \alpha}\}N^\alpha_\gamma + F_\beta = 0 . \tag{10.3}$$

For $\beta = 1$ this equation has the expanded form

$$N^1_{1,1} + N^2_{1,2} + \{^{\ 2}_{1\ 2}\}N^1_1 + \{^{\ 1}_{2\ 2}\}N^2_1 - \{^{\ 2}_{1\ 1}\}N^1_2 - \{^{\ 2}_{1\ 2}\}N^2_2 + F_1 = 0 , \tag{10.4}$$

and for $\beta = 2$ we get the corresponding equation by interchanging indices 1 and 2.

We shall specialize the equations to orthogonal coordinates, i.e., such that $a_{12} = a_{21} = 0$ everywhere. Since

$$\{^{\ 2}_{1\ 2}\} = \tfrac{1}{2}a^{22}a_{22,1} = \frac{a_{22,1}}{2a_{22}} ,$$

we can write

$$N^1_{1,1} + \{^{\ 2}_{1\ 2}\}N^1_1 = \frac{1}{\sqrt{a_{22}}}(\sqrt{a_{22}}N^1_1)_{,1} . \tag{10.5}$$

Let us now introduce the shear force,

$$S = \sqrt{a}\, N^{12} , \tag{10.6}$$

which like N^1_1 and N^2_2 has the physical dimension force per unit length in all coordinate systems.

Differentiating S with respect to u^2 we get

$$S_{,2} = \sqrt{a}\, N^{12}_{,2} + \frac{1}{2}\frac{1}{\sqrt{a}}(a_{11}a_{22,2} + a_{11,2}a_{22})N^{12},$$

and after some trivial computation we find that

$$a_{11,2}\frac{S}{\sqrt{a}} + \sqrt{\frac{a_{11}}{a_{22}}}\, S_{,2} = N^2_{1,2} + \{{}^{\,2}_{2\,2}\}N^2_1 - \{{}^{\,2}_{1\,1}\}N^1_2. \qquad (10.7)$$

The left-hand side of this equation can also be written as

$$(\sqrt{a})^{-1}(a_{11}S)_{,2}.$$

Substituting (10.5) and (10.7) into (10.4) and multiplying through by $\sqrt{a_{22}}$ we get

$$(\sqrt{a_{22}}\, N^1_1)_{,1} + \frac{1}{\sqrt{a_{11}}}(a_{11}S)_{,2} - (\sqrt{a_{22}})_{,1}N^2_2 + \sqrt{a_{22}}\, F_1 = 0. \qquad (10.8)$$

The second equation of equilibrium, corresponding to $\beta = 2$, is obtained by interchanging indices 1 and 2,

$$(\sqrt{a_{11}}\, N^2_2)_{,2} + \frac{1}{\sqrt{a_{22}}}(a_{22}S)_{,1} - (\sqrt{a_{11}})_{,2}N^1_1 + \sqrt{a_{11}}\, F_2 = 0. \qquad (10.9)$$

The third equation (10.2) can be written directly as

$$d^1_1 N^1_1 + 2\sqrt{a}\, d^{12} S + d^2_2 N^2_2 + p = 0, \qquad (10.10)$$

and, provided that d^1_1 and d^2_2 are not both zero, we can use (10.10) to eliminate either N^1_1 or N^2_2, so that we are left with a system of two first order differential equations for N^1_1 and S or N^2_2 and S, as the case may be.

3. Axisymmetrical shells

We take the x^3-axis to coincide with the axis of the shell, and ϕ to be the angle between the x^1-axis and an arbitrary half-plane bounded by the x^3-axis. Such a plane intersects the middle-surface along a *meridian*, while a plane perpendicular to the x^3-axis intersects the middle-surface along a *parallel circle*. Any point on the middle-surface defines uniquely one meridian ϕ, and one parallel circle t; t being the distance in arc-length from an arbitrarily selected parallel (see Fig. 14).

With the coordinates t, ϕ for u^1, u^2, the parametric equations of the middle-surface are

$$x^1 = r(t) \cos \phi, \qquad x^2 = r(t) \sin \phi, \qquad x^3 = q(t) \qquad (10.11)$$

where r and q are arbitrary, but sufficiently smooth functions of t. Since t is an arc length, we have

$$\left(\frac{dr}{dt}\right)^2 + \left(\frac{dq}{dt}\right)^2 = 1$$

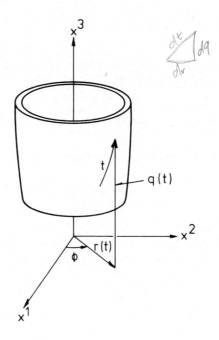

Fig. 14.

or

$$(r')^2 + (q')^2 = 1, \tag{10.12}$$

where ' denotes the derivative with respect to t. By differentiating (10.12) we obtain the important relation

$$r'r'' + q'q'' = 0. \tag{10.13}$$

From the parametric relations (10.11) we derive the matrix

$$f^i_{,\alpha} = \begin{pmatrix} r'\cos\phi & r'\sin\phi & q' \\ -r\sin\phi & r\cos\phi & 0 \end{pmatrix} \tag{10.14}$$

and hence

$$a_{\alpha\beta} = \begin{pmatrix} 1 & 0 \\ 0 & r^2 \end{pmatrix} \tag{10.15}$$

and

$$X^i = (-q'\cos\phi, -q'\sin\phi, r'). \tag{10.16}$$

Furthermore, from (10.14) we get by differentiation

$$f^i_{,11} = (r''\cos\phi, r''\sin\phi, q''),$$
$$f^i_{,12} = (-r'\sin\phi, r'\cos\phi, 0) = f^i_{,21},$$
$$f^i_{,22} = (-r\cos\phi, -r\sin\phi, 0),$$

and using the relation $d_{\alpha\beta} = X^i f^i_{,\alpha\beta}$ we get

$$d_{\alpha\beta} = \begin{pmatrix} q''/r' & 0 \\ 0 & rq' \end{pmatrix}. \tag{10.17}$$

If the tangent to the meridian at a point is parallel to the x^3-axis, we there have $r' = 0$. Then d_{11}, according to (10.17), becomes undetermined, and it may be necessary to rewrite d_{11} with the help of (10.12) and (10.13) as follows,

$$d_{11} = \frac{q''}{r'} = \frac{q''}{r'}((r')^2 + (q')^2) = q''r' - r''q'. \tag{10.18}$$

We shall recall this formula (which should be preferred in numerical work) when necessary, but retain the shorter form (10.17).

Substituting the fundamental tensors $a_{\alpha\beta}$ and $d_{\alpha\beta}$ according to (10.15) and (10.17) into the equations (10.8)–(10.10) we get

$$\frac{\partial}{\partial t}[rN_1^1] + \frac{\partial S}{\partial \phi} - \frac{\partial r}{\partial t} N_2^2 + rF_1 = 0,$$

$$\frac{\partial}{\partial \phi}[N_2^2] + \frac{1}{r}\frac{\partial}{\partial t}[r^2 S] + F_2 = 0,$$

$$\frac{q''}{r'} N_1^1 + \frac{q'}{r} N_2^2 + p = 0.$$

Let us now introduce the following notations,

$$N = N_1^1, \quad T = N_2^2, \quad (\ )^{\cdot} = \frac{\partial}{\partial t}(\ ), \quad (\ )^{\bullet} = \frac{\partial}{\partial \phi}(\ ).$$

With these notations the equations of equilibrium take the form

$$(rN)^{\cdot} + S^{\bullet} - r'T + rF_1 = 0, \tag{10.19}$$

$$T^{\bullet} + \frac{1}{r}(r^2 S)^{\cdot} + F_2 = 0, \tag{10.20}$$

$$\frac{q''}{r'} N + \frac{q'}{r} T + p = 0. \tag{10.21}$$

Solving the last equation for T, and substituting into the first two, we get, after eliminating S between them,

$$\left[\frac{r^2}{q'}(q'rN)'\right]' + \frac{rq''}{r'(q')^2}(q'rN)^{\bullet\bullet} = r\dot{F}_2 - \frac{r^2\ddot{p}}{q'} - \left[r^3 F_1 + \frac{r^3 r'}{q'} p\right]'. \tag{10.22}$$

Introducing the linear operator

$$L[\zeta] = \left[\frac{r^2}{q'}\zeta'\right]' + \frac{rq''}{r'(q')^2}\ddot{\zeta}, \qquad (10.23)$$

we can write equation (10.22) in the simple form,

$$L[q'rN] = f(t, \phi), \qquad (10.24)$$

where $f(t, \phi)$ is the right-hand side of (10.22).

Working out the derivatives in (10.22), and dividing through by r^3, we get

$$N'' + \frac{q''}{rr'q'}\ddot{N} + \cdots = 0, \qquad (10.25)$$

where the terms indicated by the dots are linear in N and its first derivatives N' and $\dot{N}$. Since the Gaussian curvature K can be written as

$$K = \frac{d}{a} = \frac{q''q'}{rr'}, \qquad (10.26)$$

we get

$$(q')^2 N'' + K\ddot{N} + \cdots = 0. \qquad (10.27)$$

The membrane force N is therefore the solution of a linear second-order partial differential equation. This equation is *elliptic* if $K > 0$, it is *hyperbolic* if $K < 0$ and *parabolic* if $K = 0$.

Depending on the sign of the Gaussian curvature of the shell, we are lead to mathematical problems that are entirely different. In this context the important difference between the three types of differential equations lies in the way the auxiliary conditions are—or rather can be—given. Let $f(u^1, u^2)$ denote an integral of the differential equation, and let it be represented by the integral surface $z = f(x, y)$.

In the elliptic case the function f can be prescribed everywhere on the boundary, and this determines the integral surface uniquely.

Alternatively, we can prescribe the normal derivative $\partial f/\partial n$ everywhere on the boundary, and that would determine the integral surface uniquely up to an arbitrary additive constant. There is also the possibility of prescribing a linear function T of f and its normal derivative all around the boundary,

$$T(s) = a(s)f(s) + b(s)\frac{\partial f}{\partial n}(s),$$

where $T(s)$, $a(s)$ and $b(s)$ are given functions. Clearly, this makes it possible to prescribe either the normal membrane force or the tangential membrane force everywhere on the boundary (but not both!), and in each case we are led to a unique solution of the membrane state. Therefore, at the boundary of a membrane there is an enforced relation between the normal and the tangential membrane stresses. If the boundary conditions for a real shell structure are such that they cannot comply with this relation, bending moments and shear forces must necessarily occur to secure the equilibrium of the shell.

In the hyperbolic case (negative Gaussian curvature) the situation is entirely different. Whereas the elliptic differential equation has no real characteristics, the hyperbolic equation has two families of mutually intersecting characteristic curves. Auxiliary conditions can be enforced in two different ways. Firstly, we may prescribe f on two intersecting characteristics (that do not have a common tangent at the point of intersection). Secondly, we may prescribe f and its normal derivative along any portion of the boundary, provided that no characteristic intersects that part more than once. In the hyperbolic case it is therefore possible to prescribe the normal as well as the tangential membrane force, but only along part of the boundary, and this would determine the solution uniquely in the smallest region bounded by the characteristics passing through the end-points of the curve. It would *not* be possible to prescribe freely either f or its normal derivative $\partial f/\partial n$ all around a closed boundary.

Finally, in the parabolic case there is just one family of characteristics. The function f and its normal derivative can be prescribed along any curve that no characteristic intersects more than once, and this would determine a unique solution in a strip between the two characteristics that pass through the end-points of the curve.

Due to the fundamentally different ways in which boundary conditions are accepted in the elliptic case on one hand and the hyperbolic and parabolic cases on the other, shells of positive Gaussian curvature behave very differently from shells of negative and zero Gaussian curvature.

In the case of positive Gaussian curvature (elliptic differential equation) a 'disturbance' (i.e., a small change) of the boundary values at any point of the boundary attenuates quickly and has in practice only local consequences. In the cases of zero and negative Gaussian curvature, however, a disturbance propagates along the characteristics and has an effect also at the boundary on the 'opposite' side.

From a physical point of view the elliptic case is the natural one for problems in static elasticity, and it is, for instance, easily shown that the three-dimensional equations of (linear) elasticity are elliptic. The fact that the membrane state leads to a hyperbolic equation when the Gaussian curvature of a shell is negative, indicates that one essential feature of the elastic structure is misrepresented, and that the results therefore should be interpreted with caution. But it also indicates that shells of negative Gaussian curvature might behave in a way that is unusual for elastic bodies. Actually, shells of nonpositive Gaussian curvature are generally speaking weak structures, sensitive to disturbances at the boundary, which tend to penetrate deep into the structure. Also, they have no possibility of supporting concentrated forces in the membrane state, in contrast with shells of positive Gaussian curvature, where the membrane state will provide an approximately correct result at sufficient distance from the point where a concentrated force is applied.

4. Drop container

The free surface of any liquid has the tendency to contract, and behaves in this respect like an almost ideal membrane with a uniform *surface-tension* $\mathcal{T}$, that for a given liquid and a given temperature is a constant. The surface-tension determines the shape of liquid drops.

Their form is of considerable technical interest, since a container of uniform thickness made in the shape of a liquid drop resting on a nonwetting surface will be *optimal* in the sense of minimum material consumption for a given volume.

To determine the shape of such a container we take our coordinates so that the x^3-axis points vertically upwards, and introduce the uniform membrane forces

$$N = T = \mathcal{T} \quad \text{and} \quad S = 0. \tag{10.28}$$

The liquid inside the container acts on the walls with a normal pressure,

$$p = p_0 - \gamma q > 0, \tag{10.29}$$

where γ is the specific weight of the liquid, and p_0 a given constant, representing the hydrostatic pressure at the top $q = 0$ of the container.

It is readily checked that, with (10.28) and (10.29), two of the equations of equilibrium are identically satisfied. The third equation (10.21) yields

$$\left(\frac{q''}{r'} + \frac{q'}{r}\right)\mathcal{T} + p_0 - \gamma q = 0,$$

or, solving for q''

$$q'' = r'\left[\frac{-p_0 + \gamma q}{\mathcal{T}} - \frac{q'}{r}\right]. \tag{10.30}$$

Together with (10.12),

$$(r')^2 + (q')^2 = 1,$$

this is a system of ordinary differential equations that determine the shape $q(t)$, $r(t)$. No closed-form solution to this system is known, but it is readily integrated numerically. Let

$$c(t) = q'(t) \quad \text{and} \quad d(t) = r'(t).$$

Then the system (10.30), (10.12) can be written as the following system of first-order differential equations,

$$q' = c,$$
$$r' = \pm\sqrt{1-c^2},$$
$$c' = \pm\sqrt{1-c^2}\left[\frac{-p_0 + \gamma q}{\mathcal{T}} - \frac{c}{r}\right],$$
(10.31)

which lends itself directly to integration by the *Runge–Kutta method*.[1] As initial values we take the values at the top, i.e.,

$$c = q = r = 0,$$

and find the initial value for c' by l'Hospital's rule,

$$c' = -p_0/2\mathcal{T}.$$

The integration is performed with the positive square root until $c = -1$, thereafter it is continued with the negative square root. If we stop at $c = 0$, we get the shape shown in Fig. 15. If the bottom of the container is taken to be a circular plate of the same thickness as the rest of the container, the stresses there will obviously be the same also.

If the integration is continued passed the value $c = 0$, we get a shape like the one depicted in Fig. 16 of an overhanging drop on a circular support.

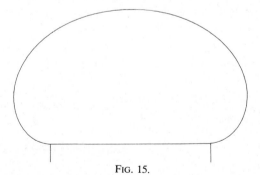

Fig. 15.

---

[1] The Runge–Kutta method is described in most textbooks on numerical analysis. This particular case has been treated by RUNGE and KÖNIG (1924), *Vorlesungen über Numerisches Rechnen*, Springer, Berlin, p. 320.

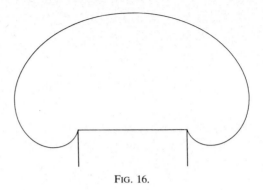

Fig. 16.

5. Domes

Although not designed in a modern sence of the word, the antique domes like the one of St. Peter's in Rome (see Fig. 17) are examples of the unique capability of curved surfaces to cover large areas without intermediate support. Modern domes of reinforced concrete will cover the same area using a shell thickness of only a few centimeters. Their safe design is based primarily upon *statically determined stresses*. Therefore, the membrane theory of shells plays an important role in the design of domes.

While in the final analysis of domes, loads due to wind, snow etc. have to be considered, their primary design is based upon the assumption that the external forces are due to gravity only.

Let the x^3-axis be pointing upwards. Then the cartesian components of the external load are

$$\bar{F}^i = -\gamma h \delta_3^i, \tag{10.32}$$

where γ is the specific weight of the material.

Multiplication by $f_{,\alpha}^i$ and X^i according to (10.14) and (10.16) yields

$$F_1 = -\gamma h q', \qquad F_2 = 0, \qquad p = -\gamma h r'. \tag{10.33}$$

In our analysis we can dispense with the assumption of uniform shell thickness for the time being, and allow h to be a function of t.

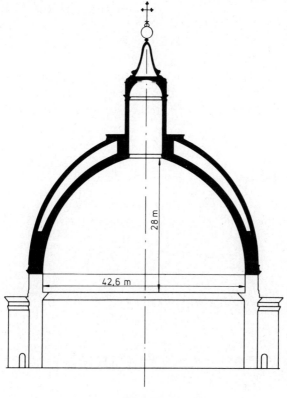

Fig. 17.

Substituting (10.33) into the equations of equilibrium (10.19)–(10.21) we get

$$(rN)' - r'T - \gamma hrq' = 0, \qquad (10.34)$$

$$\frac{q''}{r'}N + \frac{q'}{r}T - \gamma hr' = 0, \qquad (10.35)$$

and $S = 0$ everywhere.

Eliminating T we get

$$\frac{q'}{r}(rN)' + q''N - \gamma hq'^2 - \gamma hr'^2 = 0,$$

or

$$(q'rN)' = \gamma hr. \tag{10.36}$$

The equations (10.36) and (10.35) connect the geometry of the dome, given by the functions $r(t)$ and $q(t)$, with the membrane forces, the *meridional* membrane force N and the *circumferential* membrane force T.

6. Conditions at the apex

Let us assume that the dome is closed above, and take $t = q = 0$ at the apex. Expanding q and r in Taylor series at $t = 0$ we get

$$q = -at - bt^2 + \cdots, \qquad r = ct + \cdots, \tag{10.37}$$

where, due to (10.12),

$$a^2 + c^2 = 1. \tag{10.38}$$

For a pointed apex $a \neq 0$ and substitution of (10.37) into (10.36) and (10.35) yields

$$N_0 = T_0 = 0,$$

i.e., both N and T vanish.

For a flat and smooth top we must have $a = 0$ as well as $h'(0) = 0$. Then $b = \frac{1}{2}R$, where R is the radius of curvature at the apex and substitution of (10.37) into (10.36) and (10.35) yields

$$N_0 = T_0 = -\tfrac{1}{2}\gamma h_0 R \tag{10.39}$$

or

$$1/R = -\tfrac{1}{2}\gamma/\sigma_0, \tag{10.40}$$

where σ_0 is the value of the principal stresses in the tangent plane at the apex.

7. Conditions at the base

For the membrane state to prevail it is necessary to provide a base for the dome that supports the vertical *as well as the horizontal component* of the meridional force. While the vertical component presents no great problem, the horizontal one will in general require a ring-stiffener around the base. From elementary considerations of equilibrium it is clear that such a ring-stiffener will be affected by a tensile force equal to

$$P = -Nrr', \qquad (10.41)$$

where all three factors are evaluated at the base.

Since the vertical component Nq' integrated around the base must equal the total weight W of the dome, we get

$$P = \frac{1}{2\pi} W \operatorname{tg} \alpha, \qquad (10.42)$$

where $\operatorname{tg} \alpha = dr/dq$, and α therefore the angle between the meridian and the vertical at the base. To avoid unreasonably heavy ring-stiffeners in a dome like the one in Fig. 17, with a mass of approximately 10 000 000 kg, the angle α must be small, as in fact it is.

For domes loaded by their own weight only, the stress-level is independent of the thickness of the structure and the minimum thickness is choosen from other considerations (stability, wind load, geometrical imperfections, tear and wear, etc.). Furthermore, the stress level in domes of masonry is as a rule very low, typically below one per cent of the ultimate compressive stress of the material. Therefore, the influence of the shape on the size of the ring-stiffener becomes of prime importance to the designer.

8. Spherical dome of uniform thickness

The geometry of the middle-surface is given by

$$r = R \sin(t/R), \qquad q = R[\cos(t/R) - 1]. \qquad (10.43)$$

With these functions, equality (10.36) can be integrated immediately, and thereafter T may be solved from (10.35). We get

$$N = -\gamma hR \frac{1}{1+\cos(t/R)}, \tag{10.44}$$

$$T = -\gamma hR \left[\cos(t/R) - \frac{1}{1+\cos(t/R)} \right]. \tag{10.45}$$

We find that the meridional force is negative (compressive) for all values of t/R but the circumferential force changes sign when

$$\cos(t/R) = \sin^2(t/R),$$

i.e., at the angle $t/R = 51.8°$. If we are to avoid tensile forces, the spherical dome of uniform thickness must be limited to this top.

9. Dome of uniform strength

Let us determine the shape of a dome, in which the principal stresses in the tangent plane are equal and uniform. If we take $\sigma < 0$ to be this principal stress, we have

$$N = T = \sigma h(t), \tag{10.46}$$

where h is now a function of t. Substituting (10.46) into the third equation of equilibrium, we find that h cancels out, and we obtain the following differential equation for the shape,

$$q'' = \frac{\gamma}{\sigma}(r')^2 - \frac{q'r'}{r} \tag{10.47}$$

with $r' = \sqrt{1-(q')^2}$ according to (10.12). We have no closed-form solution to this equation, but again it is easy to integrate it numerically by the Runge–Kutta method.

With $\mathcal{N} = rN$ in (10.36) we find

$$(q'\mathcal{N})' = q''\mathcal{N} + q'\mathcal{N}' = \frac{\gamma}{\sigma}\mathcal{N},$$

and with q'' from (10.47) this leads to

$$\mathcal{N}' = \left(\frac{\gamma}{\sigma}q' + \frac{r'}{r}\right)\mathcal{N}$$

or

$$N' = \frac{\gamma}{\sigma}q'N$$

with the integral

$$N = N_0 e^{q\gamma/\sigma}.$$

We see that the membrane forces N and T increase exponentially with the vertical distance q from the apex, and so does the thickness, naturally,

$$h = h_0 e^{q\gamma/\sigma}, \tag{10.48}$$

where h_0 is the minimum thickness at the apex. The shape can be seen in Fig. 18.

Although, in a certain sense, this is an optimal shape, there is, in fact, very little merit to it. The ratio of maximum to minimum thickness is large, and so will the necessary ring-stiffener be. On the other hand, as was pointed out before, the stress-level is hardly worth considering in the design of domes, and therefore this shape is not of great interest and has found little use outside of textbooks on shell theory.

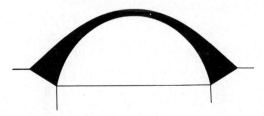

Fig. 18.

10. Domes with given circumferential stresses

Eliminating N between (10.35) and (10.36) we get

$$\left[\frac{q'r'}{q''}(\gamma hrr' - q'T)\right]' = \gamma hr,$$

from which we solve q''',

$$q''' = q'' \frac{(q'r'' + q''r')(\gamma hrr' - q'T) + q'r'(\gamma hrr' - q'T)' - rq''}{q'r'(\gamma hrr' - q'T)}.$$

(10.49)

This equation determines the shape of the dome when T and h are given functions. To prescribe T has the advantage of avoiding tensile stresses, and it is of course satisfactory to have the thickness h under control. There is nothing, for example, that prevents us from taking $h \equiv h_0$ (a constant) everywhere. However, we are not wholly free to prescribe T, since, if the dome is closed, T must equal N at the apex according to (10.39).

As an example, let us take $h \equiv h_0$ and

$$T = T_0 \, e^{-2(dq/dr)^2}.$$

(10.50)

Substitution into (10.49) and numerical integration leads to a shape as shown in Fig. 19.

In Fig. 20 the three different designs (spherical cap (a), dome of uniform strength (b), and dome with prescribed circumferential stress (c)) are compared with regards to the requirement of the ring-stiffener. On the ordinate, the dimensionless number P/P_0 is given, where P is

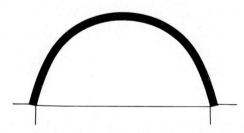

Fig. 19.

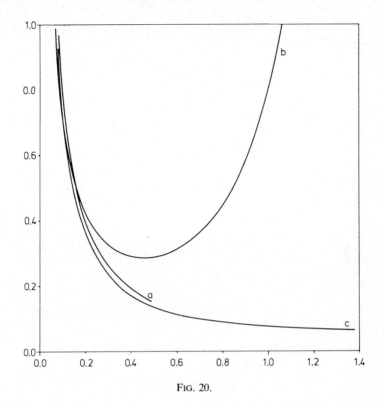

Fig. 20.

the tensile force according to (10.42), and P_0 the weight of a circular plate of the same radius as that of the dome, and of uniform thickness equal to the minimum thickness of the dome. The numbers on the abscissa represent the ratio of the height over the radius.

11. Deformation of axisymmetrical shells in the membrane state

When the membrane forces N, T and S have been determined, we can calculate the components of the strain tensor from Hooke's law (7.12),

$$E_1^1 = \frac{1}{Eh}(N - \nu T), \qquad E_2^2 = \frac{1}{Eh}(T - \nu N),$$

$$E_{12} = \frac{1}{Eh}(1 + \nu)N_{12} = \frac{1}{Eh}(1 + \nu)rS. \tag{10.51}$$

The strain tensor can be expressed in terms of the displacements according to (3.11),

$$E_{\alpha\beta} = \tfrac{1}{2}(v_{\alpha,\beta} + v_{\beta,\alpha}) - \{{}^{\gamma}_{\alpha\beta}\}v_{\gamma} - d_{\alpha\beta}w. \tag{10.52}$$

From (10.15) we find the Christoffel symbols

$$\{{}^{1}_{22}\} = -rr', \qquad \{{}^{2}_{12}\} = \{{}^{2}_{21}\} = \frac{r'}{r},$$

and hence

$$E^1_1 = u' - \frac{q''}{r'}w,$$

$$E^2_2 = \frac{1}{r}v\dot{} + \frac{r'}{r}u - \frac{q'}{r}w, \tag{10.53}$$

$$E_{12} = \frac{1}{2}\left[u\dot{} + r^2\left(\frac{v}{r}\right)'\right],$$

where the displacements u, v are taken to be

$$u = v_1, \qquad v = v_2/r \tag{10.54}$$

and hence of the physical dimension length. Equations (10.51) and (10.53) yield a system of three differential equations for the three unknown functions u, v and w.

Eliminating w between the two first ones we get

$$\frac{q'}{r}u' - \frac{q''}{r'}\left(\frac{1}{r}v\dot{} + \frac{r'}{r}u\right) = \frac{1}{Eh}\left(\frac{q'}{r}N - \nu\frac{q'}{r}T - \frac{q''}{r'}T + \nu\frac{q''}{r'}N\right)$$

or

$$\frac{(q')^2}{r}\left[\frac{u}{q'}\right]' - \frac{q''}{r'}\left[\frac{v}{r}\right]\dot{} = \frac{1}{Eh}\left(\frac{q'}{r}N - \nu\frac{q'}{r}T - \frac{q''}{r'}T + \nu\frac{q''}{r'}N\right). \tag{10.55}$$

Then u/q' is eliminated between this equation and the remaining differential equation. We find

$$\left[\frac{r^2}{q'}\left[\frac{v}{r}\right]'\right]' + \frac{q''r}{(q')^2 r'}\left[\frac{v}{r}\right]^{\cdot\cdot} = F(t,\phi), \tag{10.56}$$

where

$$F(t,\phi) = \frac{1}{Eh}\left[\frac{1}{q'}N^{\cdot} - \frac{v}{q'}T^{\cdot} - \frac{q''r}{(q')^2 r'}T + \frac{vq''r}{(q')^2 r'}N^{\cdot} + 2(1+v)\left[\frac{rS}{q'}\right]'\right].$$

The equation can be written as

$$L\left[\frac{v}{r}\right] = F(t,\phi), \tag{10.57}$$

using the linear operator (10.23). The mathematical problem of determining the displacement field is therefore equivalent to the problem (10.24) of determining the membrane forces N, T and S. Thus, if we can find the membrane forces, we can, presumably, also find the corresponding displacements.

12. General solution of the membrane state in axisymmetrical shells

Let us assume that the shell is bounded by the two parallel circles $t = t_1$ and $t = t_2 > t_1$. The external forces F_α and p are supposed to be given by periodic functions of ϕ with period 2π. Expanding them in Fourier series, we get

$$F_1 = \sum_{k=0}^{\infty}(F_{1k}\cos k\phi + \bar{F}_{1k}\sin k\phi),$$

$$F_2 = \sum_{k=0}^{\infty}(\bar{F}_{2k}\cos k\phi + F_{2k}\sin k\phi), \tag{10.58}$$

$$p = \sum_{k=0}^{\infty}(p_k\cos k\phi + \bar{p}_k\sin k\phi),$$

where F_{1k}, $\bar{F}_{1k}$, $\bar{F}_{2k}$, F_{2k}, p_k and $\bar{p}_k$ are given functions of t.

Since the problem is linear, we find the solution as the sum of solutions for $k = 0, 1, \ldots$ to the problem with

$$F_1 = F_{1k} \cos k\phi, \qquad F_2 = F_{2k} \sin k\phi, \qquad p = p_k \cos k\phi. \tag{10.59}$$

Substituting (10.59) into (10.22) we get with

$$N = N_k \cos k\phi, \qquad T = T_k \cos k\phi, \qquad S = S_k \sin k\phi \tag{10.60}$$

the equation

$$\left[\frac{r^2}{q'}(q'rN_k)'\right]' - \frac{rq''k^2}{r'(q')^2}(q'rN_k) = rkF_{2k} + \frac{r^2k^2p_k}{q'} - \left[r^3F_{1k} + \frac{r^3r'}{q'}p_k\right]', \tag{10.61}$$

which is a second-order ordinary differential equation for $N_k(t)$. When N_k is solved, we find T_k from (10.21),

$$T_k = \frac{r}{q'}\left(\frac{q''}{r'}N_k - p_k\right), \tag{10.62}$$

and S_k from (10.19),

$$S_k = r'T_k - rF_{1k} - (rN_k)'. \tag{10.63}$$

Since the differential equation is of second order, there will be two arbitrary constants in the solution of N_k. Substitution of N_k, T_k and S_k into (10.56) yields an ordinary second order differential equation for v_k. Therefore, the solution will contain four arbitrary constants that must be determined from the boundary conditions. Obviously, at least two of them must be in terms of displacements. If the shell has only one free boundary, like the hemisphere for instance, only two boundary conditions are required, and at least one of them must be in terms of displacements. The remaining two conditions, necessary to determine the constants, will be conditions of regularity at the top of the shell.

13. Axisymmetrical loading

For $k = 0$, equation (10.61) can be integrated immediately. We get

$$N_0 = -\frac{1}{q'r}\int (r'p_0 + q'F_{10})r\,dt + \frac{C_1}{q'r}, \qquad (10.64)$$

$$T_0 = \frac{q''}{r'(q')^2}\int (r'p_0 + q'F_{10})r\,dt - \frac{C_1 q''}{r'(q')^2} - \frac{rp_0}{q'}, \qquad (10.65)$$

$$S_0 = -\frac{1}{r^2}\int F_2 r\,dt + \frac{C_2}{r^2}, \qquad (10.66)$$

where C_1 and C_2 are arbitrary constants. Domes, loaded by their own weight, treated above, are a special case of this.

14. Wind load

When domes and other axisymmetrical shells are subject to wind pressure, the leading term in the Fourier expansion of the load is given by (10.59) with $k = 1$. Equation (10.61) will then take the form

$$\left(\frac{r^2}{q'}y'\right)' - \frac{q''r}{(q')^2 r'}y = F, \qquad (10.67)$$

where

$$y = q'rN_1 \qquad (10.68)$$

and

$$F = rF_{21} + \frac{r^2 p_1}{q'} - \left(r^3 F_{11} + \frac{r^3 r'}{q'}p_1\right)'. \qquad (10.69)$$

With $Y = ry$, equation (10.67) can be written as

$$\left(\frac{r}{q'}Y' - \frac{r'}{q'}Y\right)' - \frac{q''Y}{(q')^2 r'} = F$$

or

$$\left(\frac{r}{q'}Y'\right)' - \frac{r'}{q'}Y' - \frac{q'r'' - r'q''}{(q')^2}Y - \frac{q''Y}{(q')^2 r'} = F.$$

Due to equation (10.13), the two last terms on the left-hand side cancel, and therefore

$$\left(\frac{r}{q'}Y'\right)' - \frac{r'}{q'}Y' = F$$

or

$$r\left[\frac{Y'}{q'}\right]' = F, \tag{10.70}$$

which can be integrated directly. We find

$$N_1 = \frac{1}{q'r^2}\left[\int q'\left(C_1 + \int \frac{F}{r}dt\right)dt + C_2\right], \tag{10.71}$$

where C_1 and C_2 are arbitrary constants. T_1 and S_1 are now readily found from (10.62) and (10.63).

The displacements are found by integrating (10.57) in precisely the same way. Thus, for $k = 0$ and $k = 1$ we can write the solution in closed form. For $k \geq 2$ we will, in general, have to rely on numerical integration. Only for special geometries we can obtain a solution in closed form when $k \geq 2$.

15. The membrane state of a spherical shell

The analysis of the membrane state of spherical shells becomes particularly simple in *isometric coordinates*, i.e., in coordinates for which $a_{11} = a_{22}$ and $a_{12} = 0$ everywhere. There are many such coordinate systems, but we shall find two of them particularly useful, the *Mercator projection* and the *stereographic projection*, both well known in cartography.

A spherical shell is represented by

$$r = R\sin\theta, \quad q = R\cos\theta, \quad \theta = \frac{t}{R}, \tag{10.72}$$

where R is the radius of the sphere. The coordinates (θ, ϕ) will be called *geographical* (although not quite in accordance with the conventional geographic coordinates) and below we shall freely use geographical terms connected with these such as the *north pole* for $\theta = 0$, etc.

With (10.72) the homogeneous equations of equilibrium (10.19)–(10.21) take the form

$$\frac{\partial}{\partial \theta}(N \sin \theta) + \frac{\partial S}{\partial \phi} - T \cos \theta = 0,$$

$$\frac{\partial T}{\partial \phi} + \frac{1}{\sin \theta}\frac{\partial}{\partial \theta}(S \sin^2 \theta) = 0, \quad (10.73)$$

$$N + T = 0.$$

Using the last equation for eliminating N we get

$$\frac{\partial}{\partial \theta}(T \sin^2 \theta) - \frac{\partial S}{\partial \phi}\sin \theta = 0,$$

$$\frac{\partial}{\partial \theta}(S \sin^2 \theta) + \frac{\partial T}{\partial \phi}\sin \theta = 0.$$

With the help of the auxiliary quantities

$$N_1 = T \sin^2 \theta, \qquad N_2 = S \sin^2 \theta, \quad (10.74)$$

the equations take the form

$$\sin \theta \frac{\partial N_1}{\partial \theta} - \frac{\partial N_2}{\partial \phi} = 0, \qquad \sin \theta \frac{\partial N_2}{\partial \theta} + \frac{\partial N_1}{\partial \phi} = 0.$$

Let us now apply the Mercator projection by introducing the new variable

$$\psi = \log \tan \tfrac{1}{2}\theta, \qquad d\psi = \frac{d\theta}{\sin \theta}, \quad (10.75)$$

and proceed with (ψ, ϕ) as new independent variables. The equations simply become

$$\frac{\partial N_1}{\partial \psi} - \frac{\partial N_2}{\partial \phi} = 0, \qquad \frac{\partial N_2}{\partial \psi} + \frac{\partial N_1}{\partial \phi} = 0, \tag{10.76}$$

and in this form coincide with the *Cauchy–Riemann equations* for the real and imaginary part of an analytic function of a complex variable. We can interpret this fact in the following manner.

Let

$$\mathcal{N} = N_1 + iN_2 = (T + iS)\sin^2\theta \tag{10.77}$$

be any analytic function of the complex variable [2]

$$\gamma = \psi + i\phi. \tag{10.78}$$

Then T and S according to (10.77) and $N = -T$ will satisfy the homogeneous equations of equilibrium of a spherical membrane. The functions N_1 and N_2 are *harmonic* functions and the rich literature on the solution of Laplace's equation is therefore available for solving problems regarding the membrane state of spherical shells.

The surface of the sphere is mapped onto the infinite strip $0 \leq \phi \leq 2\pi$ in the γ-plane, however, the mapping is not a one-to-one map, since the meridian $\phi = 0$ coincides with $\phi = 2\pi$ on the sphere, though not in the γ-plane. The functions in the γ-plane must therefore be periodic with period 2π in the imaginary direction. We can easily get rid of this complication by changing the independent variable γ to

$$\zeta = \xi + i\eta = e^\gamma = e^\psi(\cos\phi + i\sin\phi), \tag{10.79}$$

which makes ζ periodic in ϕ. With ζ, the stress-function will be

$$\mathcal{T}(\zeta) = \mathcal{N}(\log \zeta) = \mathcal{N}(\gamma). \tag{10.80}$$

The new complex variable ζ is the stereographic projection of the

---

[2] In this section, $i = \sqrt{-1}$ and this letter is not used as an index.

spherical surface. It maps any point on the sphere with the help of a ray from the south-pole through the point and onto the tangent plane at the north-pole. The origin of the ζ-plane is the north-pole, the meridians are polar rays and the parallels are concentric circles. This mapping is conformal and one-to-one between the spherical surface and the whole plane. The south-pole is the point 'at infinity'.

Since

$$\frac{1}{\sin^2\theta} = \cosh^2\psi = \tfrac{1}{4}(e^{2\psi} + 2 + e^{-2\psi}) = \tfrac{1}{4}(\zeta\bar{\zeta} + 2 + 1/(\zeta\bar{\zeta})),$$

we have, by (10.74),

$$T = -N = \tfrac{1}{8}(\zeta\bar{\zeta} + 2 + 1/(\zeta\bar{\zeta}))(\mathcal{T} + \bar{\mathcal{T}}),$$
$$S = \tfrac{1}{8}(\zeta\bar{\zeta} + 2 + 1/(\zeta\bar{\zeta}))(\mathcal{T} - \bar{\mathcal{T}}). \tag{10.81}$$

Thus, if the membrane forces T, N and S are to be bounded at $\zeta = 0$, the function $\mathcal{T}(\zeta)$ must have a zero of second order there. The same holds good for $\zeta = \infty$. But according to the *theorem of Liouville* the only function that is analytic in the whole plane is the constant function. Therefore, if the sphere is complete, there is no regular solution to the homogeneous equations of equilibrium except the trivial one $T = N = S = 0$ everywhere. All nontrivial solutions are singular, and the singularities correspond as a rule to concentrated forces and moments applied to the membrane.[3]

We shall study this problem in some detail below.

Let the curve C on the spherical surface define a boundary, or a part of a boundary, of a proper element of a spherical shell. The membrane stresses acting on C have a resultant force with the cartesian components

$$P_j = \int_C N^{\alpha\beta} n_\alpha f^i_{,\beta}\, ds, \tag{10.82}$$

where n_α is the unit normal to C and the functions f^j given by

---

[3] A singularity can also be self-equilibrated.

$$f^1 = R \frac{\cos \phi}{\cosh \psi}, \qquad f^2 = R \frac{\sin \phi}{\cosh \psi}, \qquad f^3 = -R \tanh \psi,$$
(10.83)

since $\sin \theta = 1/\cosh \psi$. Further, we observe that in the isometric coordinates ψ, ϕ we have

$$a_{11} = a_{22} = \frac{R^2}{\cosh^2 \psi}, \qquad a_{12} = 0.$$

With

$$n_\alpha = \varepsilon_{\alpha\gamma} \frac{du^\gamma}{ds} = \sqrt{a}\, e_{\alpha\gamma} \frac{du^\gamma}{ds},$$

equation (10.82) can be written as

$$P_j = \int_C N^{\alpha\beta} e_{\alpha\gamma} f^j_{,\beta} \sqrt{a}\, du^\gamma.$$
(10.84)

We now calculate the cartesian components one by one, starting with $j = 1$. Expanding the summations we get

$$P_1 = \int_C [(N^{11} f^1_{,1} + N^{12} f^1_{,2}) \sqrt{a}\, du^2 + (N^{21} f^1_{,1} + N^{22} f^1_{,2}) \sqrt{a}\, du^1]$$

$$= \int_C R[(N_1 \sinh \psi \cos \phi - N_2 \cosh \psi \sin \phi)\, d\phi$$

$$+ (N_1 \cosh \psi \sin \phi + N_2 \sinh \psi \cos \phi)\, d\psi]$$

and therefore

$$P_1 = R\, \mathrm{Im}\left\{ \int_C \mathcal{N}(\gamma) \sinh \gamma\, d\gamma \right\}.$$
(10.85)

For $j = 2$ and $j = 3$ we find similarly

$$P_2 = -R \operatorname{Re}\left\{\int_C \mathcal{N}(\gamma) \cosh \gamma \, d\gamma\right\} \tag{10.86}$$

and

$$P_3 = R \operatorname{Im}\left\{\int_C \mathcal{N}(\gamma) \, d\gamma\right\}. \tag{10.87}$$

The membrane stresses on C also have a resultant moment. Reducing the system to the origin of (the three-dimensional) cartesian coordinate system, i.e., to the centre of the sphere, we get

$$M_j = e_{jkl} \int_C N^{\alpha\beta} n_\alpha f^k f^l_{,\beta} \, ds \tag{10.88}$$

or

$$M_j = e_{jkl} \int_C N^{\alpha\beta} e_{\alpha\gamma} f^k f^l_{,\beta} \sqrt{a} \, du^\gamma.$$

Taking first $j = 1$ we get

$$M_1 = \int_C N^{\alpha\beta} e_{\alpha\gamma} (f^2 f^3_{,\beta} - f^3 f^2_{,\beta}) \sqrt{a} \, du^\gamma$$

$$= \int_C [(N^{11} \, du^2 - N^{21} \, du^1)(f^2 f^3_{,1} - f^3 f^2_{,1}) \sqrt{a}$$

$$+ (N^{12} \, du^2 - N^{22} \, du^1)(f^2 f^3_{,2} - f^3 f^2_{,2}) \sqrt{a}].$$

But

$$f^2 f^3_{,1} - f^3 f^2_{,1} = -R^2 \frac{\sin \phi}{\cosh \psi},$$

$$f^2 f^3_{,2} - f^3 f^2_{,2} = R^2 \frac{\sinh \psi \cos \phi}{\cosh^2 \psi}$$

and after substitution we find

$$M_1 = R^2 \operatorname{Re}\left\{\int_C \mathcal{N}(\gamma) \sinh \gamma \, d\gamma\right\} \tag{10.89}$$

and similarly

$$M_2 = R^2 \operatorname{Im}\left\{\int_C \mathcal{N}(\gamma) \cosh \gamma \, d\gamma\right\} \tag{10.90}$$

and

$$M_3 = R^2 \operatorname{Re}\left\{\int_C \mathcal{N}(\gamma) \, d\gamma\right\}. \tag{10.91}$$

Combining the forces and moments in pairs we can write

$$\frac{M_1}{R^2} + i\frac{P_1}{R} = \int_C \mathcal{N}(\gamma) \sinh \gamma \, d\gamma,$$

$$\frac{M_2}{R^2} + i\frac{P_2}{R} = -i \int_C \mathcal{N}(\gamma) \cosh \gamma \, d\gamma, \tag{10.92}$$

$$\frac{M_3}{R^2} + i\frac{P_3}{R} = \int_C \mathcal{N}(\gamma) \, d\gamma.$$

Replacing γ by ζ, the formulas take the form

$$\frac{M_1}{R^2} + i\frac{P_1}{R} = \frac{1}{2}\int_C \mathcal{T}(\zeta) \frac{\zeta^2 - 1}{\zeta^2} \, d\zeta,$$

$$\frac{M_2}{R^2} + i\frac{P_2}{R} = -\frac{i}{2}\int_C \mathcal{T}(\zeta) \frac{\zeta^2 + 1}{\zeta^2} \, d\zeta, \tag{10.93}$$

$$\frac{M_3}{R^2} + i\frac{P_3}{R} = \int_C \mathcal{T}(\zeta) \frac{1}{\zeta} \, d\zeta.$$

When the curve C is the boundary of a simply connected region D, the integrals on the right-hand side of (10.93) are particularly simple to evaluate with the help of *Cauchy's integral formula*

$$f^{(n)}(z) = \frac{n!}{2\pi i} \oint_C \frac{f(\zeta)}{(\zeta - z)^{n+1}} \, d\zeta. \tag{10.94}$$

Take for instance a region containing the interior point $\zeta = 0$, and assume that $\mathcal{T}(\zeta)/\zeta^2$ is regular in the neighbourhood of $\zeta = 0$ but not at the point $\zeta = 0$ itself, where it has a singularity. Expanding the function in a *Laurent series*, we get

$$\mathcal{T}(\zeta)/\zeta^2 = \cdots + \frac{a_{-3}}{\zeta^3} + \frac{a_{-2}}{\zeta^2} + \frac{a_{-1}}{\zeta} + a_0 + a_1\zeta + \cdots \tag{10.95}$$

and with (10.94) the integrals (10.93) are evaluated with very little effort,

$$\frac{M_1}{R^2} + i\frac{P_1}{R} = \pi i(a_{-3} - a_{-1}),$$

$$\frac{M_2}{R^2} + i\frac{P_2}{R} = \pi(a_{-3} + a_{-1}), \tag{10.96}$$

$$\frac{M_3}{R^2} + i\frac{P_3}{R} = 2\pi i a_{-2}.$$

The three complex constants are uniquely determined by the six components P_j, M_j and vice versa. Let us determine the stresses due to concentrated forces and moments at $\zeta = 0$. We shall select four typical cases:

(a) a concentrated force in radial direction.
(b) a concentrated moment in radial direction.
(c) a concentrated force in tangential direction.
(d) a concentrated moment in tangential direction.

Case (a) Take $a_{-1} = a_{-3} = 0$ and $a_{-2} = P_3/(2\pi R)$. Then

$$\mathcal{T}(\zeta) = (T + iS)\sin^2\theta = \frac{P_3}{2\pi R}$$

and hence

$$T = -N = \frac{P_3}{2\pi R \sin^2 \theta}, \qquad S = 0.$$

Case (b) Take $a_{-1} = a_{-3} = 0$ and $a_{-2} = -iM_3/(2\pi R^2)$. Then

$$\mathcal{T}(\zeta) = (T + iS)\sin^2 \theta = -i\frac{M_3}{2\pi R^2}$$

and hence

$$T = -N = 0, \qquad S = -\frac{M_3}{2\pi R^2 \sin^2 \theta}.$$

Case (c) We take $P_2 \neq 0$ and to balance its moment we take $M_1 = P_2 R$ so that the net result is a tangential force in the x^2-direction at the north-pole. Now

$$\frac{P_2}{R} = \pi i(a_{-3} - a_{-1}), \qquad i\frac{P_2}{R} = \pi(a_{-3} + a_{-1}), \qquad a_{-2} = 0$$

and hence

$$a_{-1} = \frac{iP_2}{\pi R}, \qquad a_{-2} = a_{-3} = 0.$$

This leads to the following stress-function,

$$(T + iS)\sin^2 \theta = \frac{iP_2}{\pi R}\zeta = \frac{iP_2}{\pi R}e^\gamma = \frac{iP_2}{\pi R}\tan\tfrac{1}{2}\theta(\cos\phi + i\sin\phi),$$

so that

$$T = -N = -\frac{P_2}{\pi R}\tan\tfrac{1}{2}\theta\frac{\sin\phi}{\sin^2\theta}, \qquad S = \frac{P_2}{\pi R}\tan\tfrac{1}{2}\theta\frac{\cos\phi}{\sin^2\theta}.$$

Case (d) To get $M_2 \neq 0$ and all other components zero, we take

$$a_{-1} = a_{-3} = \frac{M_2}{2\pi R^2}, \qquad a_{-2} = 0.$$

This leads to

$$(T + iS) \sin^2 \theta = \frac{M_2}{2\pi R^2} \left(\frac{1}{\zeta} + \zeta\right) = \frac{M_2}{\pi R^2} \cosh \gamma$$

$$= \frac{M_2}{\pi R^2} (\cosh \psi \cos \phi + i \sinh \psi \sin \phi)$$

and hence

$$T = -N = \frac{M_2}{2\pi R^2} (\tan \tfrac{1}{2}\theta + \cot \tfrac{1}{2}\theta) \frac{\cos \phi}{\sin^2 \theta},$$

$$S = \frac{M_2}{2\pi R^2} (\tan \tfrac{1}{2}\theta - \cot \tfrac{1}{2}\theta) \frac{\sin \phi}{\sin^2 \theta}.$$

The stress-trajectories corresponding to Cases (a)–(d) are shown in Fig. 21. The forces and moments applied at the north-pole are balanced by corresponding forces and moments at the south-pole since all other external forces are assumed to be zero.[4]

The singular solutions obtained are of considerable technical interest. Forces and moments acting on a spherical container, due to tubes or supports connected to it, are balanced by membrane forces that at a reasonable distance are equivalent to membrane forces due to point-singularities.

With the help of the integral formulas (10.93) it is easy to find the membrane state corresponding to forces and moments applied to any point ζ_0. In that case the complex stress-function will have a pole at ζ_0 and in the neighbourhood of ζ_0 we have the Laurent expansion

$$\mathcal{T}(\zeta) = \cdots + \frac{a_{-3}}{(\zeta - \zeta_0)^3} + \frac{a_{-2}}{(\zeta - \zeta_0)^2} + \frac{a_{-1}}{\zeta - \zeta_0} + \cdots . \qquad (10.97)$$

---

[4] See also p. 302.

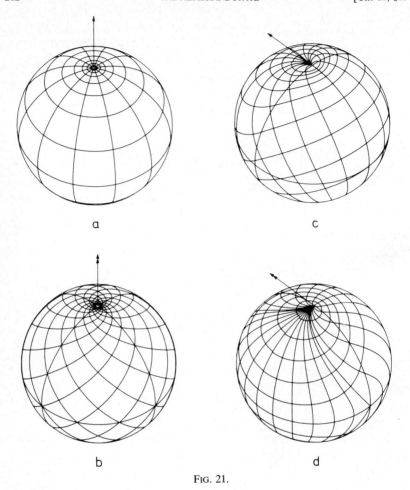

Fig. 21.

Introducing the function $\mathcal{T}$ into (10.93) and utilizing (10.94) we find, again without much effort,

$$\frac{M_1}{R^2} + i\frac{P_1}{R} = i\pi\left(a_{-1}\frac{\zeta_0^2 - 1}{\zeta_0^2} + a_{-2}\frac{2}{\zeta_0^3} - a_{-3}\frac{3}{\zeta_0^4}\right),$$

$$\frac{M_2}{R^2} + i\frac{P_2}{R} = \pi\left(a_{-1}\frac{\zeta_0^2 + 1}{\zeta_0^2} - a_{-2}\frac{2}{\zeta_0^3} + a_{-3}\frac{3}{\zeta_0^4}\right), \qquad (10.98)$$

$$\frac{M_3}{R^2} + i\frac{P_3}{R} = 2i\pi\left(\frac{a_{-1}}{\zeta_0} - \frac{a_{-2}}{\zeta_0^2} + \frac{a_{-3}}{\zeta_0^3}\right).$$

It is now possible to solve this system of equations for the unknown coefficients a_{-1}, a_{-2} and a_{-3} and thereby to determine a stress-function $\mathcal{F}$ corresponding to a given force P_j and given moment M_j at ζ_0.

This method for analyzing the membrane state of a spherical shell can be generalized to cover all second-degree surfaces of positive Gaussian curvature, i.e., the ellipsoid, the two-sheeted hyperboloid and the elliptical paraboloid.

The complex function method for the analysis of the membrane state of second-degree surfaces is discussed in detail by V.Z. VLASOV.[5]

16. Membrane state of the toroidal shell

Let us determine the membrane forces and the displacements of a toroidal shell under internal pressure (see Fig. 22).

The geometry of the middle surface is defined by the relations

$$r = R + a \cos \theta, \qquad q = a \sin \theta, \qquad \theta = \frac{t}{a}, \qquad (10.99)$$

and we find the meridional membrane force N directly from equation (10.64),

$$N = \frac{1}{r \cos \theta} \left[-p \int r \sin \theta \, dt - C_1 \right]$$

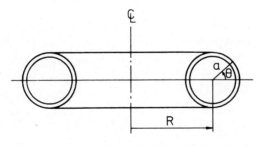

FIG. 22.

[5] VLASOV, V.Z. (1949), *General Theory of Shells*, Nauka, Moscow (in Russian); German translation: *Allgemeine Schalentheorie und ihre Anwendung in der Technik*, by KROMM, A., Springer, Berlin, 1958.

or

$$N = \frac{1}{r \cos \theta} [paR \cos \theta + \tfrac{1}{2}pa^2 \cos^2 \theta - C_1].$$

If N is to be regular at $\theta = \tfrac{1}{2}\pi$ and $\theta = \tfrac{3}{2}\pi$ (at which values of θ the Gaussian curvature of the middle surface changes sign) we must take $C_1 = 0$. Hence

$$N = pa \frac{R + \tfrac{1}{2}a \cos \theta}{R + a \cos \theta}, \qquad (10.100)$$

which is a regular function of θ, provided that $a < R$.

The circumferential membrane force T can be found in a similar way from (10.65), but it is easier to get T directly from the third equation of equilibrium (10.21),

$$T = \tfrac{1}{2}pa. \qquad (10.101)$$

From (10.66) we find the shear force

$$S = C_2/r^2,$$

which means that S does not depend on the load p. Therefore, if the shell is stress-free when $p = 0$, we must have $C_2 = 0$ and S will vanish identically.

The membrane states N, T and S thus found appear to be perfectly respectable, seemingly having all the qualities one could expect. But whether, in fact, it is a good first approximation of the true state of a toroidal shell under internal pressure or not we cannot know for sure unless we check the magnitude of the terms left out from the equations of equilibrium according to the procedure outlined in the introduction of this chapter. Let us do that.

We begin with the displacement v. Obviously, $v \equiv 0$ satisfies equation (10.56) identically. It may be checked that all nontrivial solutions of (10.56) correspond to rigid-body motions, and we shall therefore take $v \equiv 0$ to be our solution. Then u must satisfy

$$\cos^2\theta\left(\frac{u}{\cos\theta}\right)' = \frac{pa}{Eh}\left[\frac{\cos\theta}{r}(R+\tfrac{1}{2}a\cos\theta) - \tfrac{1}{2}\nu\cos\theta\right.$$
$$\left. - \frac{1}{2}\frac{r}{a} + \frac{\nu}{a}(R+\tfrac{1}{2}a\cos\theta)\right],$$

which we derive from equation (10.55) with $v \equiv 0$ for the toroidal shell. Dividing through by $\cos^2\theta$ and rearranging the terms, we get

$$\left(\frac{u}{\cos\theta}\right)' = \frac{pR}{2Eh}\left[-\frac{a^2}{R}\frac{1}{R+a\cos\theta} + \frac{a}{R\cos\theta} - \frac{1-2\nu}{\cos^2\theta}\right].$$

This expression can be integrated by using the elementary formulas

$$\int_0^\theta \frac{d\theta}{R+a\cos\theta} = \frac{2}{\sqrt{R^2-a^2}}\arctan\frac{\sqrt{R^2-a^2}\tan\tfrac{1}{2}\theta}{R+a},$$

$$\int_0^\theta \frac{d\theta}{\cos\theta} = \tfrac{1}{2}\log\frac{1+\sin\theta}{1-\sin\theta},$$

$$\int_0^\theta \frac{d\theta}{\cos^2\theta} = \tan\theta.$$

We find

$$u = \frac{pRa}{2Eh}\left[\frac{a}{2R}\log\frac{1+\sin\theta}{1-\sin\theta} - \frac{2a^2}{R\sqrt{R^2-a^2}}\arctan\frac{\sqrt{R^2-a^2}\tan\tfrac{1}{2}\theta}{R+a}\right.$$
$$\left. - (1-2\nu)\tan\theta + C\right]\cos\theta. \tag{10.102}$$

Taking $u(0) = 0$ we get $C = 0$. The terms within the square brackets have singularities at $\theta = \tfrac{1}{2}\pi$ and $\theta = \tfrac{3}{2}\pi$, but due to the factor $\cos\theta$ outside the brackets, the function u is bounded for all θ. At $\theta = \pi$, u has a discontinuity since the function arctan jumps from $\tfrac{1}{2}\pi$ to $-\tfrac{1}{2}\pi$ there.

This jump may of course be avoided if we proceed on the next branch of the multi-valued function arc tan, but then $u(2\pi)$ will not vanish, i.e., u will not be periodic.

The normal displacement w can finally be solved from (10.51) and (10.53) by eliminating E_1^1. We get

$$w = au' - \frac{a}{Eh}(N - \nu T). \qquad (10.103)$$

From (10.102) we see that u' is singular at $\theta = \tfrac{1}{2}\pi$ and $\theta = \tfrac{3}{2}\pi$ and w is unbounded as θ approaches these values.

Usually, the membrane state is an approximation that gets better the thinner the shell is. In this case, however, it is not so. We have just found that there is no regular displacement field associated with the membrane state, quite irrespective of the thickness of the shell.

Our conclusion must be that we cannot neglect the bending moments and shear forces completely in the case of a toroidal shell under internal pressure.

Although the membrane solution that we have obtained may be a reasonably good approximation almost everywhere, there is a region close to the top $\theta = \tfrac{1}{2}\pi$ and one close to the bottom $\theta = \tfrac{3}{2}\pi$ where bending plays a very important role.

We may ask how our findings can be reconciled with experience or our intuitive feeling, that a *toroidal membrane*, nevertheless, can sustain an internal pressure without being deformed completely out of shape. The answer and explanation lies in the fact that here *nonlinear* effects cannot be neglected. Especially, the nonlinear part of $E_{\alpha\beta}$, due to the rotation $w_{,\alpha}$, is of prime importance. Therefore, if we wish to analyze a toroidal *membrane* under internal pressure, we must perform a nonlinear analysis. This has been done by several investigators.[6]

17. The membrane state of shells having zero Gaussian curvature

On any surface the lines of principal curvature are orthogonal trajectories and on developable surfaces one family consists of straight lines. These

---

[6] See, for instance, SANDERS, J.L. and A. LEIPINS, (1963), "The toroidal membrane under internal pressure", *AIAA Launch and Space Vehicle Structures Conf.*

are called the *generators* of the surface and at any point on the surface the principal curvature is zero in the direction of the generator at that point. In this section we shall assume that the radius of curvature in the direction perpendicular to the generator is finite, so that no part of the middle-surface is flat.

Let us select coordinates (t, ϕ) so that $u^1 = t$ is a length-measuring coordinate in the direction of the generators and $u^2 = \phi$ is the coordinate along the orthogonal trajectories. Then

$$a_{11} = 1, \quad a_{12} = 0, \quad a = a_{22},$$

and

$$d_{11} = d_{12} = 0, \quad d_{22} \neq 0.$$

The equations of equilibrium (10.8)–(10.10) are now easily integrated. From (10.10) we immediately get

$$d_2^2 N_2^2 + p = 0$$

or

$$T = -pR, \tag{10.104}$$

where

$$T = N_2^2 \quad \text{and} \quad R = 1/d_2^2. \tag{10.105}$$

When the *transverse membrane force* T is known, we can determine the shear force S from (10.9),

$$\dot{T} + \frac{1}{\sqrt{a}}(aS)' + F_2 = 0,$$

by a quadrature,

$$S = -\frac{1}{a}\left[\int \sqrt{a}\,(\dot{T} + F_2)\,dt + f(\phi)\right]. \tag{10.106}$$

Here the partial derivative with respect to t is denoted by a prime (′) and with respect to ϕ by a dot (˙). $f(\phi)$ is an arbitrary function of ϕ. Finally, the *longitudinal membrane force* $N = N_1^1$ in the direction of the generators is determined from (10.8),

$$(\sqrt{a}\, N)' + S\dot{\ } - (\sqrt{a})' T + \sqrt{a}\, F_1 = 0,$$

by a single quadrature

$$N = -\frac{1}{\sqrt{a}}\left[\int\left(S\dot{\ } - \frac{a'}{2\sqrt{a}} T + \sqrt{a}\, F_1\right) dt + g(\phi)\right], \quad (10.107)$$

where $g(\phi)$ is another arbitrary function of ϕ.

The displacements corresponding to the membrane forces are found just as easily. Let us take the displacements of physical dimension length,

$$u = v_1, \qquad v = v_2/\sqrt{a}.$$

Then (10.52) yields with

$$\{{}^{\,2}_{1\,2}\} = \{{}^{\,2}_{2\,1}\} = \frac{a'}{2a}, \qquad \{{}^{\,2}_{2\,2}\} = -\frac{\dot{a}}{2a}$$

the following three equations,

$$E_1^1 = v_{1,1} = u',$$

$$E_{12} = \frac{1}{2}\left(v_{1,2} + v_{2,1} - \frac{a'}{a} v_2\right) = \frac{1}{2}\left[\dot{u} + a\left(\frac{v}{\sqrt{a}}\right)'\right],$$

$$E_2^2 = \frac{1}{a} v_{2,2} - d_2^2 w = \frac{(\sqrt{a}\, v)\dot{\ }}{a} - \frac{w}{R}.$$

Thus

$$u = \frac{1}{Eh}\int (N - \nu T)\, dt + h(\phi),$$

$$v = \sqrt{a}\left[\int \frac{1}{a}\left[\frac{2(1+\nu)}{Eh} S\sqrt{a} - \dot{u}\right] dt + k(\phi)\right], \quad (10.108)$$

$$w = R\left[\frac{(\sqrt{a}\, v)\dot{\ }}{a} - \frac{1}{Eh}(T - \nu N)\right],$$

where $h(\phi)$ and $k(\phi)$ are arbitrary functions of ϕ.

The solution that we have obtained contains four arbitrary functions of ϕ but none of t. This implies that we cannot prescribe any conditions on a boundary that coincides with a generator. On a boundary given by $t = $ const we are free to prescribe S and N (but not T, which is given everywhere by equality (10.104)). The same holds good for any smooth boundary that is nowhere tangent to a generator. This defines the membrane state in the whole region between the generators at the end-points of the boundary. It is also clear from the formulas obtained that an equilibrium loading is not attenuated far along the generators, but penetrates the whole shell. This is in contradiction with the so-called *principle of Saint-Venant*, and is due to the fact that for shells of zero Gaussian curvature the elliptic differential equations of the true state are 'approximated' by a parabolic system for the membrane state.

There are three kinds of developable surfaces, e.g., *cylindrical*, *conical* and *tangent surfaces*.

Any cylindrical surface can be represented by the following parametric equations,

$$f^1 = f^1(u^2), \qquad f^2 = f^2(u^2), \qquad f^3 = u^1, \qquad (10.109)$$

where f^1 and f^2 are arbitrary functions of $u^2 = \phi$. With a suitable change of variable u^2 we can always make $a_{22} = 1$, i.e., making u^2 arc-measuring. With $a = a_{22} = 1$, the equations (10.106)–(10.108) are simplified and specialized for a cylindrical shell,

$$S = -\int (T^{\cdot} + F_2)\, dt + f(\phi),$$

$$N = -\int (S^{\cdot} + F_1)\, dt + g(\phi), \qquad (10.110)$$

$$u = \frac{1}{Eh} \int (N - \nu T)\, dt + h(\phi),$$

$$v = \int \left(\frac{2(1+\nu)}{Eh} S - u^{\cdot}\right) dt + k(\phi), \qquad (10.111)$$

$$w = R\left[v^{\cdot} - \frac{1}{Eh}(T - \nu N)\right].$$

A conical surface can be described by the angle θ between a generator and the x^3-axis as a function of ϕ, the angle between the x^1-axis and the projection of the generator on the x^1, x^2-plane. Thus any conical surface can be given in the form

$$f^1 = t \sin\theta \cos\phi, \qquad f^2 = t \sin\theta \sin\phi, \qquad f^3 = t \cos\theta,$$
(10.112)

where t is measured along a generator from the vertex and where $\theta(\phi)$ determines the geometrical form of the base. From the parametric equations (10.112) we get

$$a_{11} = 1, \qquad a_{12} = a_{21} = 0, \qquad a_{22} = t^2(\sin^2\theta + \dot\theta^2)$$

and

$$d_1^1 = d_2^1 = d_1^2 = 0,$$

$$d_2^2 = \frac{\cos\theta \sin^2\theta + 2\cos\theta \dot\theta^2 - \sin\theta \ddot\theta}{t(\sin^2\theta + \dot\theta^2)^{3/2}} = \frac{1}{R}.$$

For the circular cone, θ is a constant equal to half the top angle of the cone. In that case

$$a = t^2 \sin^2\theta, \qquad R = t \tan\theta,$$

and the equations (10.104) and (10.106)–(10.108) reduce to

$$T = -pt \tan\theta,$$

$$S = -\frac{1}{t^2 \sin\theta} \left[\int (T\dot{} + F_2)t \, dt + f(\phi) \right], \qquad (10.113)$$

$$N = -\frac{1}{t} \left[\int \left(\frac{\dot S}{\sin\theta} - T + F_1 t \right) dt + g(\phi) \right]$$

and

$$u = \frac{1}{Eh} \int (N - \nu T) \, dt + h(\phi),$$

$$v = t \left[\int \left(\frac{2(1+\nu)}{Eh} St - \frac{\dot u}{\sin\theta} \right) \frac{dt}{t^2} + k(\phi) \right], \qquad (10.114)$$

$$w = \frac{\cos\theta}{t^2} \dot v - \frac{\sin\theta}{Eht} (T - \nu N).$$

The remaining class of shells of zero Gaussian curvature are those that have a middle-surface generated by the tangents to a space curve. It is generally advisable to refer such shells to coordinates of principal curvature making formulas (10.104)–(10.108) applicable.

18. Discussion

In analysing shells as membranes, bending moments and transverse shear forces are neglected. This leads to a simplified theory that has some great advantages. In many cases it gives an accurate and clear description of the state of stress and deformation. In many cases, however, bending moments and shear forces play an important role. Then, of course, the membrane theory cannot provide any answer to the problem.

In particular, this happens at boundaries and in regions close to boundaries where transverse shear is applied, but also at discontinuities of compound shells such as the joints in a pressure vessel between the cylindrical tube and its head and/or bottom. In fact, at any boundary where we cannot comply with the boundary conditions implied by the membrane state, bending plays an important role. We have also seen that the membrane state is not applicable to a toroidal shell with internal pressure. Here the 'boundaries' are the circles $\theta = \frac{1}{2}\pi$ and $\theta = \frac{3}{2}\pi$ which divide the shell in two parts, one with positive and one with negative Gaussian curvature.

We now turn our attention to the bending theory of shells.

Bibliography

TIMOSHENKO, S. (1959), *Theory of Plates and Shells*, 2nd ed., McGraw-Hill, New York, Chapter 14.

FLÜGGE, W. (1960), *Stresses in Shells*, Springer, Berlin, Chapters 2–4.

GOL'DENVEIZER, A.L. (1961), *Theory of Elastic Thin Shells*, Pergamon, Oxford, Chapters 5–9 (English translation).

WLASOW, W.S. (1958), *Allgemeine Schalentheorie und ihre Anwendung in der Technik*, Akademie-Verlag, Berlin, Chapters I–IV.

NOVOZHILOV, V.V. (1959), *The Theory of Thin Shells*, Noordhoff, Groningen, Chapter II (English translation).

SEIDE, P. (1975), *Small Elastic Deformations of Thin Shells*, Noordhoff, Leiden, Chapter 3.8.

GOULD, P.L. (1977), *Static Analysis of Shells*, Lexington Books, Lexington, MA.

CHAPTER 11

BENDING OF CIRCULAR CYLINDRICAL SHELLS

1. Introduction 225
2. Basic equations 225
3. Boundary conditions 231
4. Solution of the mathematical problem 234
5. Bending due to axisymmetrical ring-loads 244
6. Non-axisymmetric bending of cylindrical shells 250
7. Free vibrations of cylindrical shells 259
8. Cylindrical panels 261
   Bibliography 262

CHAPTER 11

BENDING OF CIRCULAR CYLINDRICAL SHELLS

1. Introduction

Not only because they represent the simplest possible case in shell theory, but also because of their technical importance, circular cylindrical shells have been the most intensively studied shell structures of all. The shortcomings of the membrane theory are particularly tangible in shells of nonpositive Gaussian curvature and a bending theory for cylindrical shells is necessary for numerous applications. This chapter is devoted to the classical linear theory of bending of cylindrical shells in coordinates of principal curvature.

2. Basic equations

Let us describe the middle-surface according to Fig. 23 in cartesian coordinates of principal curvature $(x, \phi) = (u^1, u^2)$, x being the arc-length along a generator and ϕ the arc-length in peripheral direction, perpendicular to the generators.

The parametric equations of the middle-surface become

$$x^1 = x, \qquad x^2 = R \sin \frac{\phi}{R}, \qquad x^3 = R \cos \frac{\phi}{R}. \tag{11.1}$$

With this choice of coordinates (u^1, u^2), the unit normal X^i will have the direction of the radius away from the axis of the cylinder.

From (11.1) we easily find

$$a_{11} = a_{22} = 1, \qquad a_{12} = a_{21} = 0 \tag{11.2}$$

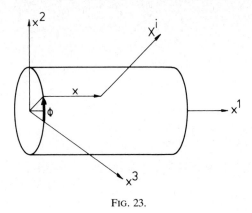

Fig. 23.

and

$$d_{11} = d_{12} = d_{21} = 0, \qquad d_{22} = -\frac{1}{R}. \tag{11.3}$$

Since the coordinates are cartesian, all Christoffel symbols vanish, and covariant derivatives reduce to partial derivatives.

We shall find it convenient to use the displacement components u, v instead of v^1, v^2, i.e.,

$$u = v^1, \qquad v = v^2. \tag{11.4}$$

With the help of the new notations we find the strain tensor from (3.11) to be

$$E_{11} = \frac{\partial u}{\partial x},$$

$$E_{12} = \frac{1}{2}\left(\frac{\partial u}{\partial \phi} + \frac{\partial v}{\partial x}\right), \tag{11.5}$$

$$E_{22} = \frac{\partial v}{\partial \phi} + \frac{w}{R},$$

and the components of the bending tensor from (3.37),

$$K_{11} = \frac{\partial^2 w}{\partial x^2},$$

$$K_{12} = \frac{\partial^2 w}{\partial x \partial \phi} - \frac{1}{R}\frac{\partial v}{\partial x}, \qquad (11.6)$$

$$K_{22} = \frac{\partial^2 w}{\partial \phi^2} - \frac{2}{R}\frac{\partial v}{\partial \phi} - \frac{w}{R^2}.$$

The equations of equilibrium (5.22)–(5.23) take the following form,

$$N^{11}_{,1} + N^{12}_{,2} + F^1 = 0,$$

$$N^{12}_{,1} + N^{22}_{,2} - \frac{2}{R}(M^{12}_{,1} + M^{22}_{,2}) + F^2 = 0, \qquad (11.7)$$

$$M^{11}_{,11} + 2M^{12}_{,12} + M^{22}_{,22} - \frac{1}{R^2}M^{22} + \frac{1}{R}N^{22} - p = 0.$$

From these equations we shall derive a set of three partial differential equations for the displacements u, v and w. To do this we must express the membrane force tensor $N^{\alpha\beta}$ and the moment tensor $M^{\alpha\beta}$ in terms of $E_{\alpha\beta}$ and $K_{\alpha\beta}$, and the straightforward procedure to do so would be to use the uncoupled stress–strain relations (7.11). This, however, we shall not do.

It was explained in Chapter 7 that the relative error inherent in the energy expression (7.7), from which the uncoupled stress–strain relations are derived, is of order h/R.[1] This error stems from neglecting the mixed term

$$D^{\alpha\beta\gamma\delta} E_{\alpha\beta} K_{\gamma\delta}$$

in the energy expression (7.3). Following the argument in Chapter 7 we find that $D^{\alpha\beta\gamma\delta}$ can be written in the form (where the factor $1 - \nu^2$ is taken out for convenience)

---

[1] In addition, there is also an error of relative order $(h/L)^2$, but that is of no concern here.

$$(1-\nu^2)D^{\alpha\beta\gamma\delta} = d_1\frac{h^2}{R}a^{\alpha\beta}a^{\gamma\delta} + d_2\frac{h^2}{R}a^{\alpha\gamma}a^{\beta\delta} + d_3h^2a^{\alpha\beta}d^{\gamma\delta}$$
$$+ d_4h^2(a^{\beta\delta}d^{\alpha\gamma} + a^{\alpha\delta}d^{\beta\gamma}) + d_5h^2a^{\gamma\delta}d^{\alpha\beta}, \quad (11.8)$$

where $d_1, \ldots, d_5$ are dimensionless numbers or, rather, dimensionless functions of Poisson's ratio ν.

Keeping in mind that adding or subtracting terms of this type does not influence the accuracy of the theory,[2] it seems reasonable to investigate if, by a suitable selection of such terms, *the resulting differential equations may be simplified*. We shall find that this is the case.

With the help of the tensor $D^{\alpha\beta\gamma\delta}$ we can write the constitutive equations in the following form,

$$N^{\alpha\beta} = \frac{Eh}{1-\nu^2}[(1-\nu)E^{\alpha\beta} + \nu a^{\alpha\beta}E^\gamma_\gamma] + \frac{Eh}{1-\nu^2}D^{\alpha\beta\gamma\delta}K_{\gamma\delta} \quad (11.9)$$

and

$$M^{\alpha\beta} = D[(1-\nu)K^{\alpha\beta} + \nu a^{\alpha\beta}K^\gamma_\gamma] + \frac{Eh}{1-\nu^2}D^{\gamma\delta\alpha\beta}E_{\gamma\delta}. \quad (11.10)$$

Substituting $N^{\alpha\beta}$ and $M^{\alpha\beta}$ into (11.7) we get the following three equations,

$$\frac{\partial^2 u}{\partial x^2} + \frac{1-\nu}{2}\frac{\partial^2 u}{\partial \phi^2} + \frac{1+\nu}{2}\frac{\partial^2 v}{\partial x \partial \phi} + \frac{\nu}{R}\frac{\partial w}{\partial x}$$
$$+ (d_1 + d_2)\frac{h^2}{R}\left(\frac{\partial^3 w}{\partial x \partial \phi^2} + \frac{\partial^3 w}{\partial x^3}\right) + \frac{1-\nu^2}{Eh}F_x = 0,$$

$$\frac{1+\nu}{2}\frac{\partial^2 u}{\partial x \partial \phi} + \frac{1-\nu}{2}\frac{\partial^2 v}{\partial x^2} + \frac{\partial^2 v}{\partial \phi^2} + \frac{1}{R}\frac{\partial w}{\partial \phi}$$
$$+ (d_1 + d_2 - d_5 - \tfrac{1}{6})\frac{h^2}{R}\left(\frac{\partial^3 w}{\partial x^2 \partial \phi} + \frac{\partial^3 w}{\partial \phi^3}\right) + \frac{1-\nu^2}{Eh}F_\phi = 0,$$

---

[2] Naturally, if the *correct* mixed terms are added, the theory will be improved, so that the inherent error will be of relative order $(h/R)^2 + (h/L)^2$ only. These terms have been derived (see NIORDSON, F. (1978), "A consistent refined shell theory", *Complex Analysis and its Applications* (*The Vekua Anniversary Volume*), Nauka, Moscow).

$$\frac{\nu}{R}\frac{\partial u}{\partial x} + \frac{1}{R}\frac{\partial v}{\partial \phi} + \frac{w}{R^2} + \frac{h^2}{12}\Delta^2 w + (d_1+d_2)\frac{h^2}{R}\left(\frac{\partial^3 u}{\partial x^3} + \frac{\partial^3 u}{\partial x \partial \phi^2}\right)$$

$$+ (d_1+d_2-d_5-\tfrac{1}{6})\frac{h^2}{R}\left(\frac{\partial^3 v}{\partial \phi^3} + \frac{\partial^3 v}{\partial x^2 \partial \phi}\right) + (2d_1-2d_5-\tfrac{1}{6}\nu)\frac{h^2}{R^2}\frac{\partial^2 w}{\partial x^2}$$

$$+ (2d_1+2d_2+2d_5-\tfrac{1}{6})\frac{h^2}{R^2}\frac{\partial^2 w}{\partial \phi^2} - \frac{1-\nu^2}{Eh}p = 0,$$

where

$$F_x = F^1, \qquad F_\phi = F^2, \tag{11.11}$$

and where we have put $d_3 = d_4 = 0$, since the corresponding terms do not contribute to any simplification.

In these equations, coefficients to partial derivatives containing the small number h^2/R^2 have been simplified by omitting terms of order h^2/R^2 in comparison with unity. Thus, for instance, in the first equation the coefficient of $\partial w/\partial x$ turns out to be $\nu/R - d_1 h^2/R^3$. Since the constitutive equations have a relative accuracy of h/R only, the omission of $d_1 h^2/R^3$ in comparison with ν/R is logical and detracts nothing from the accuracy of the equations. In the following we shall omit h^2/R^2 in comparison with unity whenever we find it convenient, without further comments.

The coefficients d_1, d_2 and d_5, which are at our disposal, can now be used for a considerable simplification of the equations. On inspection, it is seen immediately that by taking $d_1+d_2=0$ and $d_5=-\tfrac{1}{6}$, all derivatives of third order vanish. A further simplification is obtained if d_1 is chosen so that the coefficients of $\partial^2 w/\partial x^2$ and $\partial^2 w/\partial \phi^2$ become equal, i.e., by taking

$$d_2 = -d_1 = \tfrac{1}{12}(1-\nu).$$

With these values of d_1, d_2 and d_5 we get

$$\frac{\partial^2 u}{\partial x^2} + \tfrac{1}{2}(1-\nu)\frac{\partial^2 u}{\partial \phi^2} + \tfrac{1}{2}(1+\nu)\frac{\partial^2 v}{\partial x \partial \phi} + \frac{\nu}{R}\frac{\partial w}{\partial x} + \frac{1-\nu^2}{Eh}F_x = 0,$$

$$\tfrac{1}{2}(1+\nu)\frac{\partial^2 u}{\partial x \partial \phi} + \tfrac{1}{2}(1-\nu)\frac{\partial^2 v}{\partial x^2} + \frac{\partial^2 v}{\partial \phi^2} + \frac{1}{R}\frac{\partial w}{\partial \phi} + \frac{1-\nu^2}{Eh}F_\phi = 0, \tag{11.12}$$

$$\frac{\nu}{R}\frac{\partial u}{\partial x} + \frac{1}{R}\frac{\partial v}{\partial \phi} + \frac{w}{R^2} + k\left(R\Delta + \frac{1}{R}\right)^2 w - \frac{1-\nu^2}{Eh}p = 0,$$

where

$$k = \frac{h^2}{12R^2}. \tag{11.13}$$

The first complete set of equations of equilibrium for thin elastic shells was derived by LOVE[3] in 1888. His deduction was based on the Kirchhoff hypothesis, and it was noted almost from the start that LOVE had not been fully consistent, since certain small terms were retained and others omitted. Many attempts were made to derive more consistent or more accurate equations in the decades to follow, and in this way several different versions of the equations originated.

However, there was also a strong need for simpler equations to solve particular problems, so that computations would not be too laborious. In 1933, DONNELL[4] derived a set of equations which he used to investigate the stability of circular cylinders under torsion, obtaining acceptable agreement with experiments. Due to their simplicity, DONNELL's equations became widely used. However, it was soon found that the error obtained by using them becomes intolerable for large wavelengths of circumferential displacement. For such problems the more complex equations by FLÜGGE[5] were often used.

Seeking to improve DONNELL's equations without destroying their essential simplicity, MORLEY[6] proposed an equation corresponding to our equations (11.12) (using an entirely different approach). MORLEY found, in a number of numerical examples, that the source of error in DONNELL's approximation was removed, giving results close to those based on FLÜGGE's theory.

In 1968, KOITER[7] made a rigorous derivation of equations (11.12)

---

[3] LOVE, A.E.H. (1888), "On the small free vibrations and deformations of thin elastic shells", *Phil. Trans. Roy. Soc. A* **179**.

[4] DONNELL, L.H. (1933), "Stability of thin-walled tubes under torsion", *NACA Report* **479**.

[5] FLÜGGE, W. (1932), "Die Stabilität der Kreiszylinderschale", *Ing.-Arch.* **3**, p. 463.

[6] MORLEY, L.S.D. (1959), "An improvement on Donnell's approximation for thin-walled circular cylinders", *Quart. J. Mech. Appl. Math.* **12**, p. 89.

[7] KOITER, W.T. (1968), "Summary of equations for modified, simplest possible accurate linear theory of thin circular cylindrical shells", *Report* **442**, Lab. Techn. Mech., T.H. Delft.

from a modified expression for the strain energy and demonstrated, in this way, that these equations were accurate, in the sense that they are consistent with the Love–Kirchhoff assumptions, just like Love's and Flügge's, but not Donnell's equations.

The Morley–Koiter equations (11.12) seem to have the simplest possible form in which the equations of equilibrium for thin elastic cylindrical shells can be written, without relinquishing the accuracy of the theory.

The solution of the system of equations (11.12) will be discussed later. First, we shall derive the appropriate boundary conditions.

3. Boundary conditions

With d_1, d_2 and d_5 determined, we find the membrane force tensor

$$N^{11} = \frac{Eh}{1-\nu^2}\left[\frac{\partial u}{\partial x} + \nu\left(\frac{\partial v}{\partial \phi} + \frac{w}{R}\right) - kR(1-\nu)\frac{\partial^2 w}{\partial \phi^2}\right],$$

$$N^{12} = \frac{Eh}{2(1+\nu)}\left[\frac{\partial u}{\partial \phi} + \frac{\partial v}{\partial x} + 2kR\frac{\partial^2 w}{\partial x \partial \phi}\right], \tag{11.14}$$

$$N^{22} = \frac{Eh}{1-\nu^2}\left[\frac{\partial v}{\partial \phi} + \frac{w}{R} + \nu\frac{\partial u}{\partial x} + kR\left\{(1+\nu)\frac{\partial^2 w}{\partial x^2} + 2\frac{\partial^2 w}{\partial \phi^2}\right\}\right]$$

and the moment tensor

$$M^{11} = D\left[\frac{\partial^2 w}{\partial x^2} + \nu\frac{\partial^2 w}{\partial \phi^2} + \frac{1-\nu}{R}\frac{\partial v}{\partial \phi} + \frac{w}{R^2}\right],$$

$$M^{12} = D(1-\nu)\left[\frac{\partial^2 w}{\partial x \partial \phi} + \frac{1}{2R}\left(\frac{\partial u}{\partial \phi} - \frac{\partial v}{\partial x}\right)\right], \tag{11.15}$$

$$M^{22} = D\left[\frac{\partial^2 w}{\partial \phi^2} + \nu\frac{\partial^2 w}{\partial x^2} - \frac{1-\nu}{R}\frac{\partial u}{\partial x} + \frac{w}{R^2}\right].$$

The forces and moments at a boundary are given for the general case by formulas (6.15)–(6.17) and (6.19)–(6.20). The coordinate system selected is particularly suited for problems in which the boundary is perpendicular to the generators and/or follows a generator. In the first case, the unit normal and tangent vectors are

$$n_\alpha = (1, 0), \quad t_\alpha = (0, 1), \tag{11.16}$$

respectively, and in the second case

$$\bar{n}_\alpha = (0, 1), \quad \bar{t}_\alpha = (-1, 0), \tag{11.17}$$

respectively[8] (see Fig. 24).

In the first case the static boundary conditions concern the bending moment M_x, the (effective) shear force Q_x, the (effective) membrane normal force N_x and the (effective) membrane shear force S_x. According to (6.15) and (6.19) we get, for a boundary $x = \text{const}$,

$$\begin{aligned}
M_x &= D\left[\frac{\partial^2 w}{\partial x^2} + \nu \frac{\partial^2 w}{\partial \phi^2} + \frac{1-\nu}{R}\frac{\partial v}{\partial \phi} + \frac{w}{R^2}\right], \\
Q_x &= -D\left[\frac{\partial^3 w}{\partial x^3} + (2-\nu)\frac{\partial^3 w}{\partial x \partial \phi^2} + \frac{1-\nu}{R}\frac{\partial^2 u}{\partial \phi^2} + \frac{1}{R^2}\frac{\partial w}{\partial x}\right], \\
N_x &= \frac{Eh}{1-\nu^2}\left[\frac{\partial u}{\partial x} + \nu\left(\frac{\partial v}{\partial \phi} + \frac{w}{R}\right) - kR(1-\nu)\frac{\partial^2 w}{\partial \phi^2}\right], \\
S_x &= \frac{Eh}{2(1+\nu)}\left[\frac{\partial v}{\partial x} + \frac{\partial u}{\partial \phi} - 2kR\frac{\partial^2 w}{\partial x \partial \phi}\right]
\end{aligned} \tag{11.18}$$

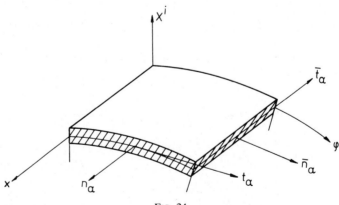

FIG. 24.

[8] Here we have adhered to the conventions for the positive direction of the vectors, given on p. 76: the normal vector points out of the material of the element, and the direction of tangent vector is taken so that n_α, t_α and X^i in that order make a right-handed system.

and, for a boundary $\phi = \text{const}$,

$$M_\phi = D\left[\frac{\partial^2 w}{\partial \phi^2} + \nu \frac{\partial^2 w}{\partial x^2} - \frac{1-\nu}{R}\frac{\partial u}{\partial x} + \frac{w}{R^2}\right],$$

$$Q_\phi = -D\left[\frac{\partial^3 w}{\partial \phi^3} + (2-\nu)\frac{\partial^3 w}{\partial x^2 \partial \phi} - \frac{1-\nu}{R}\frac{\partial^2 v}{\partial x^2} + \frac{1}{R^2}\frac{\partial w}{\partial \phi}\right],$$

$$N_\phi = \frac{Eh}{1-\nu^2}\left[\frac{\partial v}{\partial \phi} + \frac{w}{R} + \nu \frac{\partial u}{\partial x} + kR\,\Delta w\right], \qquad (11.19)$$

$$S_\phi = \frac{Eh}{2(1+\nu)}\left[\frac{\partial v}{\partial x} + \frac{\partial u}{\partial \phi} + 2kR\frac{\partial^2 w}{\partial x \partial \phi}\right].$$

The positive directions of these forces and moments are indicated in Fig. 25.

Kinematic boundary conditions are expressed in the rotation of the normal ψ and the displacements u, v and w, where

$$\psi = \frac{\partial w}{\partial x} \quad \text{for a boundary } x = \text{const} \qquad (11.20)$$

and

$$\psi = \frac{\partial w}{\partial \phi} - \frac{v}{R} \quad \text{for a boundary } \phi = \text{const.} \qquad (11.21)$$

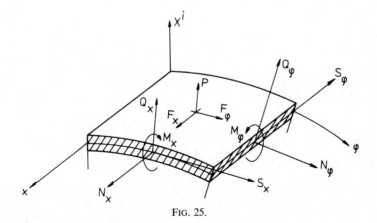

Fig. 25.

By introducing mixed terms in the energy expression, the constitutive equations become more complex. Thus, the second order derivatives of w appearing in all components of $N^{\alpha\beta}$ according to (11.14) are due to the mixed terms.

According to (11.18) and (11.19) the formulas for the effective membrane forces also include terms containing second order derivatives of w. *But such terms would appear anyhow, i.e., even if the uncoupled constitutive equations had been used*, of course only with other coefficients. Evaluating the complexity of the mathematical problem to be solved, this is what counts; the constitutive equations (11.14) act as an intermediate step only, and their form is really of no consequence. Therefore, we have no penalty to pay for simplifying the equations of equilibrium.

4. Solution of the mathematical problem

The equations to be solved constitute a system of three partial differential equations, that are linear and have constant coefficients. Using the operator method, we can easily reduce the problem to the solution of one single eight-order differential equation. For that purpose we shall write the system (11.12) in matrix-form,

$$A_{ij} V_j + F_i = 0, \tag{11.22}$$

where A_{ij} are the following elements of a symmetrical 3×3 *differential operator matrix*,

$$A_{11} = \Delta - \tfrac{1}{2}(1+\nu)\partial_\phi^2, \qquad A_{12} = A_{21} = \tfrac{1}{2}(1+\nu)\partial_x \partial_\phi,$$

$$A_{13} = A_{31} = \frac{\nu}{R}\partial_x, \qquad A_{22} = \Delta - \tfrac{1}{2}(1+\nu)\partial_x^2, \tag{11.23}$$

$$A_{23} = A_{32} = \frac{1}{R}\partial_\phi, \qquad A_{33} = \frac{1}{R^2} + k\left(R\Delta + \frac{1}{R}\right)^2,$$

and where

$$V_j = (u, v, w), \qquad F_i = \frac{1-\nu^2}{Eh}(F_x, F_y, p) \tag{11.24}$$

is the displacement 'vector' and the external load 'vector' respectively.[9] The notation ∂_x stands for the operator $\partial/\partial x$ and ∂_ϕ for $\partial/\partial\phi$. Their powers and products are the higher order derivatives, for example $\partial_x^4 = \partial^4/\partial x^4$ and $\partial_{x\phi}^2 = \partial^3/\partial x^2\partial\phi$. In particular, the Laplacian is $\Delta = \partial_x^2 + \partial_\phi^2$. Since all coefficients are constant, we can treat ∂_x and ∂_ϕ simply as factors.

Let $\mathcal{D}$ denote the determinant of A_{ij} and let $B_{ij} = \text{cofactor}(A_{ij})$, i.e., the determinant obtained by excluding the ith row and the jth column from A_{ij} and multiplying the result by $(-1)^{i+j}$. We find

$$B_{11} = \frac{1-\nu}{2R^2}\partial_x^2 + k\left\{R^2[\tfrac{1}{2}(1-\nu)\partial_x^6 + (2-\nu)\partial_x^4\partial_\phi^2 + \tfrac{1}{2}(5-\nu)\partial_x^2\partial_\phi^4 + \partial_\phi^6]\right.$$

$$\left. + (1-\nu)\partial_x^4 + (3-\nu)\partial_x^2\partial_\phi^2 + 2\partial_\phi^4 + \frac{1}{R^2}[\tfrac{1}{2}(1-\nu)\partial_x^2 + \partial_\phi^2]\right\}, \tag{11.25a}$$

$$B_{12} = B_{21} = -\frac{1-\nu}{2R^2}\partial_x\partial_\phi - \tfrac{1}{2}k(1+\nu)\left\{R^2[\partial_x^5\partial_\phi + 2\partial_x^3\partial_\phi^3 + \partial_x\partial_\phi^5]\right.$$

$$\left. + 2\partial_x^3\partial_\phi + 2\partial_x\partial_\phi^3 + \frac{1}{R^2}\partial_x\partial_\phi\right\}, \tag{11.25b}$$

$$B_{13} = B_{31} = \frac{1-\nu}{2R}(\partial_x\partial_\phi^2 - \nu\partial_x^3), \tag{11.25c}$$

$$B_{22} = \frac{1-\nu}{2R^2}[2(1+\nu)\partial_x^2 + \partial_\phi^2]$$

$$+ k\left\{R^2[\partial_x^6 + \tfrac{1}{2}(5-\nu)\partial_x^4\partial_\phi^2 + (2-\nu)\partial_x^2\partial_\phi^4 + \tfrac{1}{2}(1-\nu)\partial_\phi^6]\right.$$

$$\left. + 2\partial_x^4 + (3-\nu)\partial_x^2\partial_\phi^2 + (1-\nu)\partial_\phi^4 + \frac{1}{R^2}[\partial_x^2 + \tfrac{1}{2}(1-\nu)\partial_\phi^2]\right\}, \tag{11.25d}$$

$$B_{23} = B_{32} = -\frac{1-\nu}{2R}[(2+\nu)\partial_x^2\partial_\phi + \partial_\phi^3], \tag{11.25e}$$

$$B_{33} = \tfrac{1}{2}(1-\nu)(\partial_x^4 + 2\partial_x^2\partial_\phi^2 + \partial_\phi^4). \tag{11.25f}$$

---

[9] The word 'vector' is used here to denote a 1×3 or a 3×1 matrix.

Furthermore, let

$$V_j = B_{kj}\kappa_k(x, \phi), \qquad (11.26)$$

where κ_k are three functions of x and ϕ, so far undetermined. If (11.26) is substituted into (11.22), we find that the functions κ_k must satisfy the equations

$$\mathcal{D}\kappa_i + F_i = 0, \qquad (11.27)$$

where

$$\mathcal{D} = \tfrac{1}{2}(1-\nu)k\left\{\frac{1-\nu^2}{k}\partial_x^4 + \Delta^2(R^2\Delta+1)^2\right\}. \qquad (11.28)$$

Let κ_i^* be a particular integral of (11.27). The general solution is then given by

$$\kappa_i = \kappa_i^* + \kappa_i^{**},$$

where κ_i^{**} are the solutions of the same homogeneous equations.

Taking in particular

$$\kappa_i^{**} = \frac{2R^2}{1-\nu}(\Xi, 0, 0), \qquad (11.29)$$

we get

$$u = \frac{2R^2}{1-\nu}B_{11}\Xi, \qquad v = \frac{2R^2}{1-\nu}B_{12}\Xi, \qquad w = \frac{2R^2}{1-\nu}B_{13}\Xi, \qquad (11.30)$$

where B_{11}, B_{12} and B_{13} are given by (11.25) and where Ξ is a solution of

$$\left\{\frac{1-\nu^2}{k}\partial_x^4 + \Delta^2(R^2\Delta+1)^2\right\}\Xi = 0. \qquad (11.31)$$

The problem is thus reduced to finding solutions of one single eighth

SOLUTION OF THE MATHEMATICAL PROBLEM

order differential equation (11.31), and we may consider Ξ to be *a potential function for the integration of the homogeneous equations of equilibrium of cylindrical shells*.

Thus, any solution Ξ of (11.31) corresponds to a solution of the homogeneous system of equations of equilibrium in terms of the displacements u, v and w given by (11.30). As we shall see, the converse is not true. Not every integral of the homogeneous equations is given by the formulas (11.30).

The problem of finding particular integrals will not be considered here. For loads met in practical applications the construction of particular solutions rarely presents any difficulties and the main problem consists, generally speaking, in finding the complete solution, so that the boundary conditions can be satisfied.

For a *closed* cylindrical shell, the solution Ξ must be periodic in ϕ with period $2\pi R$, and by making a Fourier expansion in the ϕ-direction, the complete solution of the determining equation can be written as

$$\Xi = \sum_{n=0}^{\infty} \left[C_n(x) \cos \frac{n\phi}{R} + B_n(x) \sin \frac{n\phi}{R} \right] \qquad (11.32)$$

in this case.

Substituting into (11.31) and equating the coefficient of each trigonometric function to zero, we find the following set of ordinary differential equations for $n = 0, 1, 2, \ldots$, which we write in the following compact form,

$$\left\{ \frac{1-\nu^2}{k} \frac{d^4}{dx^4} + \left(\frac{d^2}{dx^2} - \frac{n^2}{R^2} \right)^2 \left(R^2 \frac{d^2}{dx^2} - n^2 + 1 \right)^2 \right\} C_n = 0 \qquad (11.33)$$

and likewise for the functions $B_n(x)$.

The solution to (11.33) can be sought in the form of an exponential function

$$e^{\alpha x/R},$$

which after substitution into (11.33) yields the following *characteristic equation* of eighth degree in α,

$$(\alpha^2 - n^2)^2(\alpha^2 - n^2 + 1)^2 + \frac{1-\nu^2}{k}\alpha^4 = 0. \tag{11.34}$$

To find the roots of this equation we introduce the dimensionless parameter

$$\rho = \sqrt[4]{3(1-\nu^2)}\sqrt{R/h}. \tag{11.35}$$

The last term of (11.34) can now be written as $4\rho^4\alpha^4$ and hence

$$(\alpha^2 - n^2)(\alpha^2 - n^2 + 1) = \pm 2i\rho^2\alpha^2.$$

The roots of the characteristic equation are therefore determined by two algebraic equations of second degree in α^2, and by solving these, we find the four roots

$$\pm\rho\left(\kappa + i(1 \pm \sqrt{1 - 2i\kappa})\right)^{1/2} \tag{11.36}$$

and their complex conjugates

$$\pm\rho\left(\kappa - i(1 \pm \sqrt{1 + 2i\kappa})\right)^{1/2}, \tag{11.37}$$

where

$$\kappa = \frac{n^2 - \frac{1}{2}}{\rho^2} = \frac{n^2 - \frac{1}{2}}{\sqrt{3(1-\nu^2)}}\frac{h}{R}. \tag{11.38}$$

Thus, the eight roots of the characteristic equation may be given in terms of four real, positive numbers p, q, r, s in the following way,

$$\alpha_1 = p + iq, \quad \alpha_2 = p - iq, \quad \alpha_3 = -p + iq, \quad \alpha_4 = -p - iq,$$
$$\alpha_5 = r + is, \quad \alpha_6 = r - is, \quad \alpha_7 = -r + is, \quad \alpha_8 = -r - is, \tag{11.39}$$

where

$$p = \frac{\rho}{\sqrt{2}}\left((1 + \omega + \kappa^2 - \sqrt{2}\sqrt{\omega + 1} - \kappa\sqrt{2}\sqrt{\omega - 1})^{1/2} + \kappa - \sqrt{\tfrac{1}{2}(\omega - 1)}\right)^{1/2}, \tag{11.40a}$$

$$q = \frac{\rho}{\sqrt{2}}\left((1+\omega+\kappa^2-\sqrt{2}\sqrt{\omega+1}-\kappa\sqrt{2}\sqrt{\omega-1})^{1/2}-\kappa+\sqrt{\tfrac{1}{2}(\omega-1)}\right)^{1/2},$$
(11.40b)

$$r = \frac{\rho}{\sqrt{2}}\left((1+\omega+\kappa^2+\sqrt{2}\sqrt{\omega+1}+\kappa\sqrt{2}\sqrt{\omega-1})^{1/2}+\kappa+\sqrt{\tfrac{1}{2}(\omega-1)}\right)^{1/2},$$
(11.40c)

$$s = \frac{\rho}{\sqrt{2}}\left((1+\omega+\kappa^2+\sqrt{2}\sqrt{\omega+1}+\kappa\sqrt{2}\sqrt{\omega-1})^{1/2}-\kappa-\sqrt{\tfrac{1}{2}(\omega-1)}\right)^{1/2}$$
(11.40d)

with

$$\omega = \sqrt{1+4\kappa^2} \qquad (11.41)$$

are the real and imaginary parts of (11.36) and (11.37).

The solution to (11.33) is therefore in the general case $(n > 1)$

$$C_n = \left(A_1 \cos\frac{qx}{R} + A_2 \sin\frac{qx}{R}\right)e^{px/R} + \left(A_3 \cos\frac{qx}{R} + A_4 \sin\frac{qx}{R}\right)e^{-px/R}$$
$$+ \left(A_5 \cos\frac{sx}{R} + A_6 \sin\frac{sx}{R}\right)e^{rx/R} + \left(A_7 \cos\frac{sx}{R} + A_8 \sin\frac{sx}{R}\right)e^{-rx/R}.$$
(11.42)

For $n = 0$ and $n = 1$, and only for these values of n, four of the eight roots of the characteristic equation are zero. In that case $p = q = 0$ and the fundamental solutions multiplying A_2, A_3 and A_4 vanish. In their place we have a polynomial of third degree, which clearly satisfies (11.33), i.e., for $n = 0$ and $n = 1$ we have

$$C_n = A_1 + A_2 x + A_3 x^2 + A_4 x^3 + \left(A_5 \cos\frac{sx}{R} + A_6 \sin\frac{sx}{R}\right)e^{rx/R}$$
$$+ \left(A_7 \cos\frac{sx}{R} + A_8 \sin\frac{sx}{R}\right)e^{-rx/R}.$$

The first terms of the expansion (11.32) will therefore be

$$\Xi^* = A_1 + A_2 x + A_3 x^2 + A_4 x^3 + (\bar{A}_1 + \bar{A}_2 x + \bar{A}_3 x^2 + \bar{A}_4 x^3)\cos\frac{\phi}{R}$$
$$+ (\bar{\bar{A}}_1 + \bar{\bar{A}}_2 x + \bar{\bar{A}}_3 x^2 + \bar{\bar{A}}_4 x^3)\sin\frac{\phi}{R}$$

and the displacements, corresponding to Ξ^* are, according to (11.30),

$$u = 2A_3 + 6A_4x + (2\bar{A}_3 + 6\bar{A}_4x)\cos\frac{\phi}{R} + (2\bar{\bar{A}}_3 + 6\bar{\bar{A}}_4x)\sin\frac{\phi}{R}, \quad (11.43a)$$

$$v = \frac{1}{R}(\bar{A}_2 + 2\bar{A}_3x + 3\bar{A}_4x^2)\sin\frac{\phi}{R} - \frac{1}{R}(\bar{\bar{A}}_2 + 2\bar{\bar{A}}_3x + 3\bar{\bar{A}}_4x^2)\cos\frac{\phi}{R},$$
$$(11.43b)$$

$$w = -6\nu R A_4 - \frac{1}{R}(\bar{A}_2 + 6\nu R^2\bar{A}_4 + 2\bar{A}_3x + 3\bar{A}_4x^2)\cos\frac{\phi}{R}$$
$$-\frac{1}{R}(\bar{\bar{A}}_2 + 6\nu R^2\bar{\bar{A}}_4 + 2\bar{\bar{A}}_3x + 3\bar{\bar{A}}_4x^2)\sin\frac{\phi}{R}. \quad (11.43c)$$

Due to the differentiation, four of the twelve arbitrary coefficients, namely A_1, A_2, $\bar{A}_1$ and $\bar{\bar{A}}_1$ disappeared in the expressions (11.43) for u, v and w. This means that four linearly independent states of displacement are missing in the expansion, and shows that not all integrals of (11.12) are given by the formulas (11.30).

To find the missing states, we return to (11.29). It should not cause any surprise to find that the states were lost when we arbitrarily put $\kappa_2^{**} = \kappa_3^{**} = 0$. In fact, taking

$$\kappa_i^{**} = \frac{2R^2}{1-\nu}(0, \Xi^*, 0) \quad (11.44)$$

instead of (11.29), the formulas corresponding to (11.30) would read

$$u = \frac{2R^2}{1-\nu}B_{21}\Xi^*, \quad v = \frac{2R^2}{1-\nu}B_{22}\Xi^*, \quad w = \frac{2R^2}{1-\nu}B_{23}\Xi^*,$$
$$(11.45)$$

and Ξ^* would still have to satisfy (11.31), i.e., the same determining equation as Ξ satisfies.

Taking

$$\Xi^* = B_1 + B_2x + B_3x^2 + B_4x^3 + (\bar{B}_1 + \bar{B}_2x + \bar{B}_3x^2 + \bar{B}_4x^3)\cos\frac{\phi}{R}$$
$$+ (\bar{\bar{B}}_1 + \bar{\bar{B}}_2x + \bar{\bar{B}}_3x^2 + \bar{\bar{B}}_4x^3)\sin\frac{\phi}{R}, \quad (11.46)$$

we find, in fact, the four missing states, corresponding to B_3, B_4, $\bar{B}_4$, $\bar{\bar{B}}_4$, i.e.,

$$u = \bar{B}_4 \frac{3x^2}{R} \sin\frac{\phi}{R} - \bar{\bar{B}}_4 \frac{3x^2}{R} \cos\frac{\phi}{R}, \tag{11.47a}$$

$$v = 4(1+\nu)B_3 + 12(1+\nu)B_4 x + \bar{B}_4\left[12(1+\nu)x - \frac{x^3}{R^2}\right]\cos\frac{\phi}{R}$$

$$+ \bar{\bar{B}}_4\left[12(1+\nu)x - \frac{x^3}{R^2}\right]\sin\frac{\phi}{R}, \tag{11.47b}$$

$$w = \bar{B}_4\left[12(1+\tfrac{1}{2}\nu)x - \frac{x^3}{R^2}\right]\sin\frac{\phi}{R} + \bar{\bar{B}}_4\left[12(1+\tfrac{1}{2}\nu)x - \frac{x^3}{R^2}\right]\cos\frac{\phi}{R}, \tag{11.47c}$$

whereas the remaining coefficients present nothing new. We have now in (11.43) together with (11.47) the twelve linearly independent states. Six of them correspond to rigid-body motion. Thus A_3, $\bar{A}_2$ and $\bar{\bar{A}}_2$ correspond to rigid-body translations in the directions of the x^1-, x^2- and x^3-axes respectively, and B_3, $\bar{A}_3$ and $\bar{\bar{A}}_3$ correspond to rigid-body rotations around the x^1-, x^2- and x^3-axes respectively. The six remaining states, i.e.,

$$u = 6A_4 x + \left(6\bar{A}_4 x - 3\bar{B}_4\frac{x^2}{R}\right)\cos\frac{\phi}{R} + \left(6\bar{\bar{A}}_4 x + 3\bar{\bar{B}}_4\frac{x^2}{R}\right)\sin\frac{\phi}{R}, \tag{11.48a}$$

$$v = 12(1+\nu)B_4 x + \left[-3\bar{\bar{A}}_4\frac{x^2}{R} + 12(1+\nu)\bar{B}_4 x - \bar{B}_4\frac{x^3}{R^2}\right]\cos\frac{\phi}{R}$$

$$+ \left[3\bar{A}_4\frac{x^2}{R} + 12(1+\nu)\bar{\bar{B}}_4 x - \bar{\bar{B}}_4\frac{x^3}{R^2}\right]\sin\frac{\phi}{R}, \tag{11.48b}$$

$$w = -6\nu R A_4 - \left[6\nu R\bar{A}_4 + 3\bar{\bar{A}}_4\frac{x^2}{R} + 12(1+\tfrac{1}{2}\nu)\bar{B}_4 x - \bar{B}_4\frac{x^3}{R^2}\right]\cos\frac{\phi}{R}$$

$$- \left[6\nu R\bar{\bar{A}}_4 + 3\bar{A}_4\frac{x^2}{R} - 12(1+\tfrac{1}{2}\nu)\bar{\bar{B}}_4 x + \bar{\bar{B}}_4\frac{x^3}{R^2}\right]\sin\frac{\phi}{R}, \tag{11.48c}$$

represent the 'beam'-states of a right circular cylindrical shell under end-loads.

Thus,

- A_4 represents uniaxial tension,
- B_4 represents torsion,
- $\bar{A}_4$ represents bending in the x^1–x^2-plane under constant moment,
- $\bar{\bar{A}}_4$ represents bending in the x^1–x^3-plane under constant moment,
- $\bar{B}_4$ represents bending in the x^1–x^2-plane under constant shear force,
- $\bar{\bar{B}}_4$ represents bending in the x^1–x^3-plane under constant shear force.

The result can be summarized as follows. For $n = 0$ and $n = 1$, four of the roots of the characteristic equation (11.34) vanish. There are twelve linearly independent integrals to the equations of equilibrium, corresponding to the zero roots. Six of the states describe all possible rigid-body motions of the shell, and six describe the shell as a beam, loaded by forces and moments at its ends.

The four nonvanishing roots of (11.34) for $n = 0$ and $n = 1$ are the solutions to

$$(\alpha^2 \pm 1)^2 + \frac{1-\nu^2}{k} = 0. \tag{11.49}$$

Omitting terms of relative order k, and reinstating k from (11.13), we find the following roots,

$$\alpha = \pm(1 \pm i)\rho, \tag{11.50}$$

where ρ is given by (11.35).

An important case is $n = 0$. This corresponds to rotational symmetry and has the solution

$$\Xi = \left(A_5 \cos\frac{\rho x}{R} + A_6 \sin\frac{\rho x}{R}\right) e^{\rho x/R} + \left(A_7 \cos\frac{\rho x}{R} + A_8 \sin\frac{\rho x}{R}\right) e^{-\rho x/R}. \tag{11.51}$$

The complete axisymmetric solution of the equations of equilibrium (11.12) is found when (11.51) is substituted into (11.30) and the

SOLUTION OF THE MATHEMATICAL PROBLEM

appropriate part[10] of the solution corresponding to the vanishing roots α is added, i.e.,

$$u = 2A_3 + 6A_4 x + 2\frac{\rho^2}{R^2}\left[\left(A_6\cos\frac{\rho x}{R} - A_5\sin\frac{\rho x}{R}\right)e^{\rho x/R}\right.$$

$$\left. + \left(A_7\sin\frac{\rho x}{R} - A_8\cos\frac{\rho x}{R}\right)e^{-\rho x/R}\right], \tag{11.52a}$$

$$v = 4(1+\nu)B_3 + 12(1+\nu)B_4 x, \tag{11.52b}$$

$$w = -6\nu R A_4 + 2\nu\frac{\rho^3}{R^2}\left\{\left[(A_5 - A_6)\cos\frac{\rho x}{R} + (A_5 + A_6)\sin\frac{\rho x}{R}\right]e^{\rho x/R}\right.$$

$$\left. + \left[(A_7 - A_8)\sin\frac{\rho x}{R} - (A_7 + A_8)\cos\frac{\rho x}{R}\right]e^{-\rho x/R}\right\}. \tag{11.52c}$$

From the displacement w we find by (11.20) the rotation of the normal at $x = \text{const}$,

$$\psi = 4\nu\frac{\rho^4}{R^3}\left[\left(A_5\cos\frac{\rho x}{R} + A_6\sin\frac{\rho x}{R}\right)e^{\rho x/R}\right.$$

$$\left. + \left(A_7\cos\frac{\rho x}{R} + A_8\sin\frac{\rho x}{R}\right)e^{-\rho x/R}\right], \tag{11.53}$$

and, by (11.18), the forces and moments

$$M_x = 4\nu D\frac{\rho^5}{R^4}\left\{\left[(A_5 + A_6)\cos\frac{\rho x}{R} + (A_6 - A_5)\sin\frac{\rho x}{R}\right]e^{\rho x/R}\right.$$

$$\left. + \left[(A_8 - A_7)\cos\frac{\rho x}{R} - (A_7 + A_8)\sin\frac{\rho x}{R}\right]e^{-\rho x/R}\right\}, \tag{11.54a}$$

$$Q_x = 8\nu D\frac{\rho^6}{R^5}\left\{\left[A_5\sin\frac{\rho x}{R} - A_6\cos\frac{\rho x}{R}\right]e^{\rho x/R}\right.$$

$$\left. + \left[A_8\cos\frac{\rho x}{R} - A_7\sin\frac{\rho x}{R}\right]e^{-\rho x/R}\right\}, \tag{11.54b}$$

---

[10] That is the ϕ-independent part of (11.43) and (11.48).

$$N_x = 6EhA_4 + 2Eh\frac{\rho^3}{R^3}\left\{\left[(A_6 - A_5)\cos\frac{\rho x}{R} - (A_5 + A_6)\sin\frac{\rho x}{R}\right]e^{\rho x/R}\right.$$

$$\left. + \left[(A_7 + A_8)\cos\frac{\rho x}{R} + (A_8 - A_7)\sin\frac{\rho x}{R}\right]e^{-\rho x/R}\right\}, \tag{11.54c}$$

$$S_x = 6EhB_4. \tag{11.54d}$$

There are eight coefficients to be determined from the boundary conditions at $x = x_1$ and $x = x_2$ ($x_2 > x_1$). If the conditions are kinematic, μ, v, w and ψ are prescribed at the boundaries, and there will be eight equations for the eight unknowns. For static boundary conditions, overall equilibrium requires N_x to be the same at both boundaries and likewise for S_x. Thus, only one condition on u and only one on v can be replaced by a static condition. But a condition on w can be replaced by one on Q_x anywhere and, likewise, one on ψ can be replaced by one on M_x at either boundary or both.

The exponentially decreasing part of the solution can be interpreted as an *attenuated* bending field, emanating from the left boundary $x = x_1$, and the exponentially increasing part as an *attenuated* field, emanating from the right boundary. If the shell is long, the interaction between the two bending fields becomes negligible, each being confined to a border zone. In this case, the constants A_5 and A_6 are determined solely from conditions at the right boundary, and A_7, A_8 from the left.

From the solution we see that $\sqrt{Rh}$ is the *characteristic length of the deformation pattern* and a measure of the width of the border zone of bending.

5. Bending due to axisymmetrical ring-loads

Let us apply the results of the preceding section to a cylindrical shell, loaded at the end $x = 0$ with a shear-force Q and a bending moment M, and free at the end $x = l$ (see Fig. 26).

We apply the first two equations of (11.54) for the bending moment and the shear force at both ends, and get

$$A_5 + A_6 - A_7 + A_8 = \frac{MR^4}{4\nu\rho^5 D}, \tag{11.55a}$$

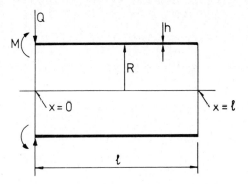

Fig. 26.

$$-A_6 + A_8 = \frac{QR^5}{8\nu\rho^6 D}, \qquad (11.55b)$$

$$[(A_5 + A_6)\cos\lambda + (A_6 - A_5)\sin\lambda]\,e^\lambda$$
$$+ [(A_8 - A_7)\cos\lambda - (A_7 + A_8)\sin\lambda]\,e^{-\lambda} = 0, \qquad (11.55c)$$

$$(A_5\sin\lambda - A_6\cos\lambda)\,e^\lambda + (A_8\cos\lambda - A_7\sin\lambda)\,e^{-\lambda} = 0, \qquad (11.55d)$$

where

$$\lambda = \frac{\rho l}{R} = \sqrt[4]{3(1-\nu^2)}\,\frac{l}{\sqrt{Rh}}. \qquad (11.56)$$

There are four equations (11.55a)–(11.55d) for the four unknown coefficients A_5, A_6, A_7 and A_8.[11] To proceed, we solve these equations and substitute the result into (11.52a)–(11.52c) and (11.53) to get the deflection w and the slope ψ at both ends.

For $x = 0$ we find

$$w_0 = \frac{MR^2}{2\rho^2 D}\frac{\sinh^2\lambda + \sin^2\lambda}{\sinh^2\lambda - \sin^2\lambda} - \frac{QR^3}{2\rho^3 D}\frac{\sinh\lambda\cosh\lambda - \sin\lambda\cos\lambda}{\sinh^2\lambda - \sin^2\lambda} \qquad (11.57)$$

---

[11] The remaining four coefficients are of no interest in this connection, A_3 and B_3 merely expressing rigid-body motion, and A_4, B_4 superimposed states of uniaxial tension and torsion respectively.

and

$$\psi_0 = -\frac{MR}{\rho D}\frac{\sinh \lambda \cosh \lambda + \sin \lambda \cos \lambda}{\sinh^2 \lambda - \sin^2 \lambda} + \frac{QR^2}{2\rho^2 D}\frac{\sinh^2 \lambda + \sin^2 \lambda}{\sinh^2 \lambda - \sin^2 \lambda}.$$

(11.58)

For $x = l$ the corresponding formulas are

$$w_l = -\frac{MR^2}{\rho^2 D}\frac{\sinh \lambda \sin \lambda}{\sinh^2 \lambda - \sin^2 \lambda} + \frac{QR^3}{2\rho^3 D}\frac{\cosh \lambda \sin \lambda - \sinh \lambda \cos \lambda}{\sinh^2 \lambda - \sin^2 \lambda}$$

(11.59)

and

$$\psi_l = -\frac{MR}{\rho D}\frac{\cosh \lambda \sin \lambda + \sinh \lambda \cos \lambda}{\sinh^2 \lambda - \sin^2 \lambda} + \frac{QR^2}{\rho^2 D}\frac{\sinh \lambda \sin \lambda}{\sinh^2 \lambda - \sin^2 \lambda}.$$

(11.60)

The formulas (11.57)–(11.60) apply to a set of *elementary cases*, that can be combined for solving a variety of problems.

As a first example, let us determine the radial displacement w_c under a ring-load Q_c, acting at the central section of a cylindrical shell (see Fig. 27).

Taking the length of the shell to be $2l$, we can apply formulas

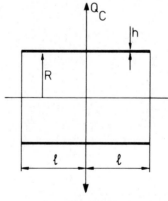

Fig. 27.

(11.57)–(11.60) to the right half of the shell with λ being given as before by (11.56).

Due to symmetry, we must have $\psi_0 = 0$ at the centre and $Q = -\tfrac{1}{2}Q_c$. Putting $\psi_0 = 0$ in (11.58) we get the following relation between the bending moment M and the shear force Q at the centre,

$$M = \frac{QR}{2\rho} \frac{\sinh^2 \lambda + \sin^2 \lambda}{\sinh \lambda \cosh \lambda + \sin \lambda \cos \lambda}. \tag{11.61}$$

Taking $Q = -\tfrac{1}{2}Q_c$ and substituting into (11.57) we find, after replacing D and ρ,

$$w_c = \frac{Q_c R^2}{2Ehl} \frac{2\lambda(\cosh^2 \lambda + \cos^2 \lambda)}{\sinh 2\lambda + \sin 2\lambda}. \tag{11.62}$$

For a very short shell ($l \ll \sqrt{Rh}$), λ is a small number, and the hyperbolic and trigonometric sine functions can be replaced by their arguments, the cosines by 1. Then

$$w_c = \frac{Q_c R^2}{2Ehl} \quad (l \ll \sqrt{Rh}) \tag{11.63}$$

and the shell acts as ring of area hl in uniform tension.

For a very long shell ($l \gg \sqrt{Rh}$), on the other hand, we get in the limit

$$w_c = \frac{Q_c R \rho}{2Eh} \quad (l \gg \sqrt{Rh}). \tag{11.64}$$

This implies that the *radial stiffness* of a long shell for a concentrated ring-load is the same as that of a ring of area bh, where b is the *effective width* of the shell, given by

$$b = \frac{2\sqrt{Rh}}{\sqrt[4]{3(1-\nu^2)}}. \tag{11.65}$$

As another example, let us consider a long cylindrical shell, reinforced by equally spaced ring-stiffeners (see Fig. 28). Such structures

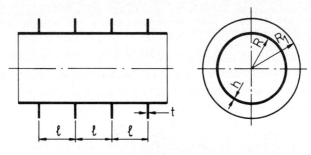

FIG. 28.

are used in pipelines, pressure vessels, submarines, airplane fuselages and elsewhere.

If the spacing l is much greater than $\sqrt{Rh}$, the interference between neighbouring stiffeners will be negligible, and the problem can be analysed as in the previous example. But if the spacing is comparable in magnitude to $\sqrt{Rh}$, we must take the interaction between the stiffeners into account.

In this problem, the bending moment is a continuous function, and we shall assign it the value M at the stiffeners. The shear-force, on the other hand, is discontinuous, making a jump at each stiffener equal to the ring-load from the stiffener. But since the magnitude of the shear-force is the same on both sides of a stiffener, the jump is due to a change of sign only. We shall assign to the shear-force immediately to the right of the stiffeners the value Q and hence the value to the left of any stiffener is $-Q$. The ring-load is then $2Q$.

To determine M and Q we shall apply formulas (11.57)–(11.60). These, however, were derived under the assumption that the end $x = l$ was unloaded. In this case, the end $x = l$ carries the load $M, -Q$ and we must therefore superimpose a state with reversed conditions, i.e., a state with the load $M, -Q$ at $x = l$ and no load at $x = 0$. For such a state, formulas (11.57)–(11.60) can be used again; we must, however, interchange the indices 0 and l. Since the direction of the x-axis is reversed, the signs of ψ and $-Q$ must also be reversed. This double change of sign for Q leaves Q unchanged; therefore, only ψ changes sign.

Thus, we get for the slope at a stiffener

$$\psi_s = \psi_0 - \psi_l \qquad (11.66)$$

and for the deflection at the same section

$$w_s = w_0 + w_l. \tag{11.67}$$

From the condition $\psi_s = 0$ we find from (11.58) and (11.60) without much difficulty

$$M = \frac{QR}{2\rho} \frac{\sinh \lambda - \sin \lambda}{\cosh \lambda - \cos \lambda}. \tag{11.68}$$

Introducing this moment into (11.57) and (11.59) we find, from (11.67),

$$w_s = -\frac{QR^3}{4\rho^3 D} \frac{\sinh \lambda + \sin \lambda}{\cosh \lambda - \cos \lambda}. \tag{11.69}$$

Let us assume that the shell is subjected to an internal pressure p and no axial force. Without stiffeners the pressure would produce the radial displacement [12]

$$\frac{pR^2}{Eh}$$

and the total radial displacement at a stiffener is therefore

$$\frac{pR^2}{Eh} + w_s.$$

Since the stiffener must suffer the same radial displacement, we get (remembering that the load is $2Q$)

$$\frac{pR^2}{Eh} + w_s = \frac{2QR^2}{EA}, \tag{11.70}$$

where $A = t(R_1 - R)$ is the area of the stiffener.[13] From this equation

---

[12] This is easily found from elementary considerations, but may of course also be obtained from the equations of equilibrium (11.12).

[13] A tacit assumption is that $R_1 - R$ is not too large in comparison with t and much smaller than R, so that the stiffener can be treated as a ring in uniform tension.

we can solve Q and find

$$Q = \frac{pR}{\rho(\sinh \lambda + \sin \lambda)/(\cosh \lambda - \cos \lambda) + 2Rh/A}. \quad (11.71)$$

With Q and M determined we are able to find all other relevant quantities at the position of a stiffener.

If we want the displacement or bending moment at the mid-section between two stiffeners, it is more convenient to apply formulas (11.57)–(11.60) to one half of the shell between two stiffeners. From the conditions of zero slope there and at the stiffeners we get the two equations needed to determine the bending moments M_0 and M_1 in terms of Q. (The shear force Q_1 at the mid-section must vanish.) However, we shall not pursue this further.

6. Non-axisymmetric bending of cylindrical shells

Let us now consider the non-axisymmetric bending of a cylindrical shell, loaded at one end. To reduce the complexity of the problem, we shall assume that it is long enough to permit us to treat it as semi-infinite.

Of the Fourier expansion (11.32) we shall consider a typical term

$$\Xi = C_n(x) \cos \frac{n\phi}{R} \quad (11.72)$$

with $n > 1$. Regularity at $x = \infty$ requires A_1, A_2, A_5 and A_6 in expression (11.42) for C_n to be zero, and therefore

$$\Xi = \left[\left(A_3 \cos \frac{qx}{R} + A_4 \sin \frac{qx}{R} \right) e^{-px/R} \right.$$
$$\left. + \left(A_7 \cos \frac{sx}{R} + A_8 \sin \frac{sx}{R} \right) e^{-rx/R} \right] \cos \frac{n\phi}{R}, \quad (11.73)$$

where p, q, r and s are given by (11.40).

We shall find it convenient to use the following four functions,

$$g_1(x) = \left(A_3 \sin\frac{qx}{R} - A_4 \cos\frac{qx}{R}\right) e^{-px/R},$$

$$g_2(x) = \left(A_3 \cos\frac{qx}{R} + A_4 \sin\frac{qx}{R}\right) e^{-px/R},$$

$$g_3(x) = \left(A_7 \sin\frac{sx}{R} - A_8 \cos\frac{sx}{R}\right) e^{-rx/R},$$

$$g_4(x) = \left(A_7 \cos\frac{sx}{R} + A_8 \sin\frac{sx}{R}\right) e^{-rx/R},$$

(11.74)

into which the so far undetermined coefficients A_3, A_4, A_7 and A_8 have been absorbed. Since

$$g_1' = (qg_2 - pg_1)/R, \qquad g_3' = (sg_4 - rg_3)/R,$$
$$g_2' = -(qg_1 + pg_2)/R, \qquad g_4' = -(sg_3 + rg_4)/R,$$

(11.75)

any linear differential operator with constant coefficients, operating on any linear combination of the functions $g_i(x)$, is itself a linear combination of the same functions.

We can now write

$$\Xi = [g_2(x) + g_4(x)] \cos\frac{n\phi}{R} \qquad (11.76)$$

and determine the displacements u, v and w by substituting (11.76) into (11.30) or (11.45). The third possibility is to use the formulas

$$u = \frac{2R^2}{1-\nu} B_{31}\Xi, \qquad v = \frac{2R^2}{1-\nu} B_{32}\Xi, \qquad w = \frac{2R^2}{1-\nu} B_{33}\Xi,$$

(11.77)

which have the advantage of utilizing the simplest expressions according to (11.25).

Substituting (11.76) into (11.77) we find

$$u = \sum_{i=1}^{4} c_{ui} g_i(x) \cos \frac{n\phi}{R},$$

$$v = \sum_{i=1}^{4} c_{vi} g_i(x) \sin \frac{n\phi}{R}, \qquad (11.78)$$

$$w = \sum_{i=1}^{4} c_{wi} g_i(x) \cos \frac{n\phi}{R},$$

where

$$\begin{aligned} c_{u1} &= q[n^2 - \nu(q^2 - 3p^2)], \\ c_{u2} &= p[n^2 - \nu(3q^2 - p^2)], \\ c_{u3} &= s[n^2 - \nu(s^2 - 3r^2)], \\ c_{u4} &= r[n^2 - \nu(3s^2 - r^2)] \end{aligned} \qquad (11.79)$$

and

$$\begin{aligned} c_{v1} &= 2(2+\nu)pqn, \\ c_{v2} &= -n[(2+\nu)(q^2 - p^2) + n^2], \\ c_{v3} &= 2(2+\nu)rsn, \\ c_{v4} &= -n[(2+\nu)(s^2 - r^2) + n^2] \end{aligned} \qquad (11.80)$$

and

$$\begin{aligned} c_{w1} &= 4pq(p^2 - q^2 - n^2), \\ c_{w2} &= (n^2 + q^2)^2 + p^2(p^2 - 6q^2 - 2n^2), \\ c_{w3} &= 4rs(r^2 - s^2 - n^2), \\ c_{w4} &= (n^2 + s^2)^2 + r^2(r^2 - 6s^2 - 2n^2). \end{aligned} \qquad (11.81)$$

Similarly, we find the rotation of the normal ψ from (11.20) to be

$$\psi = \sum_{i=1}^{4} c_{\psi i} g_i(x) \cos \frac{n\phi}{R}, \qquad (11.82)$$

where

$$c_{\psi 1} = q[(n^2+q^2)^2 + p^2(5p^2 - 10q^2 - 6n^2)]/R,$$
$$c_{\psi 2} = p[(n^2-p^2)^2 + q^2(5q^2 - 10p^2 + 6n^2)]/R,$$
$$c_{\psi 3} = s[(n^2+s^2)^2 + r^2(5r^2 - 10s^2 - 6n^2)]/R,$$
$$c_{\psi 4} = r[(n^2-r^2)^2 + s^2(5s^2 - 10r^2 + 6n^2)]/R.$$
(11.83)

From the displacement functions we can determine the boundary forces and moments according to (11.18). We find

$$M_x = D \sum_{i=1}^{4} c_{Mi} g_i(x) \cos \frac{n\phi}{R},$$
$$Q_x = D \sum_{i=1}^{4} c_{Qi} g_i(x) \cos \frac{n\phi}{R},$$
$$N_x = \frac{Eh}{1-\nu^2} \sum_{i=1}^{4} c_{Ni} g_i(x) \cos \frac{n\phi}{R},$$
$$S_x = \frac{Eh}{2(1+\nu)} \sum_{i=1}^{4} c_{Si} g_i(x) \sin \frac{n\phi}{R}.$$
(11.84)

with

$$c_{M1} = \{[p^2 - q^2 - \nu(n^2+1)]c_{w1} + 2pq c_{w2} - 2\nu n c_{v1}\}/R^2,$$
$$c_{M2} = \{[p^2 - q^2 - \nu(n^2+1)]c_{w2} - 2pq c_{w1} - 2\nu n c_{v2}\}/R^2,$$
$$c_{M3} = \{[r^2 - s^2 - \nu(n^2+1)]c_{w3} + 2rs c_{w4} - 2\nu n c_{v3}\}/R^2,$$
$$c_{M4} = \{[r^2 - s^2 - \nu(n^2+1)]c_{w4} - 2rs c_{w3} - 2\nu n c_{v4}\}/R^2$$
(11.85)

and

$$c_{Q1} = \{p[p^2 - 3q^2 - 2n^2 + \nu(n^2+1)]c_{w1}$$
$$+ q[3p^2 - q^2 - 2n^2 + \nu(n^2+1)]c_{w2} - 2n(pc_{v1} + qc_{v2})\}/R^3, \quad (11.86a)$$
$$c_{Q2} = \{p[p^2 - 3q^2 - 2n^2 + \nu(n^2+1)]c_{w2}$$
$$- q[3p^2 - q^2 - 2n^2 + \nu(n^2+1)]c_{w1} + 2n(qc_{v1} - pc_{v2})\}/R^3, \quad (11.86b)$$

$$c_{Q3} = \{r[r^2 - 3s^2 - 2n^2 + \nu(n^2 + 1)]c_{w3}$$
$$+ s[3r^2 - s^2 - 2n^2 + \nu(n^2 + 1)]c_{w4} - 2n(rc_{v3} + sc_{v4})\}/R^3, \quad (11.86c)$$
$$c_{Q4} = \{r[r^2 - 3s^2 - 2n^2 + \nu(n^2 + 1)]c_{w4}$$
$$- s[3r^2 - s^2 - 2n^2 + \nu(n^2 + 1)]c_{w3} + 2n(sc_{v3} - rc_{v4})\}/R^3 \quad (11.86d)$$

and

$$c_{N1} = [\nu c_{w1} - pc_{u1} - qc_{u2} + n\nu c_{v1} + k(1-\nu)n^2 c_{w1}]/R,$$
$$c_{N2} = [\nu c_{w2} - pc_{u2} + qc_{u1} + n\nu c_{v2} + k(1-\nu)n^2 c_{w2}]/R,$$
$$c_{N3} = [\nu c_{w3} - rc_{u3} - sc_{u4} + n\nu c_{v3} + k(1-\nu)n^2 c_{w3}]/R, \quad (11.87)$$
$$c_{N4} = [\nu c_{w4} - rc_{u4} + sc_{u4} + n\nu c_{v4} + k(1-\nu)n^2 c_{w4}]/R.$$

and

$$c_{S1} = -[nc_{u1} + pc_{v1} + qc_{v2} + 2kn(qc_{w2} + pc_{w1})]/R,$$
$$c_{S2} = -[nc_{u2} - qc_{v1} + pc_{v2} + 2kn(pc_{w2} - qc_{w1})]/R,$$
$$c_{S3} = -[nc_{u3} + rc_{v3} + sc_{v4} + 2kn(sc_{w4} + rc_{w3})]/R, \quad (11.88)$$
$$c_{S4} = -[nc_{u4} - sc_{v3} + rc_{v4} + 2kn(rc_{w4} - sc_{w3})]/R.$$

At $x = 0$ we have

$$g_1(0) = -A_4, \quad g_2(0) = A_3, \quad g_3(0) = -A_8, \quad g_4(0) = A_7,$$

and hence the boundary conditions will yield a system of linear equations for the coefficients A_3, A_4, A_7 and A_8. We can satisfy four boundary conditions and these will—generally speaking—determine the solution completely for any given value of n.

As an example, let us consider a long cylindrical shell, loaded at its central section with the external load

$$2Q \cos \frac{2\phi}{R}.$$

Now, consider the right half of the shell. Due to symmetry it will carry half of the load, and we have the following boundary conditions,

$$u(0) = 0, \quad \psi(0) = 0, \quad Q_x(0) = Q, \quad S_x(0) = 0.$$

The boundary conditions lead to the following system of equations,

$$\begin{aligned}
c_{u2}A_3 - c_{u1}A_4 + c_{u4}A_7 - c_{u3}A_8 &= 0, \\
c_{\psi2}A_3 - c_{\psi1}A_4 + c_{\psi4}A_7 - c_{\psi3}A_8 &= 0, \\
c_{Q2}A_3 - c_{Q1}A_4 + c_{Q4}A_7 - c_{Q3}A_8 &= Q/D, \\
c_{S2}A_3 - c_{S1}A_4 + c_{S4}A_7 - c_{S3}A_8 &= 0,
\end{aligned} \quad (11.89)$$

from which the coefficients A_3, A_4, A_7 and A_8 are found and the problem is solved. Taking, for instance,

$$R = 1, \quad h = 0.01, \quad \nu = 0.3, \quad n = 2,$$

we get, from (11.35) and (11.38),

$$\rho = 12.854, \quad \kappa = 0.021183,$$

and, from (11.40),

$$p = 0.13755, \quad q = 0.13467, \quad r = 12.992, \quad s = 12.719.$$

Using now formulas (11.79)–(11.88) we get the coefficients given in Table 5, from which the elements of the coefficient matrix of (11.89) can be picked.

TABLE 5. Coefficients for a cylindrical shell in mode $n = 2$ for $h/R = 0.01$ and $\nu = 0.3$.

i	1	2	3	4
c_{ui}	0.54025	0.54875	1365.7	−1181.9
c_{vi}	0.17043	−7.9964	1520.3	24.196
c_{wi}	−0.29634	15.992	1982.4	−109220
$c_{\psi i}$	−2.1130	−2.2397	−1363400	−1444100
c_{Mi}	0.89014	−14.370	-3.6079×10^7	−1397700
c_{Qi}	10.658	11.530	4.7889×10^8	-4.4886×10^8
c_{Ni}	−0.13487	−0.0024829	1202.9	28.026
c_{Si}	−0.027121	0.025310	−22744	21434

The system of linear equations (11.89) has the solution

$$A_3 = 0.89764 \times 10^{-3}, \qquad A_7 = -0.10185 \times 10^{-8},$$
$$A_4 = 0.91686 \times 10^{-3}, \qquad A_8 = -0.11324 \times 10^{-8}.$$

To find the normal displacement under the load, we substitute into (11.78) and obtain

$$w(0, \phi) = 0.01474 \frac{Q}{D} \cos 2\phi.$$

The same procedure can be repeated for any $n > 1$, provided that n is not too large. For very large values of n, i.e., when

$$6n^4 \gg \frac{1-\nu^2}{k}, \tag{11.90}$$

the characteristic equation (11.34) degenerates to

$$(\alpha^2 - n^2)^4 = 0, \tag{11.91}$$

which has only two distinct roots $\alpha = \pm n$. The physical reason for this is that for a large number of waves in the peripheral direction the curvature of the shell ceases to act as an agent for coupling the transverse and the tangential displacements. In other words, the shell acts as a plane plate. To convince ourselves that that is the case we go back to the equations of equilibrium (11.12) and take their limit as $R \to \infty$. We get,

$$\frac{\partial^2 u}{\partial x^2} + \tfrac{1}{2}(1-\nu)\frac{\partial^2 u}{\partial \phi^2} + \tfrac{1}{2}(1+\nu)\frac{\partial^2 v}{\partial x \partial \phi} + \frac{1-\nu^2}{Eh} F_x = 0,$$

$$\tfrac{1}{2}(1+\nu)\frac{\partial^2 u}{\partial x \partial \phi} + \tfrac{1}{2}(1-\nu)\frac{\partial^2 v}{\partial x^2} + \frac{\partial^2 v}{\partial \phi^2} + \frac{1-\nu^2}{Eh} F_y = 0, \tag{11.92}$$

$$\tfrac{1}{12} h^2 \Delta^2 w - \frac{1-\nu^2}{Eh} p = 0.$$

The first two equations are the equations of equilibrium for in-plane loaded plates. They are not coupled with the third equation, which is the plate equation (9.12).

Proceeding as before, we find that $A_{13} = A_{23} = 0$ and

$$A_{33} = \tfrac{1}{12}h^2(\partial_x^4 + 2\partial_x^2\partial_\phi^2 + \partial_\phi^4).$$

This leads to the following differential equation for the potential function, replacing (11.31),

$$\Delta^4 \Xi = 0, \tag{11.93}$$

which has precisely the characteristic equation (11.91). Thus, for large values of n, the shell problem degenerates to a plate problem.

To determine the elastic state in a cylindrical shell for large values of n we may therefore neglect the tangential displacements u and v and solve w from the plate equation. In our problem we take

$$w = (A + Bx)\,e^{-nx/R} \cos \frac{n\phi}{R}, \tag{11.94}$$

which clearly satisfies the plate equation, and determine the ratio between A and B from the condition $\psi = 0$ at $x = 0$. We find

$$B = nA/R$$

and we may now determine the effective shear force Q_x from equation (9.15). Putting it equal to half the external load we get

$$-\frac{2n^3 D}{R^3} A = -Q, \tag{11.95}$$

from which we get

$$w(0, \phi) = \frac{QR^3}{2n^3 D} \cos \frac{n\phi}{R}. \tag{11.96}$$

In our case the right-hand side of (11.90) is 10 9200 and n must therefore be much larger than $\sqrt[4]{10\,9200/6} = 11.6$ to satisfy the inequality. Taking $n = 15$, the solution of the complete shell equations following the scheme above, yields

$$w(0, \phi) = 1.422 \times 10^{-4} \frac{Q}{D} \cos 15\phi,$$

while the approximate plate solution according to (11.96) is

$$w(0, \phi) = 1.481 \times 10^{-4} \frac{Q}{D} \cos 15\phi$$

and thus only four per cent in error.

To solve the problem of an arbitrary load at the end, we have to consider the complete Fourier expansion (11.32). Expanding also the external load in a Fourier series, each mode n can be treated separately. For any $n > 1$, which is not too large, we use the procedure described above. This procedure is well suited for automatic computation, and since no function appearing in the procedure requires more than four numbers for its representation, the problem is easily solved on a programmable pocket calculator.

It is worthwhile to observe that all the formulas (11.79)–(11.88) have the property that any coefficient with an index 3 can be derived from the corresponding coefficient with the index 1 if p is replaced by r and q by s. The same relation holds good between coefficients with indices 4 and 2.

For a short shell we need in general all eight coefficients $A_1, A_2, \ldots, A_8$, so that four boundary conditions can be satisfied at each boundary. There will now be eight functions $g_i(x)$ and, for each displacement, force and moment, eight coefficients. The four additional coefficients can be derived from the four original ones by replacing p by $-p$ and r by $-r$. When the boundary conditions are formulated, we get a system of eight linear equations for the eight unknowns $A_1, \ldots, A_8$.

Should the shell be symmetrically loaded and supported, we take $x = 0$ to be the plane of symmetry and retain only the symmetrical part of Ξ, which can be written as

$$\Xi = \left[A_1 \cos\frac{qx}{R} \cosh\frac{px}{R} + A_2 \sin\frac{qx}{R} \sinh\frac{px}{R} \right.$$

$$\left. + A_3 \cos\frac{sx}{R} \cosh\frac{rx}{R} + A_4 \sin\frac{sx}{R} \sinh\frac{rx}{R} \right] \cos\frac{n\phi}{R}.$$

From here on we follow a scheme fully analogous with the one described above for a semi-infinite shell.

7. Free vibrations of cylindrical shells

To obtain the equations of motion of the cylindrical shell, we substitute the d'Alembert forces for F_x, F_ϕ and p in (11.12).

For a shell, the d'Alembert forces are found just as we found them for a plate in Chapter 9. We get

$$F_x = \omega^2 \gamma h u, \qquad F_\phi = \omega^2 \gamma h v, \qquad p = \omega^2 \gamma h w, \qquad (11.97)$$

when terms of relative order $(h/L)^2$ and $(h/R)^2$ are neglected.

Let us now assume the following set of displacements for the amplitudes

$$u = A \cos\frac{mx}{R} \cos\frac{n\phi}{R},$$

$$v = B \sin\frac{mx}{R} \sin\frac{n\phi}{R}, \qquad (11.98)$$

$$w = C \sin\frac{mx}{R} \cos\frac{n\phi}{R}.$$

It is easily seen that when substituted into (11.12) with (11.97) for the forces, the trigonometric functions cancel out and we obtain a set of three linear and homogeneous equations for the unknown coefficients A, B and C.

The condition for a nontrivial solution is that the determinant of the system vanishes. After some simplification, this condition can be written as

$$\begin{vmatrix} -\tfrac{1}{2}(1-\nu)n^2 - m^2 + \lambda & \tfrac{1}{2}(1+\nu)mn & \nu m \\ \tfrac{1}{2}(1+\nu)mn & -\tfrac{1}{2}(1-\nu)m^2 - n^2 + \lambda & -n \\ \nu m & -n & -1 - k(n^2 + m^2 - 1)^2 + \lambda \end{vmatrix} = 0, \quad (11.99)$$

where

$$\lambda = \frac{1-\nu^2}{E} \gamma R^2 \omega^2. \qquad (11.100)$$

When the determinant is worked out, we find the following equation of third degree in λ,

$$\lambda^3 + a_2 \lambda^2 + a_1 \lambda + a_0 = 0, \qquad (11.101)$$

where

$$a_0 = -\tfrac{1}{2}(1-\nu)[(1-\nu^2)m^4 + k(m^2+n^2)^2(m^2+n^2-1)^2],$$

$$\begin{aligned} a_1 &= \tfrac{1}{2}(1-\nu)[(m^2+n^2)^2 + n^2 + (3+2\nu)m^2] \\ &\quad + k(\tfrac{1}{2}(3-\nu))(m^2+n^2)(m^2+n^2-1)^2, \end{aligned} \qquad (11.102)$$

$$a_2 = -[1 + \tfrac{1}{2}(3-\nu)(m^2+n^2) + k(m^2+n^2-1)^2].$$

The three roots of (11.101) are real and positive and they determine therefore three (real) frequencies according to (11.100).

We have not considered any boundary conditions so far. Let us now see what conditions the functions (11.98) satisfy, assuming that the shell is complete and of length L ($0 \leq x \leq L$). Since the deformation pattern must be periodic in circumferential direction, we take n to be an integer. Furthermore, we shall take

$$m = j \frac{\pi R}{L}, \quad j = 1, 2, \ldots, \qquad (11.103)$$

where j is a positive integer.

For such values of m we get $v = w = 0$ on both boundaries $x = 0$ and $x = L$. By substituting (11.98) into (11.18) we also find that $M_x = N_x = 0$ at the boundaries. The assumed functions are therefore appropriate for the case when the cylindrical shell is closed by weightless membranes, that cannot stretch. In practice, the conditions correspond closely, but of course not precisely, to the conditions of a simply supported shell as one usually prescribes them.

The lowest root λ of (11.101), corresponding to the first fundamental frequency, is shown in Fig. 29 for a wide range of the parameters L/R and R/h. The lowest frequency is always obtained when $j = 1$, but its value of n depends on the parameters. On the line A–A the shell vibrates in the mode $j = 1$, $n = 1$, i.e., as a simply supported beam.

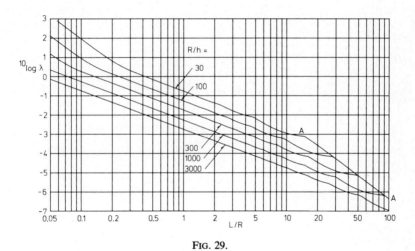

Fig. 29.

The three roots of (11.101) correspond to three different ratios $A : B : C$. In general, one of the three coefficients is considerably larger than the other two, and in such a case they can be called *longitudinal* (A is large), *torsional* (B is large) and *transverse vibrations* (C is large).

8. Cylindrical panels

In the case of a cylindrical panel, for which $0 \leq \phi \leq \chi$ and $\chi < 2\pi R$, we have to satisfy boundary conditions at all four boundaries, i.e., at the generators $\phi = 0$ and $\phi = \chi$ in addition to the meridians $x = 0$ and $x = l$.

The problem is similar to the problem of rectangular plates. When two opposite boundaries are simply supported, we can choose a Fourier expansion that satisfies this condition term by term, and it becomes possible to satisfy any conditions at the other boundaries. But when this is not the case, i.e., when no two opposite boundaries are simply supported, as in the case of a clamped plate, no simple solution is available. The same holds true for cylindrical panels.

In the case of simply supported boundaries at the generators $\phi = 0$ and $\phi = \chi$, we take instead of (11.32) the expansion

$$\Xi = \sum_{n=1}^{\infty} C_n(x) \sin \frac{n\pi\phi}{\chi} \tag{11.104}$$

and proceed as before with n replaced by $\pi n R/\chi$.

In the case of simply supported meridians $x = 0$ and $x = l$, we interchange the variables x and ϕ and write

$$\Xi = \sum_{n=1}^{\infty} C_n(\phi) \sin \frac{\pi n x}{l}. \tag{11.105}$$

With this expansion we can proceed much as before, replacing n by $\pi n R/l$ and interchanging x and ϕ.

Bibliography

TIMOSHENKO, S. (1959), *Theory of Plates and Shells*, 2nd ed., McGraw-Hill, New York, Chapter 15.
FLÜGGE, W. (1960), *Stresses in Shells*, Springer, Berlin, Chapter 5.
GOL'DENVEIZER, A.L. (1961), *Theory of Elastic Thin Shells*, Pergamon, Oxford, Chapters 10–12 (English translation).
NOVOZHILOV, V.V. (1959), *The Theory of Thin Shells*, Noordhoff, Groningen, Chapter III (English translation).
SEIDE, P. (1975), *Small Elastic Deformations of Thin Shells*, Noordhoff, Leiden, Chapter 4.
WLASSOW, W.S. (1958), *Allgemeine Schalentheorie und ihre Anwendung in der Technik*, Akademie-Verlag, Berlin, Part II, Section 12.

CHAPTER 12

AXISYMMETRICAL BENDING OF SHELLS

1. Introduction 265
2. Basic equations 265
3. Conical shells 272
4. Numerical example 276
5. Spherical shells 281
6. Toroidal shells 282
7. Approximate solution 283
    Bibliography 286

CHAPTER 12

AXISYMMETRICAL BENDING OF SHELLS

1. Introduction

The bending theory of shells of revolution is greatly simplified in the case of axisymmetric loads. Using the coordinates t, ϕ introduced in Chapter 10, the displacement v_2 in the peripheral direction vanishes, and all off-diagonal elements of $E_{\alpha\beta}$, $K_{\alpha\beta}$, $N^{\alpha\beta}$ and $M^{\alpha\beta}$ vanish too. The remaining elements are functions of one variable only, namely t.

The natural procedure to solve the problem is to express first $E_{\alpha\beta}$ and $K_{\alpha\beta}$, and then $N^{\alpha\beta}$ and $M^{\alpha\beta}$ in terms of v_1 and w. When $N^{\alpha\beta}$ and $M^{\alpha\beta}$ are substituted into the equations of equilibrium, we find one of them identically satisfied, while the other two provide a system of two ordinary differential equations for v_1 and w of total order six. This procedure was indeed followed by early authors in the field. However, in 1912, REISSNER showed[1] that, for the case of a spherical shell, a considerably simpler and clearer analysis results if the variables are suitably selected. The year after, MEISSNER extended REISSNER's results to the general case of axisymmetrical shells.[2] Here, we shall be guided by MEISSNER's approach.

2. Basic equations

Using the coordinates t and ϕ of Chapter 10, we have according to (10.11) (see also Fig. 14)

---

[1] REISSNER, H. (1912), *Spannungen in Kugelschalen* (*Kuppeln*), Müller-Breslau-Festschrift, Leipzig, p. 181.
[2] MEISSNER, E. (1913), "Das Elastizitätsproblem für dünne Schalen von Ringflächen-, Kugel- und Kegelform", *Phys. Z.* **14**, p. 343.

$$x^1 = r(t)\cos\phi, \qquad x^2 = r(t)\sin\phi, \qquad x^3 = q(t),$$

where r and q are arbitrary, smooth functions of t.

The displacement v_2 vanishes identically, and we shall use the notation u for v_1, the displacement in t-direction.

By (3.11), (10.15) and (10.17) we find the strain tensor, and after raising one index we get

$$E_1^1 = u^{\cdot} - w\frac{q^{\cdot\cdot}}{r^{\cdot}}, \qquad E_2^2 = \frac{ur^{\cdot} - wq^{\cdot}}{r} \qquad (E_2^1 = E_1^2 = 0), \qquad (12.1)$$

where the superior dot denotes differentiation with respect to t. The rotation of the normal is found to be

$$\psi = w^{\cdot} + u\frac{q^{\cdot\cdot}}{r^{\cdot}}. \qquad (12.2)$$

For the bending we shall use the measure $\hat{K}_{\alpha\beta}$ according to (8.3), which gives particularly simple formulas,

$$\hat{K}_1^1 = \psi^{\cdot}, \qquad \hat{K}_2^2 = \psi\frac{r^{\cdot}}{r} \qquad (\hat{K}_2^1 = \hat{K}_1^2 = 0). \qquad (12.3)$$

It was already mentioned in Chapter 8 that the uncoupled constitutive equations (7.11) hold with the same accuracy for $\hat{K}_{\alpha\beta}$ and $\hat{N}^{\alpha\beta}$ as for $K_{\alpha\beta}$ and $N^{\alpha\beta}$. Hence

$$E_1^1 = \frac{1}{Eh}(\hat{N}_1^1 - \nu\hat{N}_2^2) = \frac{1}{Eh}(N_1 - \nu N_2), \qquad (12.4)$$

$$E_2^2 = \frac{1}{Eh}(\hat{N}_2^2 - \nu\hat{N}_1^1) = \frac{1}{Eh}(N_2 - \nu N_1), \qquad (12.5)$$

$$M_1 = M_1^1 = D\left(\psi^{\cdot} + \nu\psi\frac{r^{\cdot}}{r}\right), \qquad (12.6)$$

$$M_2 = M_2^2 = D\left(\psi\frac{r^{\cdot}}{r} + \nu\psi^{\cdot}\right), \qquad (12.7)$$

where $\hat{N}^{\alpha\beta}$ is the membrane stress tensor defined by (8.5).

BASIC EQUATIONS

We shall find it useful to retain the shear-force vector Q_α in the equations of equilibrium. From (4.23) we get

$$Q_1 = -M_1^{\cdot} - M_1 \frac{r^{\cdot}}{r} + M_2 \frac{r^{\cdot}}{r}$$

or

$$Q_1 r = -(M_1 r)^{\cdot} + M_2 r^{\cdot} \tag{12.8}$$

and we note that the component Q_2 vanishes.

The equations of equilibrium, appropriate to $\hat{N}^{\alpha\beta}$, are (8.6) and (8.7). They yield respectively

$$(rN_1)^{\cdot} - r^{\cdot}N_2 - Q_1 r \frac{\ddot{q}}{\dot{r}} = 0, \tag{12.9}$$

$$(Q_1 r)^{\cdot} + \dot{q} N_2 + N_1 r \frac{\ddot{q}}{\dot{r}} = 0 \tag{12.10}$$

for the homogeneous case, i.e., the case when the shell is loaded at the boundaries only.

Substituting (12.6) and (12.7) into (12.8) we get

$$r\ddot{\psi} + r^{\cdot}\dot{\psi} - \frac{\dot{r}^2}{r}\psi + \nu r^{\cdot\cdot}\psi = -\frac{Q_1 r}{D}.$$

We shall write this equation in the form

$$L[\psi] + \nu \frac{\ddot{r}}{\dot{q}} \psi = -\frac{Q_1 r}{D\dot{q}}, \tag{12.11}$$

where L is the *Meissner operator*, defined by

$$L[\cdots] = \frac{1}{\dot{q}}(r[\cdots]^{\cdot})^{\cdot} - \frac{\dot{r}^2}{r\dot{q}}[\cdots]. \tag{12.12}$$

Equation (12.11) expresses Q_1 in terms of ψ. To complete the system we need another equation expressing ψ in terms of Q_1. To get this, we first eliminate N_2 between the equations of equilibrium (12.9) and (12.10) and find

$$(N_1 r\dot{q})\dot{} + (Q_1 r\dot{r})\dot{} = 0. \tag{12.13}$$

This can be integrated to

$$N_1 r\dot{q} + Q_1 r\dot{r} = \mathscr{P}, \tag{12.14}$$

where $\mathscr{P}$ is a constant of integration. It is easily seen that $P = 2\pi\mathscr{P}$ is the (constant) total axial force, and $Q_Z = \mathscr{P}/r$ is the *axial* component of the boundary force per unit length of the boundary.

The *radial* component of this force is $Q_R = \mathscr{Q}/r$, where

$$N_1 r\dot{r} - Q_1 r\dot{q} = \mathscr{Q}. \tag{12.15}$$

We shall find it convenient to use $\mathscr{Q}$ as a new independent variable instead of Q_1. Solving (12.14) and (12.15) for N_1 and Q_1 yields

$$N_1 = \frac{\mathscr{P}\dot{q} + \mathscr{Q}\dot{r}}{r} \tag{12.16}$$

and

$$Q_1 = \frac{\mathscr{P}\dot{r} - \mathscr{Q}\dot{q}}{r}. \tag{12.17}$$

Substituting (12.16) and (12.17) into (12.10) and solving for N_2 we find

$$N_2 = \dot{\mathscr{Q}}. \tag{12.18}$$

From Hooke's law (7.11) and the expressions (12.1) for the strains we get

$$\dot{u} - w\frac{\ddot{q}}{\dot{r}} = \frac{1}{Eh}(N_1 - \nu N_2) \tag{12.19}$$

and

$$u - w\frac{\dot{q}}{\dot{r}} = \frac{1}{Eh}(N_2 - \nu N_1)\frac{r}{\dot{r}}. \qquad (12.20)$$

We can now eliminate u and w between (12.2) and (12.19)–(12.20), so that ψ becomes expressed in terms of N_1 and N_2, and henceforth with the help of (12.16) and (12.18) in terms of $\mathcal{Q}$ only. Although this involves some computational effort, the process is straightforward.

Taking the derivative of (12.20) and solving for $\dot{w}$, we first eliminate $\dot{w}$, and ψ becomes expressed in u and $\dot{u}$. The derivative $\dot{u}$ is eliminated with the help of (12.19) and u with (12.20). In this last step, the terms containing w cancel, and ψ appears in terms of N_1 and N_2 only. Using (12.16) and (12.18) we finally get

$$\psi = -\frac{1}{Eh}L[\mathcal{Q}] - \frac{\nu\ddot{q}}{Eh\dot{r}}\mathcal{Q} + \frac{\mathcal{P}}{Eh}\left(\frac{\dot{r}}{r} + \frac{\nu\ddot{q}}{\dot{q}}\right). \qquad (12.21)$$

Observing that $\ddot{q}\dot{q} = -\ddot{r}\dot{r}$, we can write (12.11) and (12.21) with the help of (12.17) as the following system,

$$L[\psi] - \nu\frac{\ddot{q}}{\dot{r}}\psi - \frac{1}{D}\left(\mathcal{Q} - \frac{\dot{r}}{\dot{q}}\mathcal{P}\right) = 0 \qquad (12.22)$$

and

$$L[\mathcal{Q}] + \nu\frac{\ddot{q}}{\dot{r}}\mathcal{Q} + Eh\psi - \mathcal{P}\left(\frac{\dot{r}}{r} + \nu\frac{\ddot{q}}{\dot{q}}\right) = 0. \qquad (12.23)$$

We can now eliminate $\mathcal{Q}$ and find

$$LL[\psi] - \nu L\left[\frac{\ddot{q}}{\dot{r}}\psi\right] + \nu\frac{\ddot{q}}{\dot{r}}L[\psi] + \left[\frac{Eh}{D} - \nu^2\left(\frac{\ddot{q}}{\dot{r}}\right)^2\right]\psi = \frac{\mathcal{P}}{D}\left(\frac{\dot{r}}{r} - L\left[\frac{\dot{r}}{\dot{q}}\right]\right). \qquad (12.24)$$

This is an ordinary fourth order differential equation for ψ, and in general we shall have to resort to numerical methods for solving it. However, if the curvature $d_1^1 = \ddot{q}/\dot{r}$ in the direction of the generator, i.e., in t-direction, is independent of t, a considerable simplification occurs. This is the case for conical, spherical and toroidal shells. In such cases the second and the third terms cancel, and the coefficient of ψ in the last term on the left-hand side becomes constant. We get

$$LL[\psi] + 4\kappa^4\psi = \frac{\mathcal{P}}{D}\left(\frac{\dot{r}}{r} - L\left[\frac{\dot{r}}{\dot{q}}\right]\right), \qquad (12.25)$$

where κ is given by the formula

$$\kappa^4 = \frac{3(1-\nu^2)}{h^2} - \frac{\nu^2}{4R^2} \qquad (12.26)$$

using the notation R for the constant radius of curvature in t-direction. The second term on the right-hand side is of relative order $(h/R)^2$, and can be omitted without any loss of accuracy.

The complete solution of (12.25) is $\psi + \psi_0$, where ψ_0 is any particular integral and ψ the solution of the homogeneous equation

$$LL[\psi] + 4\kappa^2\psi = 0. \qquad (12.27)$$

By factorization, equation (12.27) can be split into two similar second order equations,

$$L[\psi] \pm 2i\kappa^2\psi = 0, \qquad (12.28)$$

and the problem is therefore reduced to solving these second order equations.

Let ζ be a solution of the first differential equation, i.e.,

$$L[\zeta] + 2i\kappa^2\zeta = 0. \qquad (12.29)$$

Then its complex conjugate $\bar{\zeta}$ is a solution of the second equation,

$$L[\bar{\zeta}] - 2i\kappa^2\bar{\zeta} = 0. \qquad (12.30)$$

But since any linear combination of ζ and $\bar{\zeta}$ satisfies (12.27), we conclude that the real imaginary parts of ζ, i.e.,

$$\zeta_1 = \tfrac{1}{2}(\bar{\zeta}+\zeta), \qquad \zeta_2 = \tfrac{1}{2}i(\bar{\zeta}-\zeta) \qquad (12.31)$$

are two real-valued integrals of (12.27).

Now, by adding and subtracting (12.29) and (12.30) we find

$$L[\zeta_1] = -2\kappa^2 \zeta_2, \qquad L[\zeta_2] = 2\kappa^2 \zeta_1, \qquad (12.32)$$

and these formulas make it particulary simple to compute other relevant quantities, as we shall see.

Since there are two linearly independent solutions (ζ_a and ζ_b, say) to the second order differential equation (12.29), we have in their real and imaginary parts all four fundamental solutions of (12.27).

Tracing our steps backwards we find, from (12.22),

$$\mathscr{Q} = D\left(L[\psi] - \nu\frac{\ddot{q}}{\dot{r}}\psi\right) + \frac{\mathscr{P}\dot{r}}{\dot{q}}. \qquad (12.33)$$

The radial displacement v_R is given by

$$v_R = u\dot{r} - w\dot{q} = \frac{r}{Eh}(N_2 - \nu N_1),$$

where (12.20) was used. Substituting N_1 and N_2 from (12.16) and (12.18) we find

$$v_R = \frac{1}{Eh}(\mathscr{Q}r - \nu\mathscr{Q}\dot{r} - \nu\mathscr{P}\dot{q}). \qquad (12.34)$$

The bending moment M_1 is given by (12.6) and the axial and radial components of the boundary force are

$$Q_Z = \mathscr{P}/r, \qquad Q_R = \mathscr{Q}/r \qquad (12.35)$$

per unit length of the boundary respectively.

The positive directions of the forces, moments, displacements and rotations are shown in Fig. 30 for positive and negative values of $\dot{q}$.

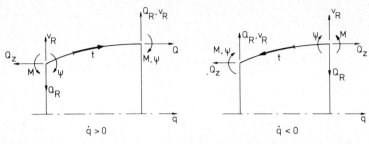

Fig. 30.

3. Conical shells

For the conical shell we have

$$r(t) = t \sin \alpha, \qquad q(t) = t \cos \alpha,$$

where t is the distance from the top and α half the top angle. The Meissner operator (12.12) becomes

$$L[\cdots] = \left\{ (t[\cdots]')' - \frac{[\cdots]}{t} \right\} \tan \alpha \qquad (12.36)$$

and the differential equations (12.28) can be written as

$$t \frac{d^2 \psi}{dt^2} + \frac{d\psi}{dt} + \left(\pm 2i\rho^2 - \frac{1}{t} \right) \psi = 0, \qquad (12.37)$$

where

$$\rho^2 = \kappa^2 \cot \alpha = \frac{\sqrt{3(1-\nu^2)}}{h} \cot \alpha. \qquad (12.38)$$

The right-hand side of equation (12.25) becomes

$$\frac{\mathscr{P}}{Dt \cos^2 \alpha}.$$

It is easily verified that $L[1/t] = 0$ and hence a particular integral of

(12.25) is

$$\psi_0 = \frac{\mathcal{P}}{Eh\, t\, \cos^2 \alpha}. \qquad (12.39)$$

Introducing the new independent variable

$$x = 2\rho\sqrt{2t}, \qquad (12.40)$$

equation (12.37) is transformed into the Bessel equations

$$x^2 \frac{d^2\psi}{dx^2} + x \frac{d\psi}{dx} - (\mp ix^2 + 4)\psi = 0. \qquad (12.41)$$

The linearly independent solutions to these equations are

$$\mathrm{ber}_2 x \mp i\, \mathrm{bei}_2 x \quad \text{and} \quad \mathrm{ker}_2 x \mp i\, \mathrm{kei}_2 x,$$

where ber_2, bei_2, ker_2 and kei_2 are the *Kelvin functions* of second order.[3]

The complete solution to the fourth order differential equation (12.27) can therefore be written as

$$\psi = \frac{C\mathcal{P}}{x^2} + A_1 \mathrm{ber}_2 x + A_2 \mathrm{bei}_2 x + B_1 \mathrm{ker}_2 x + B_2 \mathrm{kei}_2 x, \qquad (12.42)$$

where

$$C = \frac{2}{D\rho^2 \sin^2 \alpha}, \qquad (12.43)$$

and where A_1, A_2, B_1 and B_2 are arbitrary real numbers.

The bending moment M_1 is found from (12.6) to be[4]

---

[3] The Kelvin functions have been tabulated by many authors under a variety of notations. Here we shall follow the notations used by ABRAMOWITZ, M. and I.A. STEGUN (1964), *Handbook of Mathematical Functions*, Dover Publications, New York, Section 9.9. The recurrence relations used in (12.48) and the asymptotic formulas (12.60) are taken from the same reference.

[4] In order to reduce the number of formulas and their length in this section, we shall use the three dots to indicate a repetition of the previous terms, but with a replacement of A_1 and A_2 by B_1 and B_2; of ber and bei by ker and kei.

$$M_1 = -\frac{8C\mathcal{P}\rho^2 D(1-\nu)}{x^4} + \frac{4\rho^2 D}{x^2}[A_1(x\,\mathrm{ber}'_2 x + 2\nu\,\mathrm{ber}_2 x)$$
$$+ A_2(x\,\mathrm{bei}'_2 x + 2\nu\,\mathrm{bei}_2 x) + \cdots] \quad (12.44)$$

and $\mathcal{Q}$ is found from (12.33) and (12.32),

$$\mathcal{Q} = \mathcal{P}\tan\alpha - 2D\rho^2 \tan\alpha\,(A_1\mathrm{bei}_2 x - A_2\mathrm{ber}_2 x + \cdots). \quad (12.45)$$

The radial displacement v_R is found from (12.34) to be

$$v_R = -\frac{\nu\mathcal{P}}{Eh\cos\alpha} + \frac{D\rho^2\sin^2\alpha}{Eh\cos\alpha}[A_1(2\nu\,\mathrm{bei}_2 x - x\,\mathrm{bei}'_2 x)$$
$$- A_2(2\nu\,\mathrm{ber}_2 x - x\,\mathrm{ber}'_2 x) + \cdots] \quad (12.46)$$

and the radial component Q_R of the boundary force is given by (12.35). We find

$$Q_R = \frac{8\mathcal{P}\rho^2}{x^2\cos\alpha} - \frac{16D\rho^2}{x^2\cos\alpha}[A_1\mathrm{bei}_2 x - A_2\mathrm{ber}_2 x + \cdots]. \quad (12.47)$$

Kelvin functions of second order can be expressed in terms of Kelvin functions of zeroth order and their derivatives,

$$\mathrm{ber}_2 x = \frac{2}{x}\mathrm{bei}'x - \mathrm{ber}\,x,$$
$$\mathrm{bei}_2 x = -\frac{2}{x}\mathrm{ber}'x - \mathrm{bei}\,x,$$
$$\mathrm{ber}'_2 x = -\frac{4}{x^2}\mathrm{bei}'x - \mathrm{ber}'x + \frac{2}{x}\mathrm{ber}\,x, \quad (12.48)$$
$$\mathrm{bei}'_2 x = \frac{4}{x^2}\mathrm{ber}'x - \mathrm{bei}'x + \frac{2}{x}\mathrm{bei}\,x,$$
$$\cdots$$

Substituting these expressions we find our results in terms of the zeroth order Kelvin functions,

$$\psi = \frac{C\mathcal{P}}{x^2} + A_1\left(\frac{2}{x}\text{bei}'x - \text{ber } x\right) - A_2\left(\frac{2}{x}\text{ber}'x + \text{bei}\right) + \cdots,$$
(12.49)

$$M_1 = -\frac{8C\mathcal{P}\rho^2 D(1-\nu)}{x^4}$$

$$- A_1 \frac{8\rho^2 D}{x^2}\left(\frac{2(1-\nu)}{x}\text{bei}'x + \tfrac{1}{2}x\,\text{ber}'x - \tfrac{1}{2}(1-\nu)\text{ber } x\right)$$

$$+ A_2 \frac{8\rho^2 D}{x^2}\left(\frac{2(1-\nu)}{x}\text{ber}'x - \tfrac{1}{2}x\,\text{bei}'x + \tfrac{1}{2}(1-\nu)\text{bei } x\right) + \cdots,$$
(12.50)

$$v_R = -\frac{\nu C\mathcal{P}}{8}\cos\alpha + A_1\frac{\cos\alpha}{\rho^2}\left(\tfrac{1}{4}x\,\text{bei}'x - \frac{1+\nu}{x}\text{ber}'x - \tfrac{1}{2}(1+\nu)\text{bei } x\right)$$

$$- A_2\frac{\cos\alpha}{\rho^2}\left(\tfrac{1}{4}x\,\text{ber}'x + \frac{1+\nu}{x}\text{bei}'x - \tfrac{1}{2}(1+\nu)\text{ber } x\right) + \cdots,$$
(12.51)

$$Q_R = \frac{4\mathcal{P}\rho^2}{\pi x^2 \cos\alpha} + A_1\frac{16\rho^4 D}{x^2 \cos\alpha}\left(\frac{2}{x}\text{ber}'x + \text{bei } x\right)$$

$$+ A_2\frac{16\rho^4 D}{x^2 \cos\alpha}\left(\frac{2}{x}\text{bei}'x - \text{ber } x\right) + \cdots.$$
(12.52)

The four arbitrary constants A_1, A_2, B_1 and B_2 are to be determined from the boundary conditions. In addition, we can prescribe the axial force $\mathcal{P}$ and the axial displacement v_z at an arbitrary section. These are the six boundary conditions, which are necessary for a unique solution.

The Kelvin functions have an oscillatory character. The functions appearing in the A_1 and A_2 terms are such that their amplitudes increase without limit as x increases and the functions in the B_1 and B_2 terms have a singularity at $x = 0$ and decrease in amplitude as x increases.

For a cone that is closed at the top we must therefore have $B_1 = B_2 = 0$ and this condition substitutes the boundary conditions for ψ or M_1 and v_R or Q_R at $t = 0$. The axial force P may still differ from zero if we permit a concentrated force at the top.

The analysis of composite shells is perhaps best illustrated by an example.

4. Numerical example

A structure according to Fig. 31 consists of a cylindrical shell with a coaxial conical roof. We shall assume that the structure is loaded by the weight of the roof and for simplicity we shall neglect the weight of the cylindrical shell.

Let us first determine the solution corresponding to the membrane state for each shell separately. Let γh be the weight per unit area of the conical shell. From (10.113) we deduce with $\theta = \alpha$, $p = \gamma h \sin \alpha$ and $F_1 = \gamma h \cos \alpha$,

$$N_2 = -\gamma h t \frac{\sin^2 \alpha}{\cos \alpha}, \qquad N_1 = -\frac{\gamma h t}{2 \cos \alpha} + \frac{c}{t}, \qquad (12.53)$$

where c is a constant of integration. We shall take $N_1 = 0$ at the boundary between the two parts at the joint. This determines the constant c and the weight of the roof must then be supported by a concentrated force at the top.

We can now determine the deformation suffered by the conical shell in the membrane state. From (12.20) we get at the lower boundary

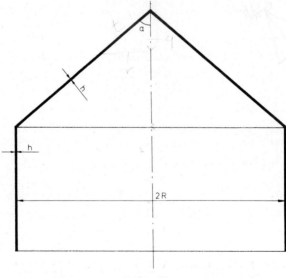

Fig. 31.

$t = t_0$,

$$v_0 = u \sin \alpha - w \cos \alpha = \frac{t_0}{Eh} N_2 \sin \alpha = -\frac{\gamma t_0^2 \sin^3 \alpha}{E \cos \alpha} \qquad (12.54)$$

and furthermore, since

$$-\psi_0 \cot \alpha = \frac{1}{Eh} [(tN_2)^\cdot - \nu(tN_1)^\cdot + \nu N_2]_{t=t_0}, \qquad (12.55)$$

we find the rotation of the normal to be

$$\psi_0 = \frac{\gamma t_0}{E} \sin \alpha \left(2 \frac{\sin^2 \alpha}{\cos \alpha} - \nu \right). \qquad (12.56)$$

Since the roof in reality is supported by the cylindrical shell, and not at the top, we shall superimpose a state of axial compression to the conical shell with a constant force P, which equals to minus the total weight, i.e.,

$$P = -\pi \gamma h t_0^2 \sin \alpha . \qquad (12.57)$$

The axial compression of the cylindrical shell will cause a radial deflection

$$v_1 = \frac{\nu \gamma t_0^2 \sin \alpha}{2E} \qquad (12.58)$$

but no rotation of the normal ($\psi_1 = 0$).

The deformations of the two parts, the conical roof and the cylindrical wall, are not compatible in the membrane state, and therefore bending will occur. At the junction there will act a radial shear-force Q, a bending moment M and a statically determined vertical force (see Fig. 32)

$$N = -\tfrac{1}{2} \gamma h t_0 . \qquad (12.59)$$

The redundant quantities Q and M are determined from the requirements of compatibility.

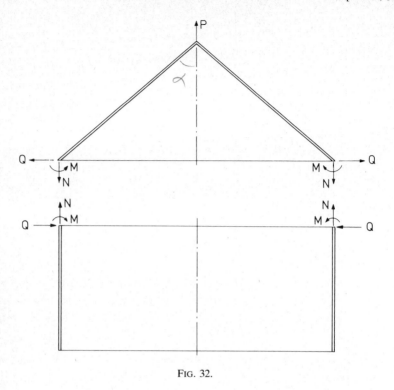

Fig. 32.

Let us assume the following numerical values,

$$h = 0.06, \quad R = 4.00, \quad \alpha = 50°, \quad \nu = 0.3,$$

and let us begin with the conical shell. We get successively $t_0 = 5.222$, $\rho = 4.807$ and $x = 31.07$. This value is far outside the region of arguments for most tables of Kelvin functions available. In fact, if the assumptions underlying our theory are to be valid, h must be much smaller than the radius of curvature $t_0 \tan \alpha$, i.e., $t_0 \gg h$, and x is therefore necessarily a large number. But for large arguments the Kelvin functions can be approximated by the first term of their asymptotic expansion,

$$\text{ber } x \approx \frac{1}{\sqrt{2\pi x}} \cos(\tfrac{1}{2} x \sqrt{2} - \tfrac{1}{8}\pi) \, e^{x/\sqrt{2}}, \qquad (12.60\text{a})$$

$$\text{bei } x \approx \frac{1}{\sqrt{2\pi x}} \sin(\tfrac{1}{2}x\sqrt{2} - \tfrac{1}{8}\pi)\, e^{x/\sqrt{2}}, \tag{12.60b}$$

$$\text{ber}' x \approx \frac{1}{\sqrt{2\pi x}} \cos(\tfrac{1}{2}x\sqrt{2} + \tfrac{1}{8}\pi)\, e^{x/\sqrt{2}}, \tag{12.60c}$$

$$\text{bei}' x \approx \frac{1}{\sqrt{2\pi x}} \sin(\tfrac{1}{2}x\sqrt{2} + \tfrac{1}{8}\pi)\, e^{x/\sqrt{2}}, \tag{12.60d}$$

$$\text{ker } x \approx \sqrt{\frac{\pi}{2x}} \cos(\tfrac{1}{2}x\sqrt{2} + \tfrac{1}{8}\pi)\, e^{-x/\sqrt{2}}, \tag{12.60e}$$

$$\text{kei } x \approx -\sqrt{\frac{\pi}{2x}} \sin(\tfrac{1}{2}x\sqrt{2} + \tfrac{1}{8}\pi)\, e^{-x/\sqrt{2}}, \tag{12.60f}$$

$$\text{ker}' x \approx -\sqrt{\frac{\pi}{2x}} \cos(\tfrac{1}{2}x\sqrt{2} - \tfrac{1}{8}\pi)\, e^{-x/\sqrt{2}}, \tag{12.60g}$$

$$\text{kei}' x \approx \sqrt{\frac{\pi}{2x}} \sin(\tfrac{1}{2}x\sqrt{2} - \tfrac{1}{8}\pi)\, e^{-x/\sqrt{2}}. \tag{12.60h}$$

For the argument $x = 31.07$ we get, by (12.60),

$$\text{ber } x \approx -2.276 \times 10^8 \quad (-2.279 \times 10^8),$$
$$\text{bei } x \approx 1.003 \times 10^8 \quad (1.012 \times 10^8),$$
$$\text{ber}' x \approx -2.318 \times 10^8 \quad (-2.291 \times 10^8),$$
$$\text{bei}' x \approx -0.8999 \times 10^8 \quad (-0.9123 \times 10^8),$$

where the values between parentheses are the accurate values.[5] We see that the approximate values have in this case a relative error of at most 1.4 per cent. Introducing these values in (12.49)–(12.52) we get

$$\psi = 2.218 \times 10^8 A_1 - 8.537 \times 10^7 A_2 - 9.575 \times 10^{-5} \gamma/D,$$
$$M = 6.600 \times 10^8 A_1 D + 2.792 \times 10^8 A_2 D + 1.284 \times 10^{-5} \gamma,$$
$$v_R = -2.099 \times 10^7 A_1 D + 4.608 \times 10^7 A_2 D + 9.641 \times 10^{-5} \gamma/D,$$
$$Q = 1.175 \times 10^9 A_1 D + 3.054 \times 10^9 A_2 D + 0.1867 \gamma. \tag{12.61}$$

---

[5] The accurate values of the Kelvin functions have been computed by a standard routine of the IMSL-library. Most computer installations have this or comparable routines. IMSL stands for "International Mathematical & Statistical Libraries, Inc.".

Eliminating the constants A_1 and A_2 we get the following relations for the conical shell,

$$\psi = 0.4609 \frac{M}{D} - 0.07009 \frac{Q}{D} - 0.01319 \frac{\gamma}{D},$$

$$v_R = -0.07009 \frac{M}{D} + 0.02150 \frac{Q}{D} + 0.004111 \frac{\gamma}{D}.$$

We note that the coefficient for Q in ψ is the same as the coefficient for M in v_R. This is a consequence of *Maxwell's reciprocity theorem*.

To determine M and Q we need the corresponding relations for the cylindrical shell. But these are already derived in Chapter 11 and are numbered (11.57) and (11.58). Assuming that the shell is not too short, i.e., assuming that λ is sufficiently large, we get, for the cylindrical shell,

$$\psi_s = -\frac{MR}{\rho_c D} - \frac{QR^2}{2\rho_c^2 D}, \qquad w_s = -\frac{MR^2}{2\rho_c^2 D} - \frac{QR^3}{2\rho_c^3 D},$$

where the signs are taken in accordance with the positive directions given in Fig. 31. The shell constant D is the same for both shells, but ρ_c for the cylindrical is given by (11.35), and is different from ρ for the conical shell. We get $\rho_c = 10.50$ in this example.

Compatibility requires

$$v_0 + v_R = v_1 + w_s, \qquad \psi_0 + \psi = \psi_s$$

or

$$0.8420 M + 0.002533 Q = 0.01307 \gamma,$$

$$0.002533 M + 0.04918 Q = -0.003672 \gamma$$

with the solution

$$\begin{aligned} M &= 0.01575\gamma \quad (0.01578\gamma), \\ Q &= -0.07547\gamma \quad (-0.07496\gamma), \end{aligned} \qquad (12.62)$$

where the values between parentheses are the corresponding values obtained by using the accurate values of the Kelvin functions. The relative error in the approximate result is below one per cent. With (12.62) substituted into (12.61) we find A_1 and A_2 to be

$$A_1 = 0.1007 \times 10^{-10} \gamma/D, \qquad A_2 = 0.3254 \times 10^{-10} \gamma/D$$

and all functions according to (12.43)–(12.52) are now determined. The distribution of the bending moment is shown in Fig. 33.

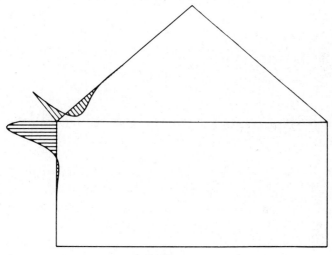

Fig. 33.

5. Spherical shells

According to (10.72) we have for a spherical shell

$$r = R \sin \theta, \qquad q = R \cos \theta, \qquad \theta = \frac{t}{R}.$$

The Meissner operator (12.12) becomes

$$L[\cdots] = -\frac{1}{R}\left(\frac{d^2}{d\theta^2} + \cot\theta \frac{d}{d\theta} - \cot^2\theta\right)[\cdots] \qquad (12.63)$$

and from (12.28) we get

$$\frac{d^2\psi}{d\theta^2} + \cot\theta \frac{d\psi}{d\theta} - \cot^2\theta\,\psi \mp 2i\rho^2\psi = 0, \tag{12.64}$$

where

$$\rho^2 = \sqrt{3(1-\nu^2)}\,R/h. \tag{12.65}$$

Using the substitution

$$x = \cos\theta \tag{12.66}$$

the equation takes the form

$$(1-x^2)\frac{d^2\psi}{dx^2} - 2x\frac{d\psi}{dx} + \left(1 + 2i\rho^2 - \frac{1}{1-x^2}\right)\psi = 0, \tag{12.67}$$

which is a special case of (13.53) of the next chapter. We shall discuss the solution of it there, in a more general context.

6. Toroidal shells

According to (10.99) we have for a toroidal shell (see Fig. 22)

$$r = R + a\cos\theta, \qquad q = a\sin\theta, \qquad \theta = \frac{t}{a}.$$

Let $\lambda = R/a$. Then $r = a(\lambda + \cos\theta)$ and the Meissner operator (12.12) becomes

$$L[\psi] = \frac{1}{a\cos\theta}\frac{d}{d\theta}\left[(\lambda + \cos\theta)\frac{d\psi}{d\theta}\right] - \frac{\sin^2\theta}{a(\lambda + \cos\theta)\cos\theta}\psi. \tag{12.68}$$

The differential equation (12.28) for ψ takes the form

$$\frac{1}{\cos\theta}\frac{d}{d\theta}\left[(\lambda + \cos\theta)\frac{d\psi}{d\theta}\right] - \frac{\sin^2\theta}{(\lambda + \cos\theta)\cos\theta}\psi \pm 2i\rho^2\psi = 0, \tag{12.69}$$

where

$$\rho^2 = \sqrt{3(1-\nu^2)}\,\frac{a}{h}. \tag{12.70}$$

Using the substitution

$$x = \cos\theta \tag{12.71}$$

the equation can be written in the form

$$(\lambda + x)^2(1-x^2)\frac{d^2\psi}{dx^2} + (\lambda + x)(2x^2 + \lambda x - 1)\frac{d\psi}{dx}$$
$$+ [1 - x^2 \pm 2i\rho^2 x(\lambda + x)]\psi = 0. \tag{12.72}$$

Let us assume that ψ can be represented by a power series in x with unknown coefficients. If this series is substituted into (12.72), this equation reduces to an identity. By equating the coefficients of each power of the independent variable separately to zero, we obtain a sequence of equations, that determine the unknown coefficients. This is a standard method, described in textbooks on differential equations.[6] We shall not enlarge upon it here, since in Chapter 14 we shall introduce a more powerful tool for the numerical solution of a much wider class of problems.

7. Approximate solution

For engineering purposes a very simple approximate solution for thin shells of revolution may sometimes prove sufficient. This approximation, which originally was introduced by GECKELER[7] is based on the fact that, for thin shells, the bending is confined to a narrow boundary zone.

---

[6] See, for instance, WHITTAKER, E.T. and G.N. WATSON (1962), *A Course of Modern Analysis*, Cambridge University Press, London, pp. 194–200. This method was applied to toroidal shells by WISSLER, H. (1916), "Festigkeitsberechnung von Ringflächenschalen", Diss., E.T.H. Zürich.

[7] GECKELER, J.W. (1926), "Ueber die Festigkeit achsensymmetrischen Schalen", *Forschg.-Arb. Ingwes.* **276**.

Let us write the Meissner operator (12.12) in the form

$$L[\cdots] = R[\cdots]'' + \frac{\dot{r}}{\dot{q}}[\cdots]' - \frac{\dot{r}^2}{r\dot{q}}[\cdots],\qquad(12.73)$$

where $R = r/\dot{q}$ is the radius of curvature in the peripheral direction. If now the bending is confined to a narrow zone, the characteristic length of the deformation pattern must necessarily be short. In such cases the contributions from the lower derivatives become negligible in comparison with the contributions from the higher ones, and consequently one may approximate the Meissner operator by the single operator-term

$$L[\cdots] \approx R[\cdots]''.\qquad(12.74)$$

Also, when the bending region is narrow, we can approximate the functions R, r, $\dot{r}$ and $\dot{q}$ by their values at the boundary and treat them as constants. We shall therefore write

$$\dot{r} = \sin\alpha,\qquad \dot{q} = \cos\alpha,\qquad r = R\cos\alpha,\qquad(12.75)$$

where α is half the top angle of the tangent cone at the boundary—and hence a constant.

With (12.74) the differential equation (12.27) for ψ takes the form

$$R^2 \frac{d^4\psi}{dt^4} + 4\kappa^2\psi = 0,\qquad(12.76)$$

which is a fourth order differential equation with constant coefficients. The solution can be sought in the form

$$e^{\mu t/R},$$

which leads to the characteristic equation

$$\mu^4 + 4\kappa^2 R^2 = 0.\qquad(12.77)$$

The roots of this equation are

$$\mu = \pm(1\pm i)\rho \tag{12.78}$$

with

$$\rho = \sqrt[4]{3(1-\nu^2)}\sqrt{R/h}\,, \tag{12.79}$$

where (12.26) has been used.

Let us now take $t=0$ at the boundary and $t>0$ in the shell. The solution to (12.76) can then be written as [8]

$$\psi = \left(A\sin\frac{\rho t}{R} + B\cos\frac{\rho t}{R}\right)e^{-\rho t/R}\,. \tag{12.80}$$

At this stage it is useful to check the approximations introduced by applying the complete Meissner operator (12.73) to the solution. When this is done, we find the relative magnitude of the three terms on the right-hand side to be

$$1:\frac{\tan\alpha}{\rho}:\left(\frac{\tan\alpha}{\rho}\right)^2.$$

Clearly, the approximation introduced in this paragraph is valid only if $(\tan\alpha)/\rho$ is small or negligible in comparison with unity, and we shall consequently assume this to be the case now.

Neglecting terms of relative order $(\tan\alpha)/\rho$ we get, by (12.6) and (12.80),

$$M_1 = D\frac{\rho}{R}\left[(A-B)\cos\frac{\rho t}{R} - (A+B)\sin\frac{\rho t}{R}\right]e^{-\rho t/R} \tag{12.81}$$

and from (12.33) we find

$$\mathcal{Q} = 2D\frac{\rho^2}{R}\left(B\sin\frac{\rho t}{R} - A\cos\frac{\rho t}{R}\right)e^{-\rho t/R}\,. \tag{12.82}$$

---

[8] The exponentially increasing part of the solution is not in concord with the assumption of a narrow bending zone at the boundary, and is therefore excluded.

Finally, by (12.34) and (12.35) we get

$$v_R = -\frac{R\cos\alpha}{2\rho}\left[(B-A)\sin\frac{\rho t}{R} - (A+B)\cos\frac{\rho t}{R}\right]e^{-\rho t/R} \quad (12.83)$$

and

$$Q_R = 2D\frac{\rho^2}{R^2\cos\alpha}\left(B\sin\frac{\rho t}{R} - A\cos\frac{\rho t}{R}\right)e^{-\rho t/R}, \quad (12.84)$$

when again terms of relative order $(\tan\alpha)/\rho$ have been omitted.

Roughly speaking, the errors introduced by this approximation are of relative order $\sqrt{h/R}\tan\alpha$, which should be compared with the corresponding relative errors of order h/R of the theory in the previous paragraphs. Thus, for all but very small angles α the relative errors are considerably greater. Nevertheless, in many engineering applications they may be acceptable.

Bibliography

TIMOSHENKO, S. (1959), *Theory of Plates and Shells*, 2nd ed., McGraw-Hill, New York, Chapter 16.
FLÜGGE, W. (1960), *Stresses in Shells*, Springer, Berlin, Chapter 6.
NOVOZHILOV, V.V. (1961), *The Theory of Thin Shells*, Pergamon, Oxford, Chapter IV (English translation).
SEIDE, P. (1975), *Small Elastic Deformations of Thin Shells*, Noordhoff, Leiden, Chapter 7.

CHAPTER 13

BENDING OF SPHERICAL SHELLS

1. Introduction 289
2. Basic relations 289
3. Solution of the mathematical problem 294
4. Boundary conditions 297
5. Spherical coordinates 298
6. Vibrations of spherical shells 312
7. Radially loaded spherical shell 334
8. Addendum 336
   Bibliography 336

CHAPTER 13

BENDING OF SPHERICAL SHELLS

1. Introduction

Spherical shells are widely used structures, and their importance in engineering is such that a detailed analysis of them seems indispensible. In Chapter 10, the membrane state of spherical shells was discussed, and, in Chapter 12, the case of axisymmetrical bending. Although this may cover a majority of problems, there are still many cases, where a general theory of (non-axisymmetrical) bending is required, a theory to which this chapter is devoted.

2. Basic relations

A spherical surface is characterized by its constant curvature. In mixed form the curvature tensor d_β^α in any coordinate system is equal to plus or minus Kronecker's delta δ_β^α divided by the radius R of the sphere. The sign will depend on our choice of surface coordinates, and so that we shall have the minus sign if the normal to the spherical surface (and hence also w) points out of the sphere. Taking this to be the case, we shall write

$$d_{\alpha\beta} = -\frac{1}{R} a_{\alpha\beta} \qquad (13.1)$$

and this formula can be used to simplify the shell equations considerably.

With (13.1) the strain tensor (3.11) takes the form

$$E_{\alpha\beta} = \tfrac{1}{2}(D_\alpha v_\beta + D_\beta v_\alpha) + \frac{1}{R} a_{\alpha\beta} w \qquad (13.2)$$

and the bending tensor (3.37) becomes

$$K_{\alpha\beta} = D_\alpha D_\beta w - \frac{1}{R}(D_\alpha v_\beta + D_\beta v_\alpha) - \frac{1}{R^2} a_{\alpha\beta} w. \qquad (13.3)$$

We shall now take advantage of the possibilities of using alternative measures of bending, as discussed in Chapter 8. In particular, we see that by taking

$$\tilde{K}_{\alpha\beta} = K_{\alpha\beta} + \frac{2}{R} E_{\alpha\beta} \qquad (13.4)$$

the bending tensor becomes a function of w only,

$$\tilde{K}_{\alpha\beta} = D_\alpha D_\beta w + \frac{1}{R^2} a_{\alpha\beta} w, \qquad (13.5)$$

and this property will prove useful in leading to simple equations.

Now the principle of virtual work will require

$$N^{\alpha\beta} \delta E_{\alpha\beta} + M^{\alpha\beta} \delta K_{\alpha\beta} = \tilde{N}^{\alpha\beta} \delta E_{\alpha\beta} + M^{\alpha\beta} \delta \tilde{K}_{\alpha\beta}$$

and when $\tilde{K}_{\alpha\beta}$ according to (13.4) is introduced, we find

$$\tilde{N}^{\alpha\beta} = N^{\alpha\beta} - \frac{2}{R} M^{\alpha\beta} \qquad (13.6)$$

to be the appropriate membrane stress tensor.

Using (13.1) and (13.6), the equations of equilibrium (5.22) and (5.23) become

$$D_\alpha \tilde{N}^{\alpha\beta} + F^\beta = 0 \qquad (13.7)$$

and

$$D_\alpha D_\beta M^{\alpha\beta} + \frac{1}{R^2} M^\alpha_\alpha + \frac{1}{R} \tilde{N}^\alpha_\alpha = p. \qquad (13.8)$$

From Hooke's law (7.11) we get

$$M^{\alpha\beta} = D\left[(1-\nu)\left(D^{\alpha}D^{\beta}w + \frac{1}{R^2}a^{\alpha\beta}w\right) + \nu a^{\alpha\beta}\left(\Delta + \frac{2}{R^2}\right)w\right] \tag{13.9}$$

and again we stress that the uncoupled equations (7.11) hold with the same accuracy for $\tilde{K}_{\alpha\beta}$ and $\tilde{N}^{\alpha\beta}$ as for $K_{\alpha\beta}$ and $N^{\alpha\beta}$.

Equation (13.7) indicates that we could formulate the problem with the help of a stress function, as in the case of in-plane loaded plates (Chapter 9), but before doing so, we recall that a spherical surface is not an Euclidean surface, and that therefore care must be taken with regards to the order of covariant differentiation. According to (2.16) and (13.1) the Riemann–Christoffel tensor for a spherical surface is

$$B_{\sigma\alpha\beta\gamma} = \frac{1}{R^2}(a_{\alpha\gamma}a_{\sigma\beta} - a_{\alpha\beta}a_{\sigma\gamma}) \tag{13.10}$$

and by (1.39) and (1.40) the rules for interchanging the order of covariant differentiation of a vector and a second order tensor are found to be

$$D_{\alpha}D_{\beta}A^{\gamma} - D_{\beta}D_{\alpha}A^{\gamma} = \frac{1}{R^2}(a_{\alpha}^{\gamma}A_{\beta} - a_{\beta}^{\gamma}A_{\alpha}) \tag{13.11}$$

and

$$D_{\alpha}D_{\beta}A_{\gamma\delta} - D_{\beta}D_{\alpha}A_{\gamma\delta} = \frac{1}{R^2}(a_{\alpha\gamma}A_{\beta\delta} - a_{\gamma\beta}A_{\alpha\delta} + a_{\delta\alpha}A_{\gamma\beta} - a_{\delta\beta}A_{\gamma\alpha}). \tag{13.12}$$

Let $\tilde{N}^{\alpha\beta} = F^{\alpha\beta}$ be a particular integral of (13.7), and let $\Phi(u^1, u^2)$ be any sufficiently smooth function. Then[1]

$$\tilde{N}^{\alpha\beta} = EhR\left(\varepsilon^{\alpha\xi}\varepsilon^{\beta\eta}D_{\xi}D_{\eta}\Phi + \frac{1}{R^2}a^{\alpha\beta}\Phi\right) + F^{\alpha\beta} \tag{13.13}$$

---

[1] The factor EhR has been added for purely aesthetical reasons, giving Φ the same physical dimension as w.

is the general solution of (13.7). To show this, we take the covariant derivative of (13.13) and get

$$D_\alpha \tilde{N}^{\alpha\beta} = Eh R \left[\varepsilon^{\alpha\xi} D_\alpha D_\xi (\varepsilon^{\beta\eta} D_\eta \Phi) + \frac{1}{R^2} D^\beta \Phi \right] - F^\beta$$

since the alternating tensor behaves like a constant in covariant differentiation. Now, the expression between square brackets vanishes, and to show this we write

$$D_\alpha D_\xi A^\beta = \tfrac{1}{2}(D_\alpha D_\xi + D_\xi D_\alpha) A^\beta + \tfrac{1}{2}(D_\alpha D_\xi - D_\xi D_\alpha) A^\beta$$

and note that the first term on the right-hand side is symmetrical with respect to the indices α and ξ while the second one may be evaluated with the help of (13.11). When multiplied by the skew-symmetrical tensor $\varepsilon^{\alpha\xi}$, the contribution from the first term is nil, and the result is

$$\varepsilon^{\alpha\xi} D_\alpha D_\xi A^\beta = \frac{1}{2R^2} (\varepsilon^{\beta\xi} A_\xi - \varepsilon^{\alpha\beta} A_\alpha) = \frac{1}{R^2} \varepsilon^{\beta\xi} A_\xi.$$

Taking $A^\beta = \varepsilon^{\beta\eta} D_\eta \Phi$, we see that $D_\alpha \tilde{N}^{\alpha\beta} = -F^\beta$, and therefore the first two equations of equilibrium (13.7) are identically satisfied whenever $\tilde{N}^{\alpha\beta}$ is derived from any function Φ according to (13.13).

The third equation of equilibrium (13.8) can now be expressed in terms of the two scalar functions w and Φ by substituting (13.9) and (13.13). Observing that, according to (13.12)

$$D_\alpha D_\beta D^\alpha D^\beta w = \Delta^2 w + \frac{1}{R^2} \Delta w,$$

we get, after some minor simplifications,

$$\left(\Delta + \frac{2}{R^2}\right)\left[\left(\Delta + \frac{1+\nu}{R^2}\right) w + \frac{1-\nu^2}{kR^2} \Phi \right] = \frac{1}{D}\left(p - \frac{1}{R} F^\alpha_\alpha \right), \tag{13.14}$$

where, as usual,

$$k = \frac{h^2}{12R^2}. \tag{13.15}$$

Although the equations of equilibrium will be satisfied by any function Φ, we must make sure that the (linearized) equations of compatibility are also satisfied, since we are not working in terms of displacements only.

By (3.48) and (3.49), using (13.1) and (13.4), we can write the equations of compatibility as follows,

$$\left(\Delta + \frac{1}{R^2}\right)E_\alpha^\alpha - D_\alpha D^\beta E_\beta^\alpha - \frac{1}{R}\tilde{K}_\alpha^\alpha = 0 \tag{13.16}$$

and

$$D_\alpha \tilde{K}_{\beta\gamma} - D_\beta \tilde{K}_{\alpha\gamma} = 0. \tag{13.17}$$

It is easily checked that (13.17) is identically satisfied. Using Hooke's law (7.12) we can write $E_{\alpha\beta}$ in terms of $\tilde{N}^{\alpha\beta}$,

$$E^{\alpha\beta} = \frac{1}{Eh}[(1+\nu)\tilde{N}^{\alpha\beta} - \nu a^{\alpha\beta}\tilde{N}_\gamma^\gamma], \tag{13.18}$$

and by introducing this expression into (13.16) we readily find that the compatibility requires

$$\left(\Delta + \frac{2}{R^2}\right)\left[\left(\Delta + \frac{1-\nu}{R^2}\right)\Phi - \frac{1}{R^2}w\right] = 0, \tag{13.19}$$

which together with (13.14) is the result. It should be stressed, that equations (13.14)–(13.19) are *accurate*, in the sense that they are derived without any approximations from the basic equations of our shell theory.

The problem of bending of spherical shells has been reduced to the problem of solving a system of two fourth order partial differential equations in two unknowns. No reference has been made to any particular coordinate system and the equations therefore hold in any conceivable system.

When R tends to infinity the two equations decouple. Equation (13.14) reduces to the plate equation (9.12) and (13.19) to equation (9.7) for Airy's stress function for flat plates.

We shall now discuss the solution of the problem.

3. Solution of the mathematical problem

We shall assume that we have a particular integral to the problem, given by the normal displacement w_1 and the membrane stress tensor $F^{\alpha\beta}$. The remaining problem, and the one on which we shall concentrate, is to find the general solution to the homogeneous problem. But before we attempt to do so we shall make the following observation.

It is readily checked that if we take $\Phi \equiv 0$ and

$$\left(\Delta + \frac{2}{R^2}\right) w^* = 0, \tag{13.20}$$

the homogeneous system (13.14)–(13.19) will be satisfied. Since Φ is zero everywhere, the membrane force tensor $\tilde{N}^{\alpha\beta}$ and hence also the strain tensor $E_{\alpha\beta}$ must vanish identically. A solution w^* to (13.20) represents therefore an *inextensional* deformation pattern.

Conversely, if we take $w \equiv 0$ and Φ a solution of

$$\left(\Delta + \frac{2}{R^2}\right) \Phi^* = 0, \tag{13.21}$$

and homogeneous system is again satisfied. With $w = 0$ the moment tensor $M^{\alpha\beta}$ vanishes identically and the solutions to (13.21) represent a class of *membrane states* that happen to be exact solutions to the shell problem.

We shall return to these solutions below.

To find the general solution of the homogeneous system (13.14)–(13.19), we write

$$W = \left(\Delta + \frac{2}{R^2}\right) w, \qquad \phi = \left(\Delta + \frac{2}{R^2}\right) \Phi, \tag{13.22}$$

and this system reduces to

$$\left(\Delta + \frac{1+\nu}{R^2}\right) W + \frac{1-\nu^2}{kR^2} \phi = 0,$$
$$-\frac{1}{R^2} W + \left(\Delta + \frac{1-\nu}{R^2}\right) \phi = 0. \tag{13.23}$$

We now proceed by taking

$$W = -\frac{1-\nu^2}{kR^2}\chi, \qquad \Phi = \left(\Delta + \frac{1+\nu}{R^2}\right)\chi, \qquad (13.24)$$

and substitute into (13.23). Clearly, the first equation of this system will be identically satisfied, while the second one will require χ to satisfy the following fourth order differential equation,

$$\left[\left(\Delta + \frac{1+\nu}{R^2}\right)\left(\Delta + \frac{1-\nu}{R^2}\right) + \frac{1-\nu^2}{kR^4}\right]\chi = 0. \qquad (13.25)$$

To complete the solution, we take

$$\chi = \left(\Delta + \frac{2}{R^2}\right)\Xi. \qquad (13.26)$$

Then

$$w = -\frac{1-\nu^2}{kR^2}\Xi + w^*, \qquad \Phi = \left(\Delta + \frac{1+\nu}{R^2}\right)\Xi, \qquad (13.27)$$

where Ξ is a solution of the sixth order differential equation

$$\left(\Delta + \frac{2}{R^2}\right)\left[\left(\Delta + \frac{1+\nu}{R^2}\right)\left(\Delta + \frac{1-\nu}{R^2}\right) + \frac{1-\nu^2}{kR^4}\right]\Xi = 0, \qquad (13.28)$$

and where w^* is the general solution of the second order equation (13.20).

It might seem that a still more general solution were obtained if Φ^*, a general solution to (13.21), was added to Φ, i.e., if we wrote

$$\Phi = \left(\Delta + \frac{1+\nu}{R^2}\right)\Xi + \Phi^*.$$

But this is not the case. By writing

$$\Xi = \Xi^* + \frac{R^2}{1-\nu}\Phi^*,$$

we see that Ξ^* will satisfy the same differential equation (13.28) as Ξ and equations (13.27) become

$$w = -\frac{1-\nu^2}{kR^2}\Xi^* - \frac{1+\nu}{k}\Phi^* + w^*, \qquad \Phi = \left(\Delta + \frac{1+\nu}{R^2}\right)\Xi^*.$$

In this way we got rid of Φ^* in the second equation. Its appearance in the first equation is of no consequence to the generality of the solution, since Φ^* and w^* are the solutions to the same equation. Hence, (13.27)–(13.28) is the complete solution to the problem.

The problem of bending of spherical shells has been reduced to finding solutions of one single sixth order differential equation (13.28) and one second order equation (13.20). Following our terminology in connection with cylindrical shells, we shall consider Ξ to be a *potential function for the integration of the homogeneous equations of spherical shells.*

We can further simplify the differential equation if we are prepared to neglect terms of relative order k. Thus, factorizing the square bracket in (13.28) and neglecting the term ν^2/R^4 in comparison with $(1-\nu^2)/(kR^4)$, we find

$$\left(\Delta + \frac{2}{R^2}\right)\left(\Delta + \frac{1+2\mathrm{i}\rho^2}{R^2}\right)\left(\Delta + \frac{1-2\mathrm{i}\rho^2}{R^2}\right)\Xi = 0, \qquad (13.29)$$

where, as before,

$$\rho^2 = \sqrt{3(1-\nu^2)}\,\frac{R}{h}. \qquad (13.30)$$

Since the operators commute, the complete solution Ξ is determined as the sum of the solutions to three second order differential equations. Two of these have complex conjugate coefficients. This implies that our problem is completely solved when we have acquired the solutions to the following *two* second order differential equations,

$$\left(\Delta + \frac{2}{R^2}\right)\Xi = 0, \qquad \left(\Delta + \frac{1+2\mathrm{i}\rho^2}{R^2}\right)\Xi = 0, \qquad (13.31)$$

since the real and imaginary parts of the second equation will separately satisfy the complete equation (13.29).

4. Boundary conditions

Let δv_α and δw be arbitrary virtual displacements and $\delta E_{\alpha\beta}$, $\delta K_{\alpha\beta}$ the corresponding tensors of strain and bending according to (13.2) and (13.3). The forces and moments acting at the boundary, corresponding to the field tensors $N^{\alpha\beta}$ and $M^{\alpha\beta}$, are derived from the principle of virtual work, which leads to the following equality,[2]

$$\iint_{\mathscr{D}} (N^{\alpha\beta}\, \delta E_{\alpha\beta} + M^{\alpha\beta}\, \delta K_{\alpha\beta})\, dA = \oint_{\mathscr{C}} \left(T^\alpha\, \delta v_\alpha + Q\, \delta w + M_B \frac{\partial}{\partial n} \delta w \right) ds,$$
(13.32)

where the surface integral is extended over a domain $\mathscr{D}$ of the middle-surface and the line integral over the closed boundary $\mathscr{C}$. Here

$$T^\alpha = (N^{\beta\alpha} + 2d^\alpha_\gamma M^{\beta\gamma}) n_\beta$$

is the membrane force vector,

$$Q = -n_\beta D_\alpha M^{\alpha\beta} - \frac{\partial}{\partial s}(M^{\alpha\beta} n_\alpha t_\beta)$$

is the effective shear force,

$$M_B = M^{\alpha\beta} n_\alpha n_\beta$$

the bending moment, and n_α, t_α the unit normal and tangent vectors respectively to the curve $\mathscr{C}$.

Using (13.4) and (13.6), equality (13.32) can be written in the form

$$\iint_{\mathscr{D}} \left(\tilde{N}^{\alpha\beta}\, \delta E_{\alpha\beta} + M^{\alpha\beta}\, \delta \tilde{K}_{\alpha\beta} \right) dA$$

$$= \oint_{\mathscr{C}} \left(N\, \delta u + S\, \delta v + Q\, \delta w + M_B \frac{\partial}{\partial n} \delta w \right) ds, \qquad (13.33)$$

---

[2] Chapter 5, equation (5.18).

where u and v are the normal and tangential displacements at the boundary,

$$N = \tilde{N}^{\alpha\beta} n_\alpha n_\beta \quad \text{and} \quad S = \tilde{N}^{\alpha\beta} n_\alpha t_\beta$$

are the normal and tangential membrane stress, respectively.

With the help of (13.6) and (13.9) we find

$$N = \tilde{N}^{\alpha\beta} n_\alpha n_\beta = Eh\, R \left(t^\alpha t^\beta D_\alpha D_\beta \Phi + \frac{1}{R^2} \Phi \right) + F^{\alpha\beta} n_\alpha n_\beta, \quad (13.34)$$

$$S = \tilde{N}^{\alpha\beta} n_\alpha t_\beta = -Eh\, R t^\alpha n^\beta D_\alpha D_\beta \Phi, \quad (13.35)$$

$$M_B = D \left[(1-\nu) n^\alpha n^\beta D_\alpha D_\beta w + \nu \Delta w + \frac{1+\nu}{R^2} w \right], \quad (13.36)$$

$$Q = -D(1-\nu) \left[n^\alpha D_\alpha \left(\Delta + \frac{2}{R^2} \right) w + t^\alpha D_\alpha (n^\beta t^\gamma D_\beta D_\gamma w) \right]. \quad (13.37)$$

The bending moment M_B and the three boundary forces N, S and Q are at our disposal and can be freely prescribed along the boundary, provided that they are in equilibrium with the external forces. The four boundary conditions and the eighth order system (13.14), (13.19) constitute a well-posed mathematical problem.

5. Spherical coordinates

The results obtained in the previous paragraphs are independent of any choice of surface coordinates. In this paragraph we shall specialize to spherical coordinates, which of course are the most commonly used for spherical shells. In such coordinates the middle-surface is represented by

$$f^1 = R \sin\theta \cos\phi, \quad f^2 = R \sin\theta \sin\phi, \quad f^3 = R \cos\theta, \quad (13.38)$$

where ϕ is the 'longitude', θ the 'co-latitude' and R the radius of the surface.

The components of the metric tensor and the curvature are given by

$$a_{11} = R^2, \quad a_{12} = a_{21} = 0, \quad a_{22} = R^2 \sin^2 \theta \quad (13.39)$$

and

$$d_{11} = -R, \quad d_{12} = d_{21} = 0, \quad d_{22} = -R \sin^2 \theta, \quad (13.40)$$

respectively. The non-vanishing Christoffel symbols are

$$\{{}_{1\,2}^{2}\} = \{{}_{2\,1}^{2}\} = \cot \theta, \quad \{{}_{2\,2}^{1}\} = -\sin \theta \cos \theta \quad (13.41)$$

and the Laplacian takes the form

$$\Delta \equiv \frac{1}{R^2}\left(\frac{\partial^2}{\partial \theta^2} + \cot \theta \frac{\partial}{\partial \theta} + \frac{1}{\sin^2 \theta}\frac{\partial^2}{\partial \phi^2}\right). \quad (13.42)$$

By introducing (13.42) in (13.29) or (13.31) we get differential equation(s) for the potential function in spherical coordinates.

For a spherical cap, bounded by $\theta = \theta_1$, or for a spherical zone bounded by two latitudes θ_1 and θ_2, the solution must be periodic in ϕ with period 2π. The function Ξ can then be expanded in a Fourier series,

$$\Xi = \sum_{m=0}^{\infty} [C_m(\theta) \cos m\phi + D_m(\theta) \sin m\phi]. \quad (13.43)$$

When Ξ is substituted into (13.29), and the coefficients of all trigonometric functions are put equal to zero, one obtains the following set of ordinary differential equations for $C_m(\theta)$,

$$\left(\frac{d^2}{d\theta^2} + \cot \theta \frac{d}{d\theta} - \frac{m^2}{\sin^2 \theta} + 2\right) C_m(\theta) = 0, \quad m = 0, 1, 2, \ldots,$$
$$(13.44)$$

$$\left(\frac{d^2}{d\theta^2} + \cot \theta \frac{d}{d\theta} - \frac{m^2}{\sin^2 \theta} + 1 + 2i\rho^2\right) C_m(\theta) = 0, \quad m = 0, 1, 2, \ldots,$$
$$(13.45)$$

and similarly for the functions $D_m(\theta)$.

The complete solution to (13.44) is

$$C_m(\theta) = A_m(m + \cos\theta)\tan^m \tfrac{1}{2}\theta + B_m(m - \cos\theta)\cot^m \tfrac{1}{2}\theta, \tag{13.46}$$

where A_m and B_m are arbitrary constants. Since $C_m(\theta)\cos m\phi$ is a solution of (13.20) as well, we see that the mth mode of inextensional deformations is given by

$$w^* = [A_m(m + \cos\theta)\tan^m \tfrac{1}{2}\theta + B_m(m - \cos\theta)\cot^m \tfrac{1}{2}\theta]\cos m\phi. \tag{13.47}$$

The normal displacement w^* must be accompanied by appropriate tangential displacements v_α in order to ensure an unstrained middle-surface. Let u and v be the displacements in longitudinal and lateral directions respectively, i.e.,

$$u = \frac{v_1}{R}, \qquad v = \frac{v_2}{R\sin\theta}. \tag{13.48}$$

In terms of these displacements the components of strain (13.2) become

$$E_{11} = \frac{\partial u}{\partial \theta} + Rw,$$

$$E_{12} = \frac{1}{2}\left(\frac{\partial u}{\partial \phi} + \frac{\partial v}{\partial \theta}\right) - v\cot\theta, \tag{13.49}$$

$$E_{22} = \frac{\partial v}{\partial \phi} + u\sin\theta\cos\theta + Rw\sin^2\theta.$$

When w is taken according to (13.47), while u and v are taken to be

$$u = (-A_m \tan^m \tfrac{1}{2}\theta + B_m \cot^m \tfrac{1}{2}\theta)\sin\theta\cos m\phi,$$
$$v = (-A_m \tan^m \tfrac{1}{2}\theta - B_m \cot^m \tfrac{1}{2}\theta)\sin\theta\sin m\phi, \tag{13.50}$$

and these functions are substituted into (13.49), we shall find that all components $E_{\alpha\beta}$ vanish.

It is easily checked that the solution for $m = 0$ corresponds to a rigid-body translation of the shell in the direction of its axis. Also, we may convince ourselves that for $m = 1$ the solution corresponds to a rigid-body rotation about a perpendicular to the axis of the shell. However, for $m > 1$ the displacements found describe a true state of bending.

If $B_m \neq 0$ and $m > 1$, the solution will be singular at $\theta = 0$. Therefore, if the shell contains the north-pole, we must have $B_m = 0$ for all $m > 1$. Similarly, if $A_m \neq 0$ and $m > 1$, the solution will be singular at $\theta = \pi$. Hence, if the shell contains the south-pole $\theta = \pi$, all coefficients A_m must vanish. This leads to the conclusion that if *the shell is complete, all coefficients for $m > 1$ must vanish, and no inextensional deformations can occur.* This, by the way, is a special case of a celebrated theorem by JELLETT,[3] which he formulated in the following words:

"If a closed oval surface be perfectly inextensible, it is also perfectly rigid".

JELLETT also showed that his theorem, which is rigorously true for an inextensible surface, is *approximately* true for a surface possessed of a small amount of extensibility. This accounts for the remarkable stiffness of such structures as for instance table tennis balls.

The function $C_m(\theta) \cos m\phi$ is also a solution of (13.21), and introducing it into (13.13) yields the following *membrane states*,

$$\tilde{N}^{11} = \frac{Ehm(1-m^2)}{R^3 \sin^2 \theta}(A_m \tan^m \tfrac{1}{2}\theta + B_m \cot^m \tfrac{1}{2}\theta) \cos m\phi,$$

$$\tilde{N}^{12} = \frac{Ehm(1-m^2)}{R^3 \sin^3 \theta}(A_m \tan^m \tfrac{1}{2}\theta + B_m \cot^m \tfrac{1}{2}\theta) \sin m\phi, \quad (13.51)$$

$$\tilde{N}^{22} = -\frac{Ehm(1-m^2)}{R^3 \sin^4 \theta}(A_m \tan^m \tfrac{1}{2}\theta + B_m \cot^m \tfrac{1}{2}\theta) \cos m\phi.$$

Although the constant factor $Ehm(1-m^2)/R^3$ could be absorbed by the arbitrary coefficients A_m and B_m, we retain it since it contains a

---

[3] JELLETT, J.H. (1855), "On the properties of inextensible surfaces", *Roy. Irish Acad. Trans.* **22**, p. 343.

warning. For a spherical cap, which has only one boundary, the solutions with $m = 0$ would have a resultant force, but since there are no surface loads, this force must be zero. Similarly, if $m = 1$, the boundary forces have a resultant moment, and again this must be zero. Therefore, quite appropriately, we get $\tilde{N}^{\alpha\beta} = 0$ for $m = 0$ and $m = 1$.

However, if the shell has two boundaries, solutions with $m = 0$ and $m = 1$ do exist, the resultant force and moment at one boundary being equilibrated at the other. In fact, all four cases (a), (b), (c) and (d) of the section on spherical shells in Chapter 10 (see Fig. 21) are the special cases of (13.51) for $m = 0$ and $m = 1$, provided that we remove the factor $m(1 - m^2)$ or incorporate it in the coefficients A_m and B_m.

Let us now proceed with the solution to the general problem. To do so we must solve (13.45). Making the substitution

$$x = \cos\theta, \tag{13.52}$$

equation (13.45) takes the form

$$(1-x^2)\frac{d^2 C_m}{dx^2} - 2x\frac{dC_m}{dx} + \left(1 + 2i\rho^2 - \frac{m^2}{1-x^2}\right)C_m = 0. \tag{13.53}$$

Before continuing we observe that for $m = 1$ this equation reduces to (12.67) of the previous chapter.

The standard form of (13.53) is[4]

$$(1-x^2)\frac{d^2 y}{dx^2} - 2x\frac{dy}{dx} + \left[\sigma(\sigma+1) - \frac{m^2}{1-x^2}\right]y = 0, \tag{13.54}$$

and we shall begin by discussing the solutions of this equation.

When $m = 0$, it reduces to *Legendres equation*, satisfied by $P_\sigma(x)$ and $Q_\sigma(x)$, the *Legendre functions* of the first and second kind respectively, σ being their degree. The Legendre functions can be expressed in terms of the *hypergeometric function*,[5] viz.

$$P_\sigma(x) = F(-\sigma, \sigma+1, 1, \tfrac{1}{2}(1-x)) \quad (-1 < x \leq 1), \tag{13.55}$$

---

[4] See WHITTAKER, E.T. and G.N. WATSON (1962), *A Course of Modern Analysis*, Cambridge University Press, London, p. 324.

[5] See ERDÉLYI, A. (1953), *Higher Transcendental Functions Vol. 1*, McGraw-Hill, New York, Chapter III. Formulas (13.55)–(13.62) can be found there.

where F is defined by the infinite series

$$F(\alpha, \beta, \gamma, x) = 1 + \frac{\alpha\beta}{1!\gamma}x + \frac{\alpha(\alpha+1)\beta(\beta+1)}{2!\gamma(\gamma+1)}x^2 + \cdots, \quad (13.56)$$

uniformly convergent in the interval $0 \leq x < 1$.

Clearly, if σ is integral, this series has only a finite number of terms, being a polynomial in x of degree $n = \sigma$. These are the *Legendre polynomials*. It can be shown that the same polynomials can be derived from *Rodriques' formula*,

$$P_n(x) = \frac{1}{2^n n!} \frac{d^n}{dx^n}[(x^2 - 1)^n], \quad (13.57)$$

and that the first few Legendre polynomials are

$$P_0(x) = 1, \quad P_1(x) = x, \quad P_2(x) = \tfrac{1}{2}(3x^2 - 1), \quad P_3(x) = \tfrac{1}{2}(5x^3 - 3x),$$
$$P_4(x) = \tfrac{1}{8}(35x^4 - 30x^2 + 3), \quad P_5(x) = \tfrac{1}{8}(63x^5 - 70x^3 + 15x).$$

We can say that (13.55) *extends* the definition of the Legendre functions from polynomials to functions of non-integral, and also complex values of σ.

When $m \neq 0$, the two linearly independent solutions to equation (13.54) are called the *associated Legendre functions* $P_\sigma^m(x)$ and $Q_\sigma^m(x)$ of first and second kind respectively. Index m is their *order*.[6]

The associated Legendre functions of the first kind and of positive integral order m, are given by the derivatives

$$P_\sigma^m(x) = (-1)^m (1 - x^2)^{m/2} \frac{d^m P_\sigma(x)}{dx^m} \quad (13.58)$$

and of negative integral order by the integrals

$$P_\sigma^{-m}(x) = (-1)^m (1 - x^2)^{-m/2} \int_1^x \cdots \int_1^x P_\sigma(x)(dx)^m. \quad (13.59)$$

---

[6] It is customary to suppress this index, when 0.

Substituting (13.55)–(13.56) into (13.59) and integrating term by term, one finds

$$P_\sigma^{-m}(x) = \frac{1}{\Gamma(m+1)} \left(\frac{1-x}{1+x}\right)^{m/2} F(-\sigma, \sigma+1, 1+m, \tfrac{1}{2}(1-x)). \tag{13.60}$$

This formula is usually taken as the definition of the associated Legendre functions of the first kind for arbitrary (complex) values of x, σ and m (except when m is a negative integer). It is a suitable expression for numerical evaluation of the functions. For negative integers m, the hypergeometric function in (13.60) breaks down. In that case we take instead

$$P_\sigma^m(x) = (-1)^m \frac{\Gamma(\sigma+m+1)}{\Gamma(\sigma-m+1)} P_\sigma^{-m}(x), \tag{13.61}$$

according to which P_σ^m is simply a multiple of $P_\sigma^{-m}(x)$.

We can also obtain the associated Legendre functions of the second kind $Q_\sigma^m(x)$ in terms of the hypergeometric function, but the formulas get considerably more complicated. Fortunately, it is not necessary. Instead, we take advantage of the following relation,

$$P_\sigma^m(-x) = P_\sigma^m(x) \cos[\pi(\sigma+m)] - \frac{2}{\pi} Q_\sigma^m(x) \sin[\pi(\sigma+m)]$$

$$(0 < x < 1). \tag{13.62}$$

Since σ is not an integer, but m is, the coefficient for Q_σ^m does not vanish, and hence $Q_\sigma^m(x)$ can be written as a linear combination of $P_\sigma^m(x)$ and $P_\sigma^m(-x)$. The functions $P_\sigma^m(x)$ and $P_\sigma^m(-x)$ are therefore two linearly independent solutions of (13.54), and hence we have the complete solution.

When σ is complex, the corresponding Legendre functions are also complex, and we write

$$P_\sigma^{-m}(x) = R_\sigma^m(\theta) + i S_\sigma^m(\theta) \quad (x = \cos\theta). \tag{13.63}$$

We shall find it convenient to resolve, not σ itself, but rather

$\sigma(\sigma+1)$ into its real and imaginary parts, viz.

$$\sigma(\sigma+1) = \xi + i\eta. \tag{13.64}$$

Due to the many parameters (ξ, η, m, θ), the functions R_σ^m and S_σ^m are not found in tables of mathematical functions, at least not in sufficient detail for our purpose. They are, however, easily computed using (13.56), (13.60) and (13.64). Thus, if the nth term of $F(\alpha, \beta, \gamma, x)$ is $a_n + ib_n$, we find

$$\frac{a_n + ib_n}{a_{n-1} + ib_{n-1}} = \frac{(\alpha+n)(\beta+n)x}{(n+1)(\gamma+n)} = \frac{n^2 + n(\alpha+\beta) + \alpha\beta}{(n+1)(n+\gamma)} x \tag{13.65}$$

and hence, when we replace $\alpha\beta$ by $-\sigma(\sigma+1)$, i.e., by $-\xi - i\eta$, $\alpha + \beta$ by 1, γ by $m+1$, and x by $\frac{1}{2}(1-x)$, we get

$$R_\sigma^m(\theta) = \frac{1}{m!} \tan^m \tfrac{1}{2}\theta \left(1 + \sum_{n=0}^\infty a_n\right),$$

$$S_\sigma^m(\theta) = \frac{1}{m!} \tan^m \tfrac{1}{2}\theta \left(\sum_{n=0}^\infty b_n\right), \tag{13.66}$$

where

$$a_0 = -\frac{\xi(1-x)}{2(m+1)}, \qquad b_0 = -\frac{\eta(1-x)}{2(m+1)} \tag{13.67}$$

and

$$a_n = [(n^2 + n - \xi)a_{n-1} + \eta b_{n-1}] \frac{1-x}{2(n+1)(n+m+1)}$$

$$b_n = [(n^2 + n - \xi)b_{n-1} - \eta a_{n-1}] \frac{1-x}{2(n+1)(n+m+1)} \tag{13.68}$$

$(x = \cos\theta)$.

The series are easily evaluated with the help of a computer or a programmable pocket calculator. For a wide range of values of practical applicability, the number of terms needed may be comparatively large, say 30–100, and formerly this did require a considerable computational effort, but not today. For values of θ close to π, however, the convergence is slow. The series diverge for $\theta = \pi$.

The functions R_σ^m and S_σ^m have an oscillatory character, and their amplitudes increase exponentially. Fig. 34(a) and (b) illustrate the behaviour of the first three of them for a thin shell ($h/R = 0.01$, $\nu = 0.3$). In the diagrams they have been scaled to the same maximum amplitude found in the range $0 \leq \theta \leq 90°$. In this scale they are indistinguishable from the abscissa for $\theta < 60°$. Fig. 34(c) shows the dependence on the relative thickness h/R of one of them, R_σ^2. Clearly, for thin spherical shells, the bending region is a narrow zone at the boundary. Furthermore, the width of the zone decreases as the shell gets thinner.

Since the real and imaginary parts of any solution to (13.53) separately satisfy the 'full' differential equation (13.29), four linearly independent solutions to (13.29) are the real and imaginary parts of $P_\sigma^{-1}(x)$ and $P_\sigma^{-1}(-x)$. The remaining are the two solutions of (13.44), already found. Since $-x = -\cos\theta = \cos(\pi - \theta)$, the complete integral can be written as

$$\Xi_m = [A_1 R_\sigma^m(\theta) + A_2 S_\sigma^m(\theta) + A_3(m + \cos\theta)\tan^m \tfrac{1}{2}\theta$$
$$+ A_5 R_\sigma^m(\pi - \theta) + A_6 S_\sigma^m(\pi - \theta) + A_7(m - \cos\theta)\cot^m \tfrac{1}{2}\theta]\cos m\phi$$
(13.69)

for the mth mode. Here, A_1, A_2, A_3, A_5, A_6, A_7 are six arbitrary coefficients.

With the potential Ξ_m in hands, we may proceed to find all relevant quantities. For that purpose we need the derivatives of the functions R_σ^m and S_σ^m.

Taking the derivative of (13.59), we find

$$(1 - x^2)\frac{d}{dx} P_\sigma^{-m} = -\sqrt{1 - x^2}\, P_\sigma^{-m+1} + mx P_\sigma^{-m}, \qquad (13.70)$$

which with (13.64) yields

$$\frac{d}{d\theta} R_\sigma^m = R_\sigma^{m-1} - m\cot\theta\, R_\sigma^m,$$
$$\frac{d}{d\theta} S_\sigma^m = S_\sigma^{m-1} - m\cot\theta\, S_\sigma^m. \qquad (13.71)$$

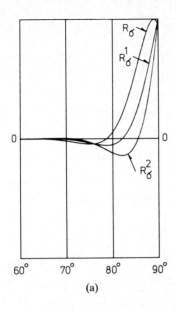
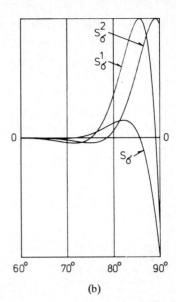
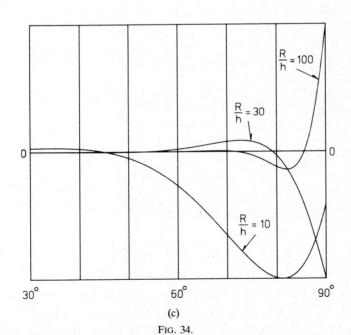

Fig. 34.

Also, since $(R_\sigma^m + iS_\sigma^m)\cos m\phi$ satisfies the differential equation

$$\left(\Delta + \frac{\xi + i\eta}{R^2}\right)(R_\sigma^m + iS_\sigma^m)\cos m\phi = 0,$$

we find the Laplacian of the functions to be

$$\Delta(R_\sigma^m \cos m\phi) = -\frac{1}{R^2}(\xi R_\sigma^m - \eta S_\sigma^m)\cos m\phi,$$

$$\Delta(S_\sigma^m \cos m\phi) = -\frac{1}{R^2}(\xi S_\sigma^m + \eta R_\sigma^m)\cos m\phi,$$
(13.72)

and from these relations and (13.71) we find

$$R_\sigma^{m-2} = 2(m-1)\cot\theta\, R_\sigma^{m-1} - (\xi - m^2 + m)R_\sigma^m + \eta S_\sigma^m,$$
$$S_\sigma^{m-2} = 2(m-1)\cot\theta\, S_\sigma^{m-1} - (\xi - m^2 + m)S_\sigma^m - \eta R_\sigma^m.$$
(13.73)

Substituting (13.69) into (13.27) and utilizing the relations (13.72), we find the complete solution

$$w = -\frac{1-\nu^2}{kR^2}[A_1 R_\sigma^m(\theta) + A_2 S_\sigma^m(\theta) + A_5 R_\sigma^m(\pi - \theta) + A_6 S_\sigma^m(\pi - \theta)$$

$$+ A_4(m + \cos\theta)\tan^m \tfrac{1}{2}\theta + A_8(m - \cos\theta)\cot^m \tfrac{1}{2}\theta]\cos m\phi,$$
(13.74)

$$\Phi = \frac{1}{R^2}[(\nu A_1 + 2\rho^2 A_2)R_\sigma^m(\theta) + (\nu A_2 - 2\rho^2 A_1)S_\sigma^m(\theta)$$

$$+ (\nu A_5 + 2\rho^2 A_6)R_\sigma^m(\pi - \theta) + (\nu A_6 - 2\rho^2 A_5)S_\sigma^m(\pi - \theta)$$

$$- (1-\nu)A_3(m + \cos\theta)\tan^m \tfrac{1}{2}\theta$$

$$- (1-\nu)A_7(m - \cos\theta)\cot^m \tfrac{1}{2}\theta]\cos m\phi.$$

The additional solution w^* according to (13.27) is included in the terms with the coefficients A_4 and A_8. The eight arbitrary coefficients ensures the possibility of satisfying four boundary conditions at two boundaries (latitudes) each. If the north-pole belongs to the shell, the coefficients A_5, A_6, A_7 and A_8 must be zero, since R_σ^m and S_σ^m are singular at π and $\cot\frac{1}{2}\theta$ at 0. The coefficients left at our disposal are sufficient in number to make it possible to satisfy four conditions at the (single) boundary. Similarly, if the south-pole is included, A_1, A_2, A_3 and A_4 must vanish. If the shell is complete, all coefficients are zero, but this result is trivial, merely stating the fact that an unloaded shell is undeformed.

We can now determine the bending moment and the boundary forces by substituting w and Φ from (13.74) into (13.34)–(13.37). In doing so we take advantage of the relations (13.72) that will permit us to express the second and third derivatives of the functions R and S in terms of the functions themselves, and their first derivative only. Consequently, only orders m and $m-1$ of the functions R and S will appear. Let us for the sake of simplicity restrict ourselves to the case of one boundary only, taking $A_5 = A_6 = A_7 = A_8 = 0$.

Then

$$M_B = \frac{Eh}{R^2}\left(\sum_{i=1}^4 c_{Mi}A_i\right)\cos m\phi,$$

$$Q = \frac{Eh}{R^3}\left(\sum_{i=1}^4 c_{Qi}A_i\right)\cos m\phi,$$

$$N = \frac{Eh}{R^3}\left(\sum_{i=1}^4 c_{Ni}A_i\right)\cos m\phi,$$

$$S = \frac{Eh}{R^3}\left(\sum_{i=1}^4 c_{Si}A_i\right)\sin m\phi,$$

(13.75)

where the functions $c_{M1}, \ldots, c_{S4}$ are given by

$$c_{M1} = \left[(1-\nu)m\left(1 - \frac{m+1}{\sin^2\theta}\right) - \nu\right]R_\sigma^m(\theta) - 2\rho^2 S_\sigma^m(\theta)$$
$$+ (1-\nu)\cot\theta\, R_\sigma^{m-1}(\theta),$$ (13.76a)

$$c_{M2} = \left[(1-\nu)m\left(1 - \frac{m+1}{\sin^2\theta}\right) - \nu\right]S_\sigma^m(\theta) + 2\rho^2 R_\sigma^m(\theta)$$
$$+ (1-\nu)\cot\theta\, S_\sigma^{m-1}(\theta),$$ (13.76b)

$$c_{M3} = 0, \tag{13.76c}$$

$$c_{M4} = (1-\nu)\frac{m(1-m^2)}{\sin^2\theta}\tan^m\tfrac{1}{2}\theta\ ; \tag{13.76d}$$

$$c_{Q1} = m\left[\frac{m(m+1)(1-\nu)}{\sin^2\theta} - 1\right]\cot\theta\, R_\sigma^m(\theta)$$

$$- 2\rho^2 m \cot\theta\, S_\sigma^m(\theta) + 2\rho^2 S_\sigma^{m-1}(\theta)$$

$$- \left[\frac{m^2(1-\nu)}{\sin^2\theta} - 1\right]R_\sigma^{m-1}(\theta), \tag{13.77a}$$

$$c_{Q2} = m\left[\frac{m(m+1)(1-\nu)}{\sin^2\theta} - 1\right]\cot\theta\, S_\sigma^m(\theta)$$

$$+ 2\rho^2 m \cot\theta\, R_\sigma^m(\theta) - 2\rho^2 R_\sigma^{m-1}(\theta)$$

$$- \left[\frac{m^2(1-\nu)}{\sin^2\theta} - 1\right]S_\sigma^{m-1}(\theta), \tag{13.77b}$$

$$c_{Q3} = 0, \tag{13.77c}$$

$$c_{Q4} = (1-\nu)\frac{m^2(1-m^2)}{\sin^3\theta}\tan^m\tfrac{1}{2}\theta\ ; \tag{13.77d}$$

$$c_{N1} = (1+m)\left(1 - \frac{m}{\sin^2\theta}\right)[\nu R_\sigma^m(\theta) + 2\rho^2 S_\sigma^m(\theta)]$$

$$+ \cot\theta[\nu R_\sigma^{m-1}(\theta) + 2\rho^2 S_\sigma^{m-1}(\theta)]\ . \tag{13.78a}$$

$$c_{N2} = (1+m)\left(1 - \frac{m}{\sin^2\theta}\right)[\nu S_\sigma^m(\theta) - 2\rho^2 R_\sigma^m(\theta)]$$

$$+ \cot\theta\,[\nu S_\sigma^{m-1}(\theta) - 2\rho^2 R_\sigma^{m-1}(\theta)], \tag{13.78b}$$

$$c_{N3} = (1-\nu)\frac{m(m^2-1)}{\sin^2\theta}\tan^m\tfrac{1}{2}\theta, \tag{13.78c}$$

$$c_{N4} = 0\ ; \tag{13.78d}$$

$$c_{S1} = -\frac{m(m+1)}{\sin\theta}\cot\theta\,[\nu R_\sigma^m(\theta) + 2\rho^2 S_\sigma^m(\theta)]$$

$$+ \frac{m}{\sin\theta}[\nu R_\sigma^{m-1}(\theta) + 2\rho^2 S_\sigma^{m-1}(\theta)], \tag{13.79a}$$

$$c_{S2} = -\frac{m(m+1)}{\sin\theta}\cot\theta\,[\nu S_\sigma^m(\theta) - 2\rho^2 R_\sigma^m(\theta)]$$

$$+ \frac{m}{\sin\theta}[\nu S_\sigma^{m-1}(\theta) - 2\rho^2 R_\sigma^{m-1}(\theta)], \tag{13.79b}$$

$$c_{S3} = (1-\nu)\frac{m(1-m^2)}{\sin^2\theta}\tan^m\tfrac{1}{2}\theta, \tag{13.79c}$$

$$c_{S4} = 0. \tag{13.79d}$$

Although there are many formulas, and although some of them are rather lengthy, the main work is done, and the application of the result requires in fact very little work. We shall illustrate that by an example.

Let us consider a thin hemispherical shell ($h/R = 0.01$, $\nu = 0.3$) loaded at the equatorial boundary by a radial load

$$P\cos 2\phi,$$

where P is constant. Thus we have $m = 2$ and we need the functions R_σ^2, R_σ^1, S_σ^2 and S_σ^1 evaluated at $\theta = \tfrac{1}{2}\pi$ for $\sigma = 12.854$. Using the series (13.66) we find

$$R_\sigma^2(\tfrac{1}{2}\pi) = 133\,740, \qquad R_\sigma^1(\tfrac{1}{2}\pi) = 2\,852\,140,$$

$$S_\sigma^2(\tfrac{1}{2}\pi) = 87\,972, \qquad S_\sigma^1(\tfrac{1}{2}\pi) = -577\,552.$$

By taking the shell to cover the northern hemisphere, we get $A_5 = A_6 = A_7 = A_8 = 0$. It is now an easy matter to evaluate the functions (13.76)–(13.79). This leads us to the following system of equations for the boundary conditions,

$$\begin{pmatrix} -8.73325\times 10^7 & 1.32507\times 10^8 & 4.2 & 0 \\ -3.79998\times 10^8 & -1.88535\times 10^9 & -4.2 & 0 \\ -2.94853\times 10^7 & 4.39226\times 10^7 & 0 & -4.2 \\ -1.95988\times 10^8 & -9.41461\times 10^7 & 0 & -8.4 \end{pmatrix}\begin{pmatrix} A_1 \\ A_2 \\ A_3 \\ A_4 \end{pmatrix}$$

$$= \left(0, 0, 0, \frac{PR^3}{D}\right)$$

with the solution

$$\begin{pmatrix} A_1 \\ A_2 \\ A_3 \\ A_4 \end{pmatrix} = \frac{PR^3}{D} \begin{pmatrix} 0.72776 \times 10^{-8} \\ -0.19403 \times 10^{-8} \\ 0.21254 \\ -0.071382 \end{pmatrix}.$$

The problem is now solved, and we can for instance find the radial displacement w under the load by substituting into (13.74). We find

$$w = 0.1420 \frac{PR^3}{D} \cos 2\phi.$$

6. Vibrations of spherical shells

When the shell performs free, harmonic vibrations of small amplitude, the state is governed by the equations of equilibrium (13.7)–(13.8) with F^β and p given by the d'Alembert forces

$$F^\beta = \omega^2 \gamma h v^\beta, \qquad p = \omega^2 \gamma h w, \tag{13.80}$$

where ω is the angular frequency of the vibrations, and γ the mass density of the material.

Equations (13.7) can no longer be solved by the stress function Φ, since we have no particular integral $F^{\alpha\beta}$. We shall therefore use an alternative formulation of these equations, namely in terms of the displacements. With the help of Hooke's law we get

$$\tilde{N}_{\alpha\beta} = \frac{Eh}{1-\nu^2}\left[(1-\nu)\left(\tfrac{1}{2}D_\alpha v_\beta + \tfrac{1}{2}D_\beta v_\alpha + \frac{1}{R}a_{\alpha\beta}w\right) + \nu a_{\alpha\beta}\left(D_\gamma v^\gamma + \frac{2w}{R}\right)\right].$$

$$\tag{13.81}$$

When (13.81), (13.9) and (13.80) are substituted into (13.7) and (13.8), we find

$$\tfrac{1}{2}(1-\nu)\left(\Delta + \frac{1}{R^2}\right)v_\beta + \tfrac{1}{2}(1+\nu)D_\beta D_\gamma v^\gamma + \frac{1+\nu}{R}D_\beta w + \lambda v_\beta = 0$$

$$\tag{13.82}$$

and

$$\left(\Delta + \frac{2}{R^2}\right)\left(\Delta + \frac{1+\nu}{R^2}\right)w + \frac{1+\nu}{kR^3}\left(D_\alpha v^\alpha + \frac{2}{R}w\right) - \frac{\lambda}{kR^2}w = 0 \tag{13.83}$$

respectively, where

$$\lambda = \omega^2(1-\nu^2)\frac{\gamma}{E}. \tag{13.84}$$

Since any vector field may be written as the sum of an irrotational and a solenoidal part, we write

$$v_\beta = D_\beta \Psi + \varepsilon_{\gamma\beta} D^\gamma \chi, \tag{13.85}$$

where Ψ and χ are two scalar functions. Taking the covariant derivative and contracting, we get

$$D_\beta v^\beta = \Delta \Psi. \tag{13.86}$$

If instead of contracting we multiply by $\varepsilon^{\alpha\beta}$, we find

$$\varepsilon^{\alpha\beta} D_\alpha v_\beta = \Delta \chi = \Theta, \tag{13.87}$$

where Θ is the *local average rotation* (see (3.30)).

By substituting (13.85) into (13.82) and observing the rules for interchanging the order of covariant differentiation, we get

$$D_\beta\left(\Delta\Psi + \frac{1-\nu}{R^2}\Psi + \lambda\Psi + \frac{1+\nu}{R}w\right)$$
$$+ \tfrac{1}{2}(1-\nu)\varepsilon_{\gamma\beta}D^\gamma\left(\Delta\chi + \frac{2}{R^2}\chi + \frac{2\lambda}{1-\nu}\chi\right) = 0. \tag{13.88}$$

When multiplying by the operator $a^{\alpha\beta}D_\alpha$, we get

$$\Delta\left(\Delta\Psi + \frac{1-\nu}{R^2}\Psi + \lambda\Psi + \frac{1+\nu}{R}w\right) = 0 \tag{13.89}$$

and hence

$$\left(\Delta + \frac{1-\nu}{R^2} + \lambda\right)\Psi + \frac{1+\nu}{R} w = \kappa, \tag{13.90}$$

where κ is a harmonic function.

If instead we multiply (13.88) by $\varepsilon^{\alpha\beta} D_\alpha$, we get

$$\Delta\left(\Delta + \frac{2}{R^2} + \frac{2\lambda}{1-\nu}\right)\chi = 0 \tag{13.91}$$

and thus

$$\left(\Delta + \frac{2}{R^2} + \frac{2\lambda}{1-\nu}\right)\chi = \mu, \tag{13.92}$$

where μ is a harmonic function.

By substituting (13.90) and (13.92) back into (13.88), we find that κ and μ must satisfy the following equations,

$$D_\beta \kappa + \frac{1-\nu}{2} \varepsilon_{\gamma\beta} D^\gamma \mu = 0, \tag{13.93}$$

which, when compared with the Cauchy–Riemann equations for the real and imaginary part of an analytic function, reveal that κ and $\frac{1}{2}(1-\nu)\mu$ are conjugate harmonic.

A particular integral of (13.90) is

$$\Psi_0 = \frac{\kappa}{(1-\nu)/R^2 + \lambda} \tag{13.94}$$

and, of (13.92),

$$\chi_0 = \frac{\mu}{2/R^2 + 2\lambda/(1-\nu)}. \tag{13.95}$$

Substituting Ψ_0 and χ_0 into (13.85) we get

$$v_\beta = \frac{D_\beta \kappa}{(1-\nu)/R^2 + \lambda} + \varepsilon_{\gamma\beta} \frac{D^\gamma \mu}{2/R^2 + 2\lambda/(1-\nu)} = 0$$

when (13.93) is utilized. Thus, conjugate harmonic functions do not contribute to the displacements, and we may therefore, without loss of generality, take $\kappa = \mu = 0$.

The first two equations of equilibrium can therefore be written in terms of the rotation and dilatation as

$$\left(\Delta + \frac{2}{R^2} + \frac{2\lambda}{1-\nu}\right)\chi = 0 \qquad (13.96)$$

and

$$\left(\Delta + \frac{1-\nu}{R^2} + \lambda\right)\Psi + \frac{1+\nu}{R} w = 0 \qquad (13.97)$$

respectively. The third equation is obtained when (13.85) is substituted into (13.83), viz.

$$\left(\Delta + \frac{2}{R^2}\right)\left(\Delta + \frac{1+\nu}{R^2}\right)w + \frac{1+\nu}{kR^3}\left(\Delta\Psi + \frac{2}{R}w\right) - \frac{\lambda}{kR^2} w = 0. \qquad (13.98)$$

The equations of equilibrium have been transformed into three differential equations for the scalar functions χ, Ψ and w. *The only differential operator appearing in the equations is the (invariant) Laplacian operator* Δ.

The equations are partly decoupled, since only χ appears in the first one, and since it only appears there. But we shall see that in general the system is coupled through the boundary conditions.

Turning to these, we find that formulas (13.36) and (13.37) for M_B and Q are still applicable, while (13.34) and (13.35) for N and S must be replaced by new ones in terms of χ, Ψ and w.

From (13.34), (13.81) and (13.85) we get

$$N = \frac{Eh}{1-\nu^2}\left[(1-\nu)(n^\alpha n^\beta D_\alpha D_\beta \Psi - n^\alpha t^\beta D_\alpha D_\beta \chi) + \nu\Delta\Psi + \frac{1+\nu}{R} w\right] \qquad (13.99)$$

and similarly

$$S = \frac{Eh}{2(1+\nu)} [2n^\alpha t^\beta D_\alpha D_\beta \Psi + (n^\alpha n^\beta - t^\alpha t^\beta) D_\alpha D_\beta \chi]. \quad (13.100)$$

The four boundary conditions for M_B, Q, N and S, and the eighth order system (13.96)–(13.98) constitute again a well-posed mathematical problem.

The static boundary conditions may be replaced by kinematic conditions. Thus, at the boundary, we may prescribe

either M_B or $\partial w/\partial n$, and
either Q or w, and
either N or $u = \partial \Psi/\partial n - \partial \chi/\partial s$, and
either S or $v = \partial \Psi/\partial s + \partial \chi/\partial n$

or any linear combination of the two.

Whether we have static or kinematic conditions at the boundary, we see that, in general, there will be a coupling between the function χ and the other two functions Ψ and w through the boundary conditions. However, when the shell is complete, there are no boundary conditions and therefore no such coupling. Then the solution χ of (13.96) is independent of Ψ and w.

Let us now turn to the solution of the mathematical problem and begin with the simple case of the complete spherical shells in spherical coordinates.

As we already know, the functions

$$P_\sigma^{-m}(\cos\theta) \begin{cases} \sin m\phi \\ \cos m\phi \end{cases}, \quad P_\sigma^{-m}(\cos(\pi-\theta)) \begin{cases} \sin m\phi \\ \cos m\phi \end{cases} \quad (13.101)$$

satisfy the differential equation

$$\left[\Delta + \frac{\sigma(\sigma+1)}{R^2}\right] y = 0 \quad (13.102)$$

and hence also (13.96), provided that σ is real and satisfies the relation

$$\frac{\sigma(\sigma+1)}{R^2} = \frac{2}{R^2} + \frac{2\lambda}{1-\nu}. \tag{13.103}$$

If σ is not integral, all functions (13.101) are singular, either at $\theta = \pi$ or at $\theta = 0$. Conversely, when $\sigma = n$ is an integer, they are regular everywhere on the sphere.

The functions

$$P_n^m(\cos\theta)\cos m\phi, \qquad P_n^m(\cos\theta)\sin m\phi \tag{13.104}$$

with integral values of n and m are called *surface harmonics*, *tesseral harmonics* for $m < n$ and *sectoral harmonics* for $m = n$.[7] These functions are periodic with respect to the angles θ and ϕ with periods π and 2π respectively. They are single-valued and continuous everywhere on the sphere and are therefore the only solutions of (13.96) when the shell is complete.

Solving (13.103) for ω^2 we find

$$\omega^2 = (n-1)(n+2)\frac{E}{2(1+\nu)\gamma R^2} \tag{13.105}$$

for the complete spherical shell.

We observe that the fundamental modes of vibration fall into two classes. In the modes of the *first class*, the motion at every point is wholly tangential, and we have $w = \Psi = 0$ everywhere. The mode is a surface harmonic, and the frequency is independent of the thickness of the shell.

In the modes of the *second class*, the motion is partly radial and partly tangential. We shall return to this class presently.

Let us first investigate the axisymmetrical modes of the first class. These are determined by the complete solution

$$\chi = AP_n(\cos\theta), \qquad \Psi = 0, \qquad w = 0, \tag{13.106}$$

from which we find the displacements according to (13.85),

---

[7] For $m > n$ the functions P_n^m vanish identically.

$$v_1 = 0, \qquad v_2 = \varepsilon_{12} a^{11} \chi_{,1} = -A \sin^2 \theta \, \frac{dP_n}{dx} \qquad (13.107)$$

or, by (13.48),

$$u = 0, \qquad v = -\frac{A}{R} \sin \theta \, \frac{dP_n}{dx}. \qquad (13.108)$$

For the first few values of n the solutions are described below.

$n = 1$: $P_1 = x$, $v = -(A/R) \sin \theta$. The frequency $\omega = 0$, and the motion is a rotation of the shell as a rigid body.

$n = 2$: $P_2 = \frac{1}{2}(3x^2 - 1)$, $v = -(3A/2R) \sin 2\theta$. The tangential displacement v is of opposite directions on the two hemispheres. The equator $\theta = 90°$ is a nodal circle.

$n = 3$: $P_3 = \frac{1}{2}(5x^3 - 3x)$, $v = -(6A/R) \sin \theta (1 - \frac{5}{4} \sin^2 \theta)$. There are two polar zones rotating in the same direction and an equatorial zone rotating in the opposite direction. The nodal circles are at the co-latitudes $\theta = 63°.4$ and $116°.6$.

$n = 4$: $P_4 = \frac{1}{8}(35x^4 - 30x^2 + 3)$, $v = -(5A/4R) \sin 2\theta \, (7 \cos^2 \theta - 3)$. There are four zones, separated by the nodal circles $\theta = 49°.1$, $90°$, $130°.9$; consecutive zones moving in opposite directions.

...

For each integer $n > 0$ there is one axisymmetric mode, and n non-axisymmetric modes, given by $m = 1, 2, \ldots, n$, corresponding to the surface harmonics of order m. The frequency ω, however, is independent of m, the wave-number in peripheral direction. This seemingly paradoxical situation can readily be explained as a consequence of spherical symmetry.

For instance, let us take the axisymmetric mode

$$\chi = P_2 = \tfrac{1}{2}(3 \cos^2 \theta - 1) \qquad (13.109)$$

and tilt the axis $90°$, rotating the mode around the x^1-axis. This will correspond to a change of coordinates (θ, ϕ) to (θ_1, ϕ_1), where

$$\cos \theta_1 = \sin \theta \sin \phi, \qquad \sin \theta_1 \sin \phi_1 = -\cos \theta_1,$$

$$\sin \theta_1 \cos \phi_1 = \sin \theta \cos \phi.$$

Clearly, the mode $\chi_1 = \frac{1}{2}(3\cos^2\theta_1 - 1)$ is the same as χ in (13.109), only tilted 90°. But

$$\chi_1 = \frac{1}{2}(3\sin^2\theta \sin^2\phi - 1) = -\frac{3}{4}(1 - \cos^2\theta)\cos 2\phi - \frac{1}{4}(3\cos^2\theta - 1)$$

or

$$\chi_1 = -6P_2^{-2}(\cos\theta)\cos 2\phi - \frac{1}{2}P_2(\cos\theta).$$

The mode $P_2^{-2}(\cos\theta)\cos 2\phi$ can therefore be obtained by superimposing two axisymmetric modes $P_2(\cos\theta)$, the one tilted 90° relative the other, and with amplitudes in the ratio 1:2. In fact, all higher modes of a given degree n can be obtained by superposition of a number of axisymmetric modes of degree n, with properly chosen axes and amplitudes. This explains why the frequency depends on n only, not on m.

We shall now extend our analysis of vibrations of the first class to a spherical shell, that is *not* complete.

Suppose then that the middle-surface is bounded by a circle of co-latitude $\theta = \alpha$, and that the pole $\theta = 0$ belongs to the middle-surface.

For axisymmetric modes of vibration, the solution to (13.96) can be written as

$$\chi = AP_\sigma(\cos\theta), \tag{13.110}$$

where σ is not integral. There is no contribution from the second solution, obtained by replacing θ by $\pi - \theta$, since that function is singular at $\theta = 0$. The constant A is the amplitude of the mode.

It is readily checked that $M_B = Q = N = 0$ at the boundary and that the remaining condition for a free boundary, $S = 0$, is equivalent to

$$\frac{d^2 P_\sigma}{d\theta^2} - \cot\theta \frac{dP_\sigma}{d\theta} = 0 \quad (\theta = \alpha).$$

But

$$\frac{d^2 P_\sigma}{d\theta^2} = R^2 \Delta P_\sigma - \cot\theta \frac{dP_\sigma}{d\theta} = -\sigma(\sigma+1)P_\sigma - \cot\theta \frac{dP_\sigma}{d\theta}.$$

Also

$$\frac{dP_\sigma}{d\theta} = -\sin\theta \frac{dP_\sigma}{dx} = P^1 = -\sigma(\sigma+1)P_\sigma^{-1}$$

and hence the condition $S = 0$ can be written as

$$P_\sigma(\cos\alpha) = 2\cot\alpha\, P_\sigma^{-1}(\cos\alpha). \tag{13.111}$$

Here, α is a given number and the roots σ of this transcendental equation determine the frequencies according to (13.105) with n replaced by σ. For the range $30° \leq \alpha \leq 150°$ the lowest root is shown in Fig. 35. The result was obtained numerically by a simple trial-and-error method.

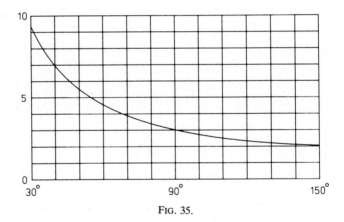

Fig. 35.

The vibrational modes of the second class are solutions of the system (13.97)–(13.98). When the shell is complete, w and Ψ are surface harmonics. Taking

$$w = AP_n^m(\cos\theta)\cos m\phi, \qquad \Psi = BRP_n^m(\cos\theta)\cos m\phi, \tag{13.112}$$

where m and n are integral, we find, after substitution, the following system of linear, homogeneous equations for A and B,

$$(1+\nu)A + [1-\nu+\lambda R^2 - n(n+1)]B = 0,$$

$$\left\{[2-n(n+1)][1+\nu-n(n+1)] + \frac{2(1+\nu)}{k} - \frac{\lambda R^2}{k}\right\} A \quad (13.113)$$

$$-\frac{1+\nu}{k} n(n+1)B = 0.$$

When $n = 0$, the second equation yields

$$\lambda = \frac{2(1+\nu)}{R^2}(1+k)$$

and, after omitting the small term k, we get, from (13.84),

$$\omega^2 = \frac{2E}{(1-\nu)\gamma R^2}. \quad (13.114)$$

The mode is $P_0 = 1$, corresponding to a uniform extension-compression of the shell in a wholly radial motion.

In the general case $(n > 0)$ the frequency is determined from the condition that the determinant of (13.113) vanishes. Working it out, we find[8]

$$\lambda^2 R^4 - \lambda R^2(n^2+n+1+3\nu) + (1-\nu^2)(n^2+n-2) = 0. \quad (13.115)$$

For each value of n there are two roots to this equation, and the corresponding modes are of quite different character.

Let us first look at the axisymmetrical modes, characterized by $m = 0$ and $v = 0$. Using (13.85) and (13.48) we find

$$u = -B \sin \theta \frac{dP_n(x)}{dx}, \quad v = 0. \quad (13.116)$$

---

[8] Here terms of relative order k are omitted, which is permissible only if n^4 can be neglected in comparison with $1/k$, i.e., for thin shells and low numbers n. In other cases these terms must be retained.

For $n = 1, 2$ the result is given below.

$n = 1$: The roots are $\lambda = 0$ and $3(1+\nu)/R^2$. The zero root corresponds to the ratio $A:B = 1$, and we get

$$w = A \cos \theta, \qquad u = -A \sin \theta,$$

which is a translation of the shell as a rigid body parallel to its axis.

In the other mode, $A:B = -2$, and hence

$$w = A \cos \theta, \qquad u = \tfrac{1}{2} A \sin \theta.$$

The poles move in the same direction along the axis of the sphere, leaving the polar diameter unchanged. The equator moves also parallel to the axis, but in opposite direction and with half the amplitude. The two hemispheres stretch and contract in opposite phase, so that one stretches while the other contracts; the centre of gravity being at rest.

$n = 2$: The roots λR^2 for different values of ν are given by the following table,

$\nu = 0.0$	$\nu = 0.1$	$\nu = 0.2$	$\nu = 0.3$	$\nu = 0.4$	$\nu = 0.5$
0.6277	0.5902	0.5442	0.4913	0.4326	0.3690
6.3723	6.7098	7.0558	7.4087	7.7674	8.1310

The ratio $A:B$ is positive for the lower root and negative for the higher. We find

$$w = \tfrac{1}{2} A (3 \cos^2 \theta - 1), \qquad u = -\tfrac{3}{2} B \sin 2\theta.$$

In this mode, the poles move in opposite directions. When the polar diameter elongates, the equator contracts and vice versa. In the mode corresponding to the lower root, the tangential motion u is towards the poles when the polar diameter is lengthening. The reverse is the case in the other mode. This explains the great difference in frequency.

For complete spherical shells the frequencies are independent of the thickness of the shell and of the wave number m. This follows from (13.105) and (13.115).

The vibrations of complete spherical shells were first discussed by LAMB,[9] who solved the problem by adapting his earlier derived results on the vibrations of elastic spheres to the case of a shell bounded by two concentric spherical surfaces.

Of course, the vibrations of an open spherical shell are in no way analogous to the vibrations of a closed shell. By the theorem due to JELLETT earlier referred to,[10] any deformation of a closed surface of positive Gaussian curvature must involve a straining of at least some part of the surface. If such a shell is sufficiently thin, the energy due to bending can always be neglected in comparison with the energy due to stretching. For an open shell the matter is very different. We have already seen that such a shell can deform inextensionally, and therefore bending may—and indeed most often will—play an important rôle. Lord RAYLEIGH[11] concluded from physical reasoning that the middle-surface of a vibrating shell remains unstretched and determined the frequencies of an open, vibrating, spherical shell using this condition. But as a matter of fact, equation (13.89) shows that if the deformation is inextensional, and hence $\Delta \Psi = 0$, then Δw must vanish everywhere. On the other hand, as we have seen, inextensional deformations are characterized by the condition $(\Delta + 2/R^2)w = 0$, and these two conditions cannot be reconciled, unless R is infinite, or w vanishes identically. Therefore, strictly speaking, unlike flat plates, spherical shells *cannot* perform free, inextensional vibrations, irrespective of the boundary conditions. But even more serious is the fact that Lord RAYLEIGH's theory cannot satisfy obvious physical requirements at the boundary.

To determine the frequencies and modes of vibration of an open spherical shell, we must solve the complete system (13.96)–(13.98) taking all four boundary conditions into account. To accomplish that, we write (13.97)–(13.98) in the form

$$A_{11}w + A_{12}\Psi = 0, \qquad A_{21}w + A_{22}\Psi = 0, \qquad (13.117)$$

where the operators are given by

---

[9] LAMB, H. (1883), "On the vibrations of a spherical shell", *London Math. Soc. Proc.* **14**, p. 50.

[10] See footnote 3, this chapter.

[11] RAYLEIGH, J.W.S. (1894), *The Theory of Sound, Vol. 1*, Dover Publications, New York, 1945, p. 395 ff.

$$A_{11} = \frac{1+\nu}{R}, \qquad A_{12} = \Delta + \frac{1-\nu}{R^2} + \lambda,$$

$$A_{21} = \left(\Delta + \frac{2}{R^2}\right)\left(\Delta + \frac{1+\nu}{R^2}\right) + \frac{2(1+\nu)}{kR^4} - \frac{\lambda}{kR^2}, \qquad A_{22} = \frac{1+\nu}{kR^3}\Delta.$$

$$\tag{13.118}$$

Taking

$$w = -A_{12}\Xi, \qquad \Psi = A_{11}\Xi \tag{13.119}$$

the system (13.117) will be satisfied whenever Ξ is a solution of the equation

$$(A_{11}A_{22} - A_{12}A_{21})\Xi = 0, \tag{13.120}$$

where Ξ is a new potential function for solving the system (13.117).

This differential equation can be factorized and written in the form

$$\left(\Delta - \frac{\beta}{R^2}\right)\left(\Delta - \frac{\xi + i\eta}{R^2}\right)\left(\Delta - \frac{\xi - i\eta}{R^2}\right)\Xi = 0, \tag{13.121}$$

where β is the real root, and $\xi \pm i\eta$ the two complex conjugate roots of the following algebraic equation of third degree in Δ,

$$\Delta^3 + c_2\Delta^2 + c_1\Delta + c_0 = 0, \tag{13.122}$$

where

$$\begin{aligned}
c_2 &= 4 + \lambda R^2, \\
c_1 &= (1 - \nu^2 - \lambda R^2)/k + 5 - \nu^2 + (3+\nu)\lambda R^2, \\
c_0 &= [2(1-\nu^2) + (1+3\nu)\lambda R^2 - \lambda^2 R^4]/k \\
&\quad + 2(1-\nu^2) + 2(1+\nu)\lambda R^2.
\end{aligned} \tag{13.123}$$

We can determine the roots by *Cardan's formula*. Thus let

$$p = \tfrac{1}{3}c_1 - \tfrac{1}{9}c_2^2, \qquad q = \tfrac{1}{27}c_2^3 - \tfrac{1}{6}c_1c_2 + \tfrac{1}{2}c_0,$$
$$\hat{u} = (-q + \sqrt{p^3 + q^2})^{1/3}, \qquad \hat{v} = (-q - \sqrt{p^3 + q^2})^{1/3}. \tag{13.124}$$

Then the roots are given by

$$\beta = \hat{u} + \hat{v} - \tfrac{1}{3}c_2, \qquad \xi = -\tfrac{1}{2}(\hat{u} + \hat{v}) - \tfrac{1}{3}c_2, \qquad \eta = \tfrac{1}{2}\sqrt{3}\,(\hat{u} - \hat{v}). \tag{13.125}$$

For small values of λ, β is close to -2, ξ close to -1, and η approximately equal to $\sqrt{(1-\nu^2)/k}$. Therefore, η is of order R/h, while β and ξ are of order unity. This means that $\hat{u} + \hat{v}$ is of order 1 while $\hat{u} - \hat{v}$ is of order R/h. Due to this, the small terms in (13.123) should *not* be omitted.

With the help of (13.121) the system (13.96)–(13.98) is reduced to four second order differential equations of standard type.

In spherical coordinates the solution for the mth mode can be written in the form

$$\begin{aligned}\Xi &= [A_1 P_\sigma^{-m}(\theta) + A_2 R_\zeta^m(\theta) + A_3 S_\zeta^m(\theta) + A_5 P_\sigma^{-m}(\pi - \theta) \\ &\quad + A_6 R_\zeta^m(\pi - \theta) + A_7 S_\zeta^m(\pi - \theta)]\cos m\phi, \\ \chi &= \frac{1}{R}[A_4 P_\tau^{-m}(\theta) + A_8 P_\tau^{-m}(\pi - \theta)]\sin m\phi,\end{aligned} \tag{13.126}$$

where

$$\beta = -\sigma(\sigma + 1), \qquad \xi + i\eta = -\zeta(\zeta + 1), \qquad \gamma = -\tau(\tau + 1). \tag{13.127}$$

We find β, ξ and η from (13.125), while γ is determined by the condition that χ given by (13.126) will satisfy (13.96), i.e.,

$$\gamma = 2 + \frac{2\lambda R^2}{1 - \nu}. \tag{13.128}$$

When the middle-surface contains the pole $\theta = 0$, we must have $A_5 = A_6 = A_7 = A_8 = 0$, and the solution in this case is

$$\begin{aligned}w = \frac{1}{R^2}\{&A_1(\beta - 1 + \nu - \lambda R^2)P_\sigma^{-m}(\theta) \\ &+ [A_2(\xi - 1 + \nu - \lambda R^2) + A_3\eta]R_\zeta^m(\theta) \\ &+ [A_3(\xi - 1 + \nu - \lambda R^2) - A_2\eta]S_\zeta^m(\theta)\}\cos m\phi, \quad (13.129a)\end{aligned}$$

$$\Psi = \frac{1+\nu}{R}[A_1 P_\sigma^{-m}(\theta) + A_2 R_\zeta^m(\theta) + A_3 S_\zeta^m(\theta)] \cos m\phi,$$

(13.129b)

$$\chi = \frac{1}{R} A_4 P_\tau^{-m}(\theta) \sin m\phi.$$

(13.129c)

When the shell is bounded by two circles, neither pole belongs to the middle-surface, and the solution is obtained by adding analogous terms with coefficients $A_5, \ldots, A_8$ instead of $A_1, \ldots, A_4$ and the argument $\pi - \theta$ instead of θ for the corresponding function. For the sake of simplicity we shall confine the analysis to the case of a single boundary $\theta = \alpha$, assuming that $\theta = 0$ belongs to the shell.

At the boundary we can prescribe four conditions, which is the necessary and sufficient number of conditions for determining the coefficients $A_1, \ldots, A_4$.

The bending moment and the boundary forces are obtained, when w, Ψ and χ are substituted into (13.34)–(13.35) and (13.99)–(13.100). The normal and tangential displacements at the boundary are $u = v_\alpha n^\alpha$ and $v = v_\alpha t^\beta$, where v_α is given by (13.85).

The result can be written in the form

$$N = \frac{Eh}{R^3}\left(\sum_{i=1}^{4} c_{Ni} A_i\right) \cos m\phi, \qquad u = \frac{1}{R^2}\left(\sum_{i=1}^{4} c_{ui} A_i\right) \cos m\phi,$$

$$S = \frac{Eh}{R^3}\left(\sum_{i=1}^{4} c_{Si} A_i\right) \sin m\phi, \qquad v = \frac{1}{R^2}\left(\sum_{i=1}^{4} c_{vi} A_i\right) \sin m\phi,$$

(13.130)

$$M_B = \frac{D}{R^4}\left(\sum_{i=1}^{4} c_{Mi} A_i\right) \cos m\phi, \qquad \frac{\partial w}{\partial n} = \frac{1}{R^3}\left(\sum_{i=1}^{4} c_{\psi i} A_i\right) \cos m\phi.$$

$$Q = \frac{D}{R^5}\left(\sum_{i=1}^{4} c_{Qi} A_i\right) \cos m\phi,$$

To be able to write the coefficients more compact, we shall use the following notations,

$$a_1 = \frac{m(m+1)}{\sin^2\theta},$$

$$a_2 = (m+1)\left(\frac{m}{\sin^2\theta} - 1\right) - \frac{\lambda R^2}{1-\nu},$$

$$a_3 = (1-\nu)(m+1)\left(\frac{m}{\sin^2\theta} - 1\right) + 2 - \beta,$$

$$a_4 = \lambda R^2 + 1 - \nu - \beta,$$

$$a_5 = (1-\nu)(m+1)\left(\frac{m}{\sin^2\theta} - 1\right) + 2 - \xi,$$

$$a_6 = \lambda R^2 + 1 - \nu - \xi, \qquad (13.131)$$

$$a_7 = (1-\nu)\frac{m(m+1)}{\sin^2\theta} - 2,$$

$$a_8 = (1-\nu)\frac{m^2}{\sin^2\theta} - 2.$$

Furthermore, let

$$\begin{aligned}
\mathcal{P} &= P_\sigma^{-m}(\theta), & \mathcal{P}_1 &= P_\sigma^{-m+1}(\theta), \\
\mathcal{Q} &= P_\tau^{-m}(\theta), & \mathcal{Q}_1 &= P_\tau^{-m+1}(\theta), \\
\mathcal{R} &= R_\zeta^m(\theta), & \mathcal{R}_1 &= R_\zeta^{m-1}(\theta), \\
\mathcal{S} &= S_\zeta^m(\theta), & \mathcal{S}_1 &= S_\zeta^{m-1}(\theta).
\end{aligned} \qquad (13.132)$$

Then

$$c_{N1} = a_2 \mathcal{P} - \cot\theta\, \mathcal{P}_1, \qquad (13.133a)$$

$$c_{N2} = a_2 \mathcal{R} - \cot\theta\, \mathcal{R}_1, \qquad (13.133b)$$

$$c_{N3} = a_2 \mathcal{S} - \cot\theta\, \mathcal{S}_1, \qquad (13.133c)$$

$$c_{N4} = \frac{m}{(1+\nu)\sin\theta}\left[(m+1)\cot\theta\, \mathcal{Q} - \mathcal{Q}_1\right]; \qquad (13.133d)$$

$$c_{S1} = a_1 \cos\theta\, \mathcal{P} - \frac{m}{\sin\theta}\mathcal{P}_1, \qquad (13.134a)$$

$$c_{S2} = a_1 \cos\theta\, \mathcal{R} - \frac{m}{\sin\theta}\mathcal{R}_1, \qquad (13.134b)$$

$$c_{S3} = a_1 \cos\theta\, \mathcal{S} - \frac{m}{\sin\theta}\mathcal{S}_1, \qquad (13.134c)$$

$$c_{S4} = (a_1 - m - \tfrac{1}{2}\gamma)\frac{\mathcal{Q}}{1+\nu} - \frac{\cot\theta}{1+\nu}\mathcal{Q}_1 ; \qquad (13.134d)$$

$$\begin{aligned}
c_{M1} &= a_4[(1-\nu)\cot\theta\,\mathcal{P}_1 - a_3\mathcal{P}], \\
c_{M2} &= (\eta^2 - a_5 a_6)\mathcal{R} - \eta(a_5 + a_6)\mathcal{S} \\
&\quad + (1-\nu)\cot\theta\,(a_6\mathcal{R}_1 + \eta\mathcal{S}_1), \\
c_{M3} &= (\eta^2 - a_5 a_6)\mathcal{S} + \eta(a_5 + a_6)\mathcal{R} \\
&\quad + (1-\nu)\cot\theta\,(a_6\mathcal{S}_1 - \eta\mathcal{R}_1), \\
c_{M4} &= 0 ;
\end{aligned} \qquad (13.135)$$

$$\begin{aligned}
c_{Q1} &= m\cot\theta\,a_4(a_7+\beta)\mathcal{P} - a_4(a_8+\beta)\mathcal{P}_1, \\
c_{Q2} &= m\cot\theta\,\eta(a_7 - a_6 + \xi)\mathcal{S} + m\cot\theta\,[\eta^2 + a_6(a_7+\xi)]\mathcal{R} \\
&\quad - [\eta^2 + a_6(a_8+\xi)]\mathcal{R}_1 - \eta(a_8 - a_6 + \xi)\mathcal{S}_1, \\
c_{Q3} &= -m\cot\theta\,\eta(a_7 - a_6 + \xi)\mathcal{R} + m\cot\theta[\eta^2 + a_6(a_7+\xi)]\mathcal{S} \\
&\quad - [\eta^2 + a_6(a_8+\xi)]\mathcal{S}_1 + \eta(a_8 - a_6 + \xi)\mathcal{R}_1, \\
c_{Q4} &= 0 ;
\end{aligned} \qquad (13.136)$$

$$\begin{aligned}
c_{u1} &= (1+\nu)(\mathcal{P}_1 - m\cot\theta\,\mathcal{P}), \\
c_{u2} &= (1+\nu)(\mathcal{R}_1 - m\cot\theta\,\mathcal{R}), \\
c_{u3} &= (1+\nu)(\mathcal{S}_1 - m\cot\theta\,\mathcal{S}), \\
c_{u4} &= -\frac{m}{\sin\theta}\mathcal{Q} ;
\end{aligned} \qquad (13.137)$$

$$\begin{aligned}
c_{v1} &= -(1+\nu)\frac{m}{\sin\theta}\mathcal{P}, \\
c_{v2} &= -(1+\nu)\frac{m}{\sin\theta}\mathcal{R}, \\
c_{v3} &= -(1+\nu)\frac{m}{\sin\theta}\mathcal{S}, \\
c_{v4} &= -m\cot\theta\,\mathcal{Q} + \mathcal{Q}_1 ;
\end{aligned} \qquad (13.138)$$

$$c_{\psi 1} = a_4(m\cot\theta\,\mathcal{P} - \mathcal{P}_1), \qquad (13.139a)$$

$$c_{\psi 2} = a_6(m \cot \theta \, \mathcal{R} - \mathcal{R}_1) + m \cot \theta \, \eta \mathcal{S}, \qquad (13.139b)$$

$$c_{\psi 3} = a_6(m \cot \theta \, \mathcal{S} - \mathcal{S}_1) - m \cot \theta \, \eta \mathcal{R}, \qquad (13.139c)$$

$$c_{\psi 4} = 0. \qquad (13.139d)$$

If the boundary is free, we have $N = S = M_B = Q = 0$ at $\theta - \alpha$. The condition for a nontrivial solution leads to the following frequency equation,

$$\begin{vmatrix} c_{N1} & c_{N2} & c_{N3} & c_{N4} \\ c_{S1} & c_{S2} & c_{S3} & c_{S4} \\ c_{M1} & c_{M2} & c_{M3} & 0 \\ c_{Q1} & c_{Q2} & c_{Q3} & 0 \end{vmatrix} = 0, \qquad (13.140)$$

which may be solved by trial-and-error. From the first root λR^2 we find the lowest frequency for given values of m, h/R, ν and α. Writing the frequency ω as

$$\omega = c \frac{h}{R^2} \sqrt{\frac{G}{\gamma}}, \qquad (13.141)$$

where $G = E/2(1 + \nu)$ is the shear modulus, the coefficient c is determined by λR^2 and equation (13.84). Table 6 gives c for different values of m, h/R and α. In Fig. 36 the shape of the first mode for $m = 6$, $h/R = 0.01$, $\nu = 0.3$ and $\alpha = 60°, 90°, 120°, 150°$ is shown.

Lord RAYLEIGH's solution of the problem is based on the assumption of inextensional deformation of the middle-surface. Since we know that the true mode cannot be inextensional, we shall regard RAYLEIGH's method as approximate, which due to its nature gives an upper bound for the frequency. Following his approach we take for the mth mode the displacements (13.47) and (13.50),

$$u = -A_m \tan^m \tfrac{1}{2}\theta \sin \theta \cos m\phi \cos \omega t,$$

$$v = -A_m \tan^m \tfrac{1}{2}\theta \sin \theta \sin m\phi \cos \omega t, \qquad (13.142)$$

$$w = A_m(m + \cos \theta) \tan^m \tfrac{1}{2}\theta \cos m\phi \cos \omega t.$$

Recalling that $m = 0$ and $m = 1$ represent rigid-body motions of the shell, we may confine our considerations to integral values $m > 1$.

TABLE 6. Frequencies of an open spherical shell with $\nu = 0.3$, coefficient c.

α	m	\multicolumn{4}{c}{h/R}	RAYLEIGH
		0.04	0.02	0.01	0.005	
60°	2	3.085	3.145	3.184	3.210	3.264
	3	7.647	7.929	8.132	8.267	8.537
	4	13.659	14.350	14.901	15.284	16.050
90°	2	2.007	2.052	2.081	2.100	2.139
	3	5.356	5.577	5.724	5.819	6.012
	4	9.834	10.411	10.817	11.086	11.619
120°	2	2.341	2.429	2.485	2.521	2.596
	3	6.608	7.025	7.300	7.477	7.830
	4	12.191	13.239	13.974	14.461	15.413
150°	2	6.315	6.989	7.429	7.710	8.257
	3	16.811	19.479	21.372	22.645	25.094
	4	28.538	34.489	38.917	42.076	48.377

For a thin shell the kinetic energy is

$$E_{\text{kin}} = \tfrac{1}{2}\gamma h \iint \left[\left(\frac{\partial u}{\partial t}\right)^2 + \left(\frac{\partial v}{\partial t}\right)^2 + \left(\frac{\partial w}{\partial t}\right)^2 \right] dA,$$

from which we get

$$E_{\text{kin}} = A_m^2 \tfrac{1}{2}\pi \gamma h \omega^2 R^2 \sin^2 \omega t$$

$$\times \int_0^\alpha [2\sin^2\theta + (m + \cos\theta)^2] \tan^{2m} \tfrac{1}{2}\theta \sin\theta \, d\theta. \quad (13.143)$$

Since the displacements (13.142) produce no strain, the potential energy is due to bending only, and hence

$$E_{\text{pot}} = \tfrac{1}{2} D \iint \left[(1-\nu) \tilde{K}^\alpha_\beta K^\beta_\alpha + \nu \tilde{K}^\alpha_\alpha \tilde{K}^\beta_\beta \right] dA.$$

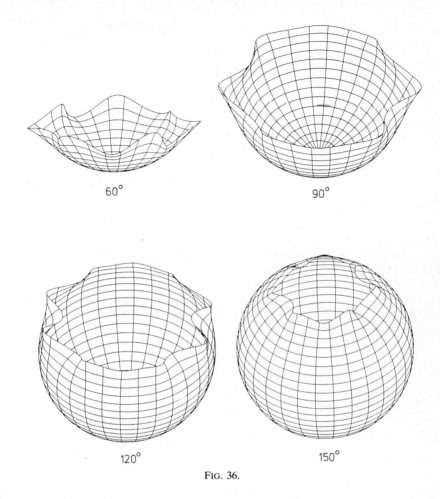

FIG. 36.

But, due to (13.20), $\tilde{K}_\alpha^\alpha = (\Delta + 2/R^2)w = 0$, so that the second term of the integrand vanishes. Substitution yields

$$E_{\text{pot}} = A_m^2 \frac{2\pi}{R^2} D(1-\nu)m^2(m^2-1)^2 \cos^2 \omega t \int_0^\alpha \tan^{2m} \tfrac{1}{2}\theta \, \frac{d\theta}{\sin^3 \theta}.$$

(13.144)

Since the total energy $E_{\text{kin}} + E_{\text{pot}}$ is constant, the coefficients for

$\sin^2 \omega t$ and $\cos^2 \omega t$ must be equal. From this condition we find

$$\omega^2 = \frac{h^2}{R^4} \frac{E}{3(1+\nu)\gamma} m^2 (m^2 - 1)^2$$

$$\times \frac{\int_0^\alpha \tan^{2m} \tfrac{1}{2}\theta \, \dfrac{d\theta}{\sin^3 \theta}}{\int_0^\alpha [2\sin^2 \theta + (m + \cos \theta)^2] \tan^{2m} \tfrac{1}{2}\theta \sin \theta \, d\theta}$$

(13.145)

which is RAYLEIGH's approximation.

When m is a positive integer greater than unity, the integration can always be performed. Thus,

$$\int_0^\alpha \tan^{2m} \tfrac{1}{2}\theta \, \frac{d\theta}{\sin^3 \theta} = \frac{1}{8} \left[\frac{\tan^{2m-2} \tfrac{1}{2}\alpha}{m-1} + 2 \frac{\tan^{2m} \tfrac{1}{2}\alpha}{m} + \frac{\tan^{2m+2} \tfrac{1}{2}\alpha}{m+1} \right].$$

For the second integral we write $\tan \tfrac{1}{2}\theta = \sin \theta / (1 + \cos \theta)$ and make the substitution $x = 1 + \cos \theta$. In this way we find

$$\int_0^\alpha [2\sin^2 \theta + (m + \cos \theta)^2] \tan^{2m} \tfrac{1}{2}\theta \sin \theta \, d\theta$$

$$= \int_{1+\cos \alpha}^{2} \frac{(2-x)^m}{x^m} [(m-1)^2 + 2(m+1)x - x^2] \, dx,$$

which can be evaluated for any integer m.

The values in the last column of Table 6 were obtained in this way. For $m = 2$, a very thin shell, and moderate angles α, the error is below a few per cent and presumably acceptable, but outside this range the error becomes excessive. For angles α close to π, RAYLEIGH's formula (13.145) gives frequencies that tend to infinity as $1/(\pi - \alpha)^2$, which is nonsense.

For $m = 0$ and $m = 1$ the lowest root λR^2 of the frequency determinant (13.140) is zero. The corresponding modes represent rigid-body motion. The second root of (13.140) for any m including the lowest non-zero root for $m = 0$ and $m = 1$ does not depend strongly on the shell thickness h, as we can see in Table 7. This indicates that the corresponding modes are mainly extensional, i.e., bending—although present—plays a minor rôle. The second fundamental mode for $m = 6$, $h/R = 0.01$, $\nu = 0.3$ is shown in Fig. 37.

When α approaches 180°, the root λR^2 approaches the lowest root of the determinant of (13.113) for $n = 2$, independently of m.

The method developed in this paragraph, although primarily intended for solving the problem of free vibrations, is immediately applicable to the case of forced vibrations and in particular to the static case ('forced vibrations' with $\lambda = 0$). Since it can cope with static as well as kinematic boundary conditions, the general problem of spheri-

TABLE 7. Second root λR^2 of frequency determinant (13.140) for $\nu = 0.3$.

α	m	h/R			
		0.04	0.02	0.01	0.005
60°	0	0.7839	0.7742	0.7701	0.7687
	1	0.8493	0.8186	0.8101	0.8076
	2	0.9788	0.8760	0.8500	0.8433
90°	0	0.6998	0.6917	0.6896	0.6890
	1	0.7078	0.7010	0.6990	0.6985
	2	0.7932	0.7715	0.7659	0.7644
120°	0	0.6161	0.6124	0.6114	0.6112
	1	0.5903	0.5878	0.5871	0.5870
	2	0.6796	0.6715	0.6695	0.6690
150°	0	0.5344	0.5327	0.5323	0.5322
	1	0.5167	0.5152	0.5148	0.5147
	2	0.5801	0.5725	0.5703	0.5698
180°		0.4931	0.4918	0.4914	0.4913

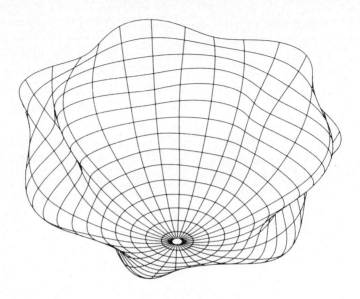

Fig. 37.

cal shells bounded by one or two parallel circles can be solved by this method, whether this problem is static or dynamic.

7. Radially loaded spherical shell

As an example of its application to a static case, let us consider the deformation of a complete spherical shell loaded by a radial load

$$2Q_m \cos m\phi$$

at the equator.

Dividing the shell at the equator into two hemispherical parts, we may take it that each part carries half the load $Q_m \cos m\phi$. For $m > 1$ we have, due to symmetry, the boundary conditions

$$u = S = \frac{\partial w}{\partial n} = 0, \qquad Q = Q_m \cos m\phi \qquad (\theta = 90°)$$

and this leads to the following system of equations,

$$\begin{pmatrix} c_{u1} & c_{u2} & c_{u3} & c_{u4} \\ c_{S1} & c_{S2} & c_{S3} & c_{S4} \\ c_{\psi 1} & c_{\psi 2} & c_{\psi 3} & 0 \\ c_{Q1} & c_{Q2} & c_{Q3} & 0 \end{pmatrix} \begin{pmatrix} A_1 \\ A_2 \\ A_3 \\ A_4 \end{pmatrix} = \frac{Q_m R^5}{D}[0,0,0,1].$$

The coefficients $A_1, \ldots, A_4$ are determined by evaluating $c_{u1}, \ldots, c_{Q3}$ for $\lambda = 0$ and solving the system of linear equations. By substituting $A_1, \ldots, A_4$ into (13.129) and (13.85) we find all displacements. Taking $h/R = 0.01$ and $\nu = 0.3$ we find

$$w = c_m \frac{Q_m R^3}{D} \cos m\phi,$$

where the Fourier coefficients c_m are given in Table 8.

TABLE 8. Radially loaded spherical shell with $h/R = 0.01$ and $\nu = 0.3$, Fourier coefficients c_m.

m	c_m
0	0.0002354
1	–
2	0.018608
3	0.007030
4	0.003786
5	0.002396
6	0.001666

For $m = 0$, the boundary condition $u = 0$ must be replaced by $N = 0$. For $m = 1$, the external forces are not in equilibrium and this case is therefore excluded.

For $m = 2$, the coefficient is 0.018608. This may be compared with the corresponding coefficient 0.1420 for an open hemispherical shell and the coefficient 0.01474 for a long cylindrical shell loaded at its mid-section, derived in Chapter 11.

8. Addendum

General differential equations for spherical shells in spherical coordinates were first derived by VAN DER NEUT [12] and HAVERS [13] for the static case and generalized to the dynamic case by FEDERHOFER.[14] Equations (13.14), (13.19) for the static case and (13.96)–(13.98) for the dynamic case are due to the present author.[15] They are more general in the sense that they are independent of coordinate system; their application to spherical coordinates is a special case. The *explicit* formulas for the boundary forces and moments (13.75)–(13.79) in the static case and for all kinematic and static boundary values (13.130)–(13.139) in the general case makes it a straight forward procedure to apply the theory to any specific problem.

Bibliography

NOVOZHILOV, V.V. (1959), *The Theory of Thin Shells*, Noordhoff, Groningen, pp. 346–358 (English translation).

SEIDE, P. (1975), *Small Elastic Deformations of Thin Shells*, Noordhoff, Leiden, pp. 333–404.

---

[12] NEUT, A. VAN DER (1932), *De Elastische Stabiliteit van den Dunwandigen Bol*, H.J. Paris Amsterdam.

[13] HAVERS, A. (1935), "Asymptotische Biege theorie der unbelastete Kugelschale", *Ing. Arch.* **6**, pp. 282–312.

[14] FEDERHOFER, K. (1937), "Zur Berechnung der Eigenschwingungen der Kugelschale", *Sitzber. Akad. Wiss. Wien, Math.-Naturw. Kl. 2A* **146**, pp. 57–69 and 505–514.

[15] NIORDSON, F.I. (1984), "Free vibrations of thin elastic spherical shells", *Internat. J. Solids & Structures* **20** (7) pp. 667–687.

CHAPTER 14

BENDING OF SHELLS OF REVOLUTION

1. Introduction 339
2. Shell equations 339
3. Boundary conditions 341
4. The differential equations for V_i and Q_i 343
5. Numerical solution 346
6. Bending of a toroidal shell under internal pressure 349

CHAPTER 14

BENDING OF SHELLS OF REVOLUTION

1. Introduction

In this chapter we shall derive the equations which govern the general case of non-axisymmetrical bending of shells of revolution. Since no solution in closed form is known to this problem, our aim will be to derive equations suitable for numerical integration.

We shall describe the middle-surface of the shell as in Chapter 10 with the functions $r(t)$ and $q(t)$ (see Fig. 14 and equations (10.11)–(10.18)). It is tacitly assumed that these two functions are sufficiently smooth.

In many applications the shell consists of a number of interconnected elementary parts, such as cylindrical, conical and spherical. It is therefore advantageous to find a formulation, in which discontinuities in r' and q' are permitted. This we shall do, but at this stage we shall restrict ourselves to smooth functions.

2. Shell equations

By expanding the external forces and the dependent functions in a Fourier series, just as we did in the case of cylindrical and spherical shells, we can describe the nth mode of the displacement field by the functions

$$v^1(t, \phi) = \boldsymbol{u}(t) \cos n\phi ,$$
$$v^2(t, \phi) = \boldsymbol{v}(t) \sin n\phi , \qquad (14.1)$$
$$w(t, \phi) = \boldsymbol{w}(t) \cos n\phi ,$$

where the boldface quantities depend on t only.

The components of the membrane stress tensor are

$$N_{11}(t, \phi) = N_{11}(t) \cos n\phi,$$
$$N_{12}(t, \phi) = N_{12}(t) \sin n\phi, \qquad (14.2)$$
$$N_{22}(t, \phi) = N_{22}(t) \cos n\phi,$$

where, according to (3.11) and (7.11),

$$N_{11} = Eh\left[u' - \frac{q''}{r'}w + \frac{\nu}{r}\left(r'u + \frac{n}{r}v - q'w\right)\right], \qquad (14.3)$$

$$N_{12} = Eh\tfrac{1}{2}(1 - \nu)\left[v' - nu - \frac{2r'}{r}v\right], \qquad (14.4)$$

$$N_{22} = Eh\left[rr'u + nv - rq'w + \nu r^2\left(u' - \frac{q''}{r'}w\right)\right]. \qquad (14.5)$$

Similarly, by using (3.37) and (7.11), we find the components of the moment tensor to be

$$M_{11}(t, \phi) = M_{11}(t) \cos n\phi,$$
$$M_{12}(t, \phi) = M_{12}(t) \sin n\phi, \qquad (14.6)$$
$$M_{22}(t, \phi) = M_{22}(t) \cos n\phi,$$

where

$$M_{11} = D\left\{\frac{1}{r'}\left(q''' - \frac{r''q''}{r'}\right)u + \frac{2q''}{r'}u' - \left(\frac{q''}{r'}\right)^2 w + w''\right.$$
$$\left. + \frac{\nu}{r}\left[\left(\frac{r'q'}{r} + q''\right)u + \frac{2nq'}{r^2}v - \frac{1}{r}((q')^2 + n^2)w + r'w'\right]\right\}, \qquad (14.7)$$

$$M_{12} = D(1 - \nu)\left\{\frac{nr'}{r}w - nw' - \frac{nq''}{r'}u + \frac{q'}{r}v' - \frac{2q'r'}{r^2}v\right\}, \qquad (14.8)$$

$$M_{22} = D\left\{(r'q' + rq'')u + \frac{2nq'}{r}v + rr'w' - ((q')^2 + n^2)w\right.$$
$$\left. + \nu r^2\left[\frac{1}{r'}\left(q''' - \frac{r''q''}{r'}\right)u + \frac{2q''}{r'}u' - \left(\frac{q''}{r'}\right)^2 w + w''\right]\right\}. \qquad (14.9)$$

The nth mode of the external forces is given by

$$F^1(t, \phi) = F_1(t) \cos n\phi,$$
$$F^2(t, \phi) = F_2(t) \sin n\phi, \qquad (14.10)$$
$$p(t, \phi) = p(t) \cos n\phi.$$

When substituted into the equations of equilibrium (5.22)–(5.23), we find

$$N'_{11} + \frac{r'}{r} N_{11} + \frac{n}{r^2} N_{12} - \frac{r'}{r^3} N_{22} + 2\frac{q''}{r'} M'_{11}$$
$$- \left(\frac{r''q''}{(r')^2} - \frac{q'''}{r'} - \frac{2q''}{r}\right) M_{11} + \frac{2nq''}{r^2 r'} M_{12} - \left(q'' + \frac{q'r'}{r}\right) M_{22} + F_1 = 0, \quad (14.11)$$

$$N'_{12} + \frac{r'}{r} N_{12} - \frac{n}{r^2} N_{22} + \frac{2q'}{r} M'_{12} + \frac{2q''}{r} M_{12} - \frac{2nq'}{r^3} M_{22} + F_2 = 0, \qquad (14.12)$$

$$M'''_{11} + \frac{2r'}{r} M''_{11} + \frac{2n}{r^2} M'_{12} - \frac{r'}{r^3} M'_{22} + \left(\frac{r''}{r} - \left(\frac{q''}{r'}\right)^2\right) M_{11}$$
$$- \left(\frac{n^2}{r} + r'' + \frac{(q')^2}{r} - \frac{2(r')^2}{r}\right) \frac{1}{r^3} M_{22} - \frac{q''}{r'} N_{11} - \frac{q'}{r^3} N_{22} - q = 0. \qquad (14.13)$$

This completes the derivation of the shell equations. By substituting (14.3)–(14.5) and (14.7)–(14.9) into (14.11)–(14.13) we get a system of three ordinary differential equations for the unknown functions u, v and w.

3. Boundary conditions

In Chapter 6 we derived the appropriate boundary conditions from the principle of virtual work. Utilizing the results, and in particular (6.15), (6.17) and (6.19) we get by (14.2) and (14.6) the bending moment M_B, the effective shear force Q, and the effective membrane force vector $\hat{T}^\alpha$ in the following form,

$$M_B = M_{11} \cos n\phi, \qquad (14.14)$$

$$Q = -\left(M'_{11} + \frac{r'}{r} M_{11} + \frac{2n}{r^2} M_{12} - \frac{r'}{r^3} M_{22}\right) \cos n\phi, \qquad (14.15)$$

$$\hat{T}^1 = \left(N_{11} + \frac{q''}{r'} M_{11}\right) \cos n\phi, \qquad (14.16)$$

$$\hat{T}^2 = \frac{1}{r^2}\left(N_{12} + \frac{2q'}{r} M_{12}\right) \sin n\phi. \qquad (14.17)$$

Instead of $\hat{T}^1$, $\hat{T}^2$ and Q we can prescribe the displacements u, v and w respectively, and instead of M_B we can prescribe the rotation of the normal $\psi \cos n\phi$, where ψ according to (6.22) is given as

$$\psi = w' + \frac{q''}{r'} u. \qquad (14.18)$$

This completes the mathematical formulation of our problem, but as already indicated, this formulation has the disadvantage of requiring smooth functions r and q. If the shell is composed of different parts, there will be conditions of continuity to satisfy at each joint. The displacements u and w, and the forces $\hat{T}^1$ and Q are related to the slope r', and since r' may be discontinuous at a joint, all these quantities will suffer a discontinuity.

This difficulty can be avoided, if we use the axial, radial and peripheral directions for reference. These directions are unique, and if the displacements and the membrane forces are resolved into the corresponding components, continuity of the functions is ensured also at the joints.

Thus, the components of the displacement vector in axial, radial and peripheral directions are

$$v_Z = q'u + r'w, \qquad v_R = r'u - q'w, \qquad v_T = rv, \qquad (14.19)$$

respectively, and similarly we can write the components of the force vector, but let that wait.

We shall find it convenient to describe the rotation of the normal ψ and the displacements by one single 'vector' with four components,[1]

$$V_i = (\psi, v_Z, v_R, v_T). \qquad (14.20)$$

---

[1] The word 'vector' is used here to denote a 1×4 or a 4×1 matrix.

The bending moment and the components of the force at the boundary will be organized in a similar way,

$$Q_i = (M, Q_Z, Q_R, Q_T) \tag{14.21}$$

and so that the scalar product[2]

$$Q_i V_i = c \int_0^{2\pi} (M\psi + Q_Z v_Z + Q_R v_R + Q_T v_T) \cos^2 n\phi \, d\phi \tag{14.22}$$

will equal the work done by the generalized forces Q_i on the generalized displacements V_i. This will obviously be the case if $c = 1/2\pi$ for $n = 0$, and $c = 1/\pi$ for $n > 0$ and

$$\begin{aligned} M &= M_B r/\cos n\phi, \\ Q_Z &= (q'\hat{T}^1 + r'Q)r/\cos n\phi, \\ Q_R &= (r'\hat{T}^1 - q'Q)r/\cos n\phi, \\ Q_T &= \hat{T}^2 r/\sin n\phi. \end{aligned} \tag{14.23}$$

The components of V_i are continuous functions of t everywhere, and so are the functions Q_i, provided that no concentrated forces or moments are applied.

4. The differential equations for V_i and Q_i

In this section we shall derive the shell equations in terms of V_i and Q_i. For the purpose of numerical integration we shall replace the equations of equilibrium (14.11)–(14.13) by an equivalent system of eight first order differential equations.

Taking V_i and Q_i as new dependent variables, the most general linear form of first order differential equations is

$$\begin{aligned} Q_i' + A_{ij}Q_j + B_{ij}V_j &= f_i, \\ V_i' + C_{ij}Q_j + D_{ij}V_j &= g_i, \end{aligned} \tag{14.24}$$

---

[2] The summation convention for latin subscripts is assumed to hold in this chapter for the range 1 to 4.

where the coefficients A_{ij}, B_{ij}, C_{ij} and D_{ij} are four 4×4 matrices, and f_i and g_i are two 4-vectors.

Our task is now to determine the 72 coefficients, which appear in the new shell equations (14.24).

To do that, we determine the derivatives V'_i and Q'_i from (14.8), (14.9) and (14.23). With the help of (14.14)–(14.17) we can express them all in terms of the following functions,

$$\begin{array}{ll} u, u' & N_{12}, N'_{12} \\ v, v' & M_{11}, M'_{11}, M''_{11} \\ w, w', w'' & M_{12}, M'_{12} \\ N_{11}, N'_{11} & M_{22}, M'_{22}. \end{array}$$

By a rather tedious elimination process, all these functions and derivatives can be expressed in terms of Q_i and V_i. Thus, u, v and w are found by solving (14.19), and w' can be solved from (14.18). Next, by inverting (14.23) we get M_B, Q, $\hat{T}^1$ and $\hat{T}^2$ and then N_{12}, M_{12} and v' are found from (14.4), (14.8) and (14.17). Then u', w'', N_{11} and M_{11} are determined from (14.3), (14.7) and (14.16). With u' and w'' in hands we can also determine N_{22} and M_{22}, etc. When the elimination process is completed, the derivatives V'_i and Q'_i are obtained in terms of Q_i and V_i and a comparison with (14.24) yields

$$A_{ij} = \frac{1}{r} \begin{pmatrix} -vr' & rr' & -q'r & 4k\dfrac{q'n}{r} \\ \dfrac{vr'n^2}{r} & 0 & 0 & q'n \\ \dfrac{vq''}{r'} - \dfrac{vq'}{r}(n^2+1) & -vq' & -vr' & r'n \\ \dfrac{vq''n}{r'} - \dfrac{2vq'n}{r} & -vq'n & -vr'n & r' \end{pmatrix},$$

(14.25)

THE DIFFERENTIAL EQUATIONS FOR V_i AND Q_i

$$B_{ij} = \frac{Eh}{r}\begin{pmatrix} -k\left[(r')^2 + \frac{2n^2}{1+\nu}\right] & \frac{kn^2}{r}\left[(r')^2 + \frac{2}{1+\nu}\right] & -\frac{kr'q'}{r}(n^2+1) & -\frac{2kr'q'n}{r} \\ \frac{kn^2}{r}\left[(r')^2 + \frac{2}{1+\nu}\right] & -\frac{kn^2}{r^2}\left[(nr')^2 + \frac{2}{1+\nu}\right] & \frac{kr'q'n^2}{r^2}(n^2+1) & \frac{2kr'q'n^3}{r^2} \\ \frac{kr'q'}{r}(n^2+1) & \frac{kr'q'n^2}{r^2}(n^2+1) & -1 & -n \\ -\frac{2kr'q'n}{r} & \frac{2kr'q'n^3}{r^2} & -n & -n^2 \end{pmatrix},$$

(14.26)

$$C_{ij} = \frac{1-\nu^2}{Ehr}\begin{pmatrix} -\frac{1}{k} & \frac{q''q'}{r'} & q'' & 0 \\ \frac{q''q'}{r'} & -(q')^2 & -r'q' & 0 \\ q'' & -r'q' & -(r')^2 & 0 \\ 0 & 0 & 0 & -\frac{2}{1-\nu} \end{pmatrix},$$

(14.27)

$$D_{ij} = -\frac{1}{r}\begin{pmatrix} -\nu r' & \frac{\nu r' n^2}{r} & \frac{\nu q''}{r'} - \frac{\nu q'}{r}(n^2+1) & \frac{\nu q'' n}{r'} - \frac{2\nu q' n}{r} \\ rr' & 0 & -\nu q' & -\nu q' n \\ -q'r & 0 & -\nu r' & -\nu r' n \\ 4k\frac{q'n}{r} & q'n & r'n & r' \end{pmatrix},$$

(14.28)

$$f_i = r(0, -q'F_1 - r'p, -r'F_1 + q'p, -rF_2),$$
$$g_i = (0, 0, 0, 0),$$

(14.29)

where

$$k = \tfrac{1}{12}h^2.$$

(14.30)

We note that the following relations hold good,

$$B_{ij} = B_{ji}, \qquad C_{ij} = C_{ji}, \qquad A_{ij} = -D_{ji}. \tag{14.31}$$

It can be shown that they follow from the fact that our elastic system is conservative.

At the boundaries we may give either Q_i or V_i, or in general any linear combination of them,

$$b_{ij}Q_j + c_{ij}V_j = d_i, \tag{14.32}$$

where the 36 numbers b_{ij}, c_{ij} and d_i are constants specific for each boundary. A boundary is called *singular* and the conditions there are said to be singular if the matrix b_{ij} is singular; otherwise, it is called *regular*. We shall assume, to start with, that the boundaries are regular.

5. Numerical solution

Mathematically, the problem is reduced to a two-point boundary value problem in ordinary first order differential equations. Starting at one boundary we may obtain a set of eight linearly independent solutions of the homogeneous system and a particular integral by a forward integration scheme, such as the Runge–Kutta method. The general solution, which is a linear combination of the fundamental solutions, will have eight arbitrary constants, which can be determined from the boundary conditions. This method is analogous to the analytical methods used in Chapter 11 for cylindrical shells and Chapter 13 for spherical shells.

Although it worked well when the fundamental system was obtained in closed form, it is easy to see that it is numerically ill-conditioned. For the cylindrical shell, for example, four fundamental solutions were exponentially increasing, and four exponentially decreasing, and hence, for a sufficiently long interval of integration, the superposition of fundamental solutions will involve small differences between large numbers. The same thing happens for other types of shells. Therefore, another approach has to be sought, and the *field method* described below offers an effective alternative.

Let r, q and h define the shell in the interval $t_0 \leq t \leq t_1$, and consider the part $t_0 \leq t \leq \tau$, where τ is a fixed number between t_0 and t_1 ($t_0 < \tau < t_1$). The differential equations (14.24), the initial conditions

(14.32) at $t = t_0$, and the given vector $V_i(\tau)$ uniquely determine the solution for that part of the shell independently of all data for $t > \tau$. In particular, the values of V_i determine Q_i at $t = \tau$, which in view of the linearity of the problem are expressed by

$$Q_i = U_{ij}V_j + W_i, \tag{14.33}$$

where U_{ij} is a (symmetrical) 4×4 stiffness matrix and W_i a 4-vector. Equation (14.33) is called the *field relation*, since it is satisfied by the 'field' of all possible solutions $V_i(\tau), Q_i(\tau)$ depending on the unused data for $t > \tau$. Correspondingly, U_{ij} and W_i are called *field functions*. They are governed by differential equations that we obtain by differentiating (14.33) with respect to t, use (14.24) to eliminate Q'_i and V'_i and (14.33) to eliminate Q_i. This gives

$$(U'_{ij} - U_{ik}C_{kl}U_{lj} + A_{ik}U_{kj} - U_{ik}D_{kj} + B_{ij})V_j$$
$$+ W'_i - U_{ij}C_{jk}W_k + A_{ij}W_j + U_{ij}g_j - f_i = 0. \tag{14.34}$$

Since this equation must be satisfied for all V_i (which depend on data at $t > \tau$, whereas U_{ij} and W_i do not), it follows that

$$U'_{ij} - U_{ik}C_{kl}U_{lj} + A_{ik}U_{kj} - U_{ik}D_{kj} + B_{ij} = 0 \tag{14.35}$$

and

$$W'_i - U_{ij}C_{jk}W_k + A_{ij}W_j + U_{ij}g_j - f_i = 0. \tag{14.36}$$

This is a system of 20 first order nonlinear differential equations for the field functions. Due to the symmetry of U_{ij} we have, however, only 14 to solve.

The corresponding initial conditions for a regular boundary are found by substituting (14.33) into (14.32). Let b_{ij}^{-1} be the inverse matrix of b_{ij} so that

$$b_{ij}^{-1}b_{jk} = \delta_{ik}, \tag{14.37}$$

where δ_{ik} is Kronecker's delta. Then

$$b_{ki}^{-1}b_{ij}(U_{jl}V_l + W_j) + b_{ki}^{-1}c_{ij}V_j = b_{ki}^{-1}d_i$$

or
$$(U_{kl} + b_{ki}^{-1}c_{il})V_l + W_k - b_{ki}^{-1}d_i = 0. \tag{14.38}$$

Since (14.38) is also an identity with respect to V_l, we get the initial values
$$\begin{aligned} U_{ij} &= -b_{ik}^{-1}c_{kj} \\ W_i &= b_{ij}^{-1}d_j \end{aligned} \quad \text{for } t = t_0. \tag{14.39}$$

With these initial conditions we can solve equations (14.35) and (14.36) by a forward integration method, for instance by Runge–Kutta's method. When this integration is performed, and we have U_{ij} and W_i at $t = t_1$, we can use (14.32) and (14.33) to find the displacement vector V_i there, by solving
$$(b_{ik}U_{kj} + c_{ij})V_j = d_i - b_{ij}W_j \quad (t = t_1). \tag{14.40}$$

With this value as an initial condition, a backward integration of (14.24) and (14.33), i.e.,
$$V_i' + (C_{ij}U_{jk} + D_{ik})V_k = g_i - C_{ij}W_j \tag{14.41}$$

is performed, and from the values so obtained Q_i is calculated from (14.33), completing the solution.

The advantage of using the field method, rather than integrating (14.24) directly, is that it is numerically stable. In general, the field functions have a narrow boundary zone of rapid variation (for values of t close to t_0) but are otherwise slowly varying.

The numerical integration technique, called the field method, which does not use elementary and particular solutions at all, but replaces the boundary-value problem by two initial-value problems to be solved in succession, was originally developed by JORDAN and SHELLY.[3] It was later formulated by MILLER[4] for a two point boundary value problem governed by an ordinary linear differential equation of even order. The presentation in this chapter is mainly due to COHEN.[5]

---

[3] JORDAN, P.F. and P.E. SHELLY (1966), "Stabilization of unstable two-point boundary-value problems", *AIAA Journal* **4** (5) pp. 923–924.
[4] MILLER, R.E. (1967), "Use of the field method for numerical integration of two-point boundary-value problems", *AIAA Journal* **5** (4) pp. 811–813.
[5] COHEN, G.A. (1974), "Numerical integration of shell equations using the field method", *Journal of Applied Mechanics Series E* **41** (1) pp. 261–266.

6. Bending of a toroidal shell under internal pressure

In Chapter 10 we found the membrane state of a toroidal shell subjected to uniform lateral pressure. We recall that although the membrane forces were statically determined, there was no regular displacement field associated with the membrane state and we concluded that bending plays an essential role in this case.

To illustrate the numerical method developed in this chapter we shall compute the state of stress and strain in a given toroidal shell with internal hydrostatic pressure and compare the result with the membrane solution.

Since the state is axisymmetric, we have $n = 0$ and $V_4 = Q_4 = 0$ everywhere. All components A_{ij}, B_{ij}, C_{ij} and D_{ij} with at least one index 4 vanish. From (10.99) and (14.25)–(14.29) we get

$$A_{ij} = \begin{pmatrix} \frac{\nu}{r}\sin\theta & -\sin\theta & -\cos\theta \\ 0 & 0 & 0 \\ -\frac{\nu}{r}\left(\frac{\cos\theta}{r}-\frac{1}{a}\right) & -\frac{\nu}{r}\cos\theta & \frac{\nu}{r}\sin\theta \end{pmatrix}, \quad (14.42)$$

$$B_{ij} = \frac{Eh}{r}\begin{pmatrix} -k\sin^2\theta & 0 & \frac{k}{r}\sin\theta\cos\theta \\ 0 & 0 & 0 \\ \frac{k}{r}\sin\theta\cos\theta & 0 & -1 \end{pmatrix}, \quad (14.43)$$

$$C_{ij} = \frac{1-\nu^2}{Ehr}\begin{pmatrix} -\frac{1}{k} & \frac{\cos\theta}{a} & -\frac{\sin\theta}{a} \\ \frac{\cos\theta}{a} & -\cos^2\theta & \sin\theta\cos\theta \\ -\frac{\sin\theta}{a} & \sin\theta\cos\theta & -\sin^2\theta \end{pmatrix}, \quad (14.44)$$

$$f_i = (0, pr\sin\theta, pr\cos\theta). \quad (14.45)$$

We perform the numerical integration of the field equations (14.35) and (14.36) taking a shell with the following data,

$$R/a = 2, \quad a/h = 40, \quad p/E = 5 \times 10^{-7}, \quad \nu = 0.3.$$

We start the forward integration at $\theta = 0$ proceeding in steps of $1°$ ($\pi/180$) unto $\theta = 2\pi$. The initial conditions at $\theta = 0$ are $\psi = v_Z = Q_R = 0$, or

$$V_1 = 0, \quad V_2 = 0, \quad Q_3 = 0. \tag{14.46}$$

A comparison with (14.32) shows that b_{ij} is singular. Roughly speaking it means that the initial conditions require U_{11} and U_{22} to be infinite, all remaining U_{ij} being zero. For the numerical solution we take arbitrary but large values, say $U_{11} = U_{22} = 10^6$ at $\theta = 0$.

After performing the forward integration by the Runge–Kutta method we reach the 'boundary' $\theta = 2\pi$ where the conditions (14.46) also hold. We can satisfy them exactly by taking

$$V_1 = 0, \quad V_2 = 0, \quad V_3 = -W_3/U_{33}. \tag{14.47}$$

With these values of V_i we start the backwards integration of (14.41) and find on reaching $\theta = 0$ that the conditions $V_1 = V_2 = 0$ are only approximately fulfilled.

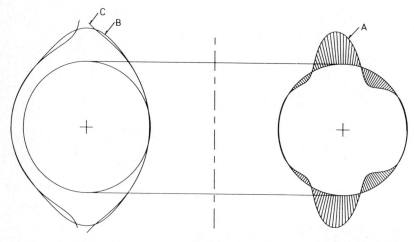

Fig. 38.

This rough solution can be greatly improved by repeating the integration scheme using the newly obtained values of Q_1 and Q_2 at $\theta = 0$ as initial values of W_1 and W_2 respectively. By repeating this we can satisfy the boundary conditions (14.46) with any desired accuracy.

The resulting bending moment is shown in Fig. 38 (curve A). As already mentioned in Chapter 10, bending plays an important role at $\theta = \frac{1}{2}\pi$ and $\frac{3}{2}\pi$. The maximum value in this particular case,

$$M_{max} = 10.39 \, pa,$$

is obtained at $\theta = 95°$ and $\theta = 265°$. The deformation of the shell is indicated by curve B in Fig. 38. The maximum radial displacement w is found to be

$$w_{max} = 77.3 \frac{pa}{E},$$

and occurs at almost exactly the same place as the maximum moment. Curve C is the (singular) solution of the membrane state (10.102)–(10.103) shown for comparison.

CHAPTER 15

THE DONNELL–MUSHTARI–VLASOV THEORY

1. Introduction 355
2. Simplified equations of equilibrium 356
3. Equilibrium in the deformed state 359
4. Summary 362
   Bibliography 362

CHAPTER 15

THE DONNELL–MUSHTARI–VLASOV THEORY

1. Introduction

The first theory of shells, founded on the Kirchhoff hypothesis that normals remain normals, was due to ARON,[1] who obtained the components of bending and twist in a form that was independent of the tangential displacements v^α. LOVE[2] found that this was not strictly correct, and derived accurate formulas, corresponding to (3.37). Much later it was discovered that in many cases ARON's approximation was justified. Thus DONNELL[3] successfully used the simplified expression

$$K_{\alpha\beta} \approx D_\alpha D_\beta w \qquad (15.1)$$

in his investigations concerning the stability of shells, and about the same time a similar theory was introduced in the U.S.S.R. by MUSHTARI[4] and VLASOV.[5] MUSHTARI could show that if the bending stresses are comparable in magnitude to (or smaller than) those due to the membrane

---

[1] ARON, H. (1874), "Das Gleichgewicht und die Bewegung einer unendlich dünnen beliebig gekrümmten, elastischen Schale", *J. f. Reine Angew. Math. (Crelle)* **78**.
[2] LOVE, A.E.H., see footnote 3 on p. 230.
[3] DONNELL, L.H. (1976), see bibliography at the end of this chapter.
[4] MUSHTARI, Kh.M. (1938), "On the stability of cylindrical shells subjected to torsion" (in Russian), *Trudy Kaz. Aviats. In-Ta.* **5** (2).
   MUSHTARI, Kh.M. (1938), "Certain generalizations of the theory of thin shells" (in Russian), *Izv. Fiz. Mat. Ob. Pri Kaz. U.* **11** (8).
[5] VLASOV, V.Z. (1944), "The fundamental differential equations of the general theory of elastic shells" (in Russian), *Prikl. Mat. Mech. Akad. Nauk SSSR* **8** (2).

forces, it is justified to neglect the tangential displacements in $K_{\alpha\beta}$, and this leads to a greatly simplified shell theory, the DONNELL–MUSHTARI–VLASOV theory, henceforth called the DMV theory.

2. Simplified equations of equilibrium

Let at any typical point ε be the absolute value of the numerically largest principal strain in the middle-surface, and, similarly, κ the absolute value of the numerically largest principal change of curvature. Then, the principal membrane force will be in the order of magnitude $Eh\varepsilon$ and the moment tensor in the order of magnitude $Eh^3\kappa$. We shall write this as

$$N^{\alpha\beta} = O(Eh\varepsilon), \qquad M^{\alpha\beta} = O(Eh^3\kappa),$$

assuming that the coordinate system is locally cartesian.

Further, let R be the smallest principal radius of curvature, and l a characteristic length of the deformation pattern.

We can now estimate the order of magnitude of the first two terms in the equations of equilibrium (5.22),

$$D_\alpha N^{\alpha\beta} = O(Eh\varepsilon/l), \qquad 2d^\beta_\gamma D_\alpha M^{\gamma\alpha} = O(Eh^3\kappa/Rl).$$

Clearly, their ratio is $\varepsilon : h^2\kappa/R$, and since we assume that $h\kappa$ is comparable in magnitude to ε, or smaller, the second term of (5.22) is at most of order h/R in comparison with the first, and may hence be neglected. It is easy to see that the same holds for the third term of (5.22), and hence we may write this equation as

$$D_\alpha N^{\alpha\beta} + F^\beta = 0, \tag{15.2}$$

omitting terms of relative order h/R.

Making a similar estimate of the relative order of magnitude of the terms in (5.23), we find that the term $d_{\alpha\gamma}d^\gamma_\beta M^{\alpha\beta}$ can be neglected in comparison with $d_{\alpha\beta}N^{\alpha\beta}$. Therefore, omitting terms of relative order h/R, we get

$$D_\alpha D_\beta M^{\alpha\beta} - d_{\alpha\beta}N^{\alpha\beta} - p = 0. \tag{15.3}$$

The equations (15.2)–(15.3) are the simplified equations of equilibrium of the DMV theory. Following the scheme used in Chapter 6, they can be derived from the principle of virtual work using the expression (15.1) for $K_{\alpha\beta}$. Thus, assuming that the bending stresses are at most of the same order of magnitude as the membrane stresses, the influence of the tangential displacements may be neglected in $K_{\alpha\beta}$.

Now, let $F^{\alpha\beta}$ be a particular integral of (15.2), and let

$$N^{\alpha\beta} = F^{\alpha\beta} + \varepsilon^{\alpha\xi}\varepsilon^{\beta\eta}D_\xi D_\eta \Phi, \qquad (15.4)$$

where $\Phi(u^1, u^2)$ is any sufficiently smooth function. Then

$$D_\alpha N^{\alpha\beta} = -F^\beta + \varepsilon^{\alpha\xi}D_\alpha D_\xi \varepsilon^{\beta\eta}D_\eta \Phi.$$

But

$$D_\alpha D_\xi \varepsilon^{\beta\eta}D_\eta \Phi = \tfrac{1}{2}(D_\alpha D_\xi + D_\xi D_\alpha)\varepsilon^{\beta\eta}\Phi_{,\eta} + \tfrac{1}{2}(D_\alpha D_\xi - D_\xi D_\alpha)\varepsilon^{\beta\eta}\Phi_{,\eta}$$

$$= \tfrac{1}{2}(D_\alpha D_\xi + D_\xi D_\alpha)\varepsilon^{\beta\eta}\Phi_{,\eta} + \tfrac{1}{2}B_{\rho\alpha\xi\zeta}a^{\zeta\beta}\varepsilon^{\rho\eta}\Phi_{,\eta},$$

where (1.39) was used.

The ratio of the order of magnitude of these two terms is $1/l^2 : K$ where K is the Gaussian curvature of the shell. If the characteristic length of the deformation pattern l is much smaller than the (numerically) *largest* principal radius of curvature, we have $1/l^2 \gg K$, and consequently we can neglect the second term and write

$$D_\alpha D_\xi \varepsilon^{\beta\eta}D_\eta \Phi = \tfrac{1}{2}(D_\alpha D_\xi + D_\xi D_\alpha)\varepsilon^{\beta\eta}\Phi_{,\eta}.$$

Multiplying by $\varepsilon^{\alpha\xi}$, the right-hand side (which is symmetrical with respect to α and ξ) vanishes, and hence expression (15.4) satisfies the equations of equilibrium (15.2); exactly for shells of zero Gaussian curvature, and approximately for any other shell, provided that $l \ll R_{\max}$, where $R_{\max}$ is the numerically largest principal radius of curvature.

The remaining equation of equilibrium (15.3) can be written in terms of the scalar functions w and Φ. Thus, expressing the moment tensor according to (7.11) and (15.1) in terms of w,

$$M_{\alpha\beta} = D[(1-\nu)D_\alpha D_\beta w + \nu a_{\alpha\beta}\Delta w] \qquad (15.5)$$

and substituting (15.1) and (15.5) into (15.3), we find

$$D\Delta^2 w - d_{\alpha\beta}\varepsilon^{\alpha\xi}\varepsilon^{\beta\eta}D_\xi D_\eta \Phi - d_{\alpha\beta}F^{\alpha\beta} - p = 0,$$

which we shall write in the form

$$D\Delta^2 w - L[\Phi] = \hat{p}, \qquad (15.6)$$

where $L[\cdots]$ is a second order linear differential operator of the form

$$L[\cdots] = d_{\alpha\beta}\varepsilon^{\alpha\xi}\varepsilon^{\beta\eta}D_\xi D_\eta[\cdots] \qquad (15.7)$$

and $\hat{p}$ the augmented lateral pressure

$$\hat{p} = d_{\alpha\beta}F^{\alpha\beta} + p. \qquad (15.8)$$

Another relation between w and Φ is found from the linearized equation of compatibility (3.48),

$$D_\alpha D^\alpha E^\beta_\beta - D_\alpha D^\beta E^\alpha_\beta + d^\alpha_\alpha K^\beta_\beta - d^\beta_\alpha K^\alpha_\beta - K E^\alpha_\alpha = 0.$$

The last term of this expression may be omitted under precisely the same circumstances $(1/l^2 \gg K)$ as earlier. Introducing E^α_β from Hooke's law (7.12), $N^{\alpha\beta}$ from (15.4), and K^α_β from (15.1), we get

$$\frac{1}{Eh}\Delta^2\Phi + L[w] = 0, \qquad (15.9)$$

where we have taken $F^{\alpha\beta} = 0$.

The two fourth order differential equations (15.6) and (15.9) are the simplified shell equations of the DMV theory.

We repeat that they are approximate, and that the approximation may be justified for shells of zero Gaussian curvature if the bending stresses are at most of the same order as the membrane stresses. For shells of nonzero Gaussian curvature an additional condition is that the characteristic length of the deformation pattern is much smaller than the largest principal radius of curvature.

3. Equilibrium in the deformed state

In a number of problems of importance we cannot ignore the fact that equilibrium is a state of the *deformed* structure. This is clearly important to take into account, in cases when the displacements are large, i.e., when the displacement vector can be compared in size with the linear dimensions of the structure. But there are other cases, including problems regarding the stability of shells, which are essentially nonlinear, although we may take all displacements, strains and rotations to be arbitrarily small. In fact, when the slope dw/ds somewhere along a given curve, although very small in comparison with unity, is much larger than the numerically largest principal strain ε, and so that the square $(dw/ds)^2$ is comparable to ε, we should take the nonlinearity of the strain tensor into account. Going back to the complete expression (3.10) for $E_{\alpha\beta}$, we shall write

$$E_{\alpha\beta} = \tfrac{1}{2}(D_\alpha v_\beta + D_\beta v_\alpha) - d_{\alpha\beta} w + \tfrac{1}{2} w_{,\alpha} w_{,\beta} \tag{15.10}$$

disregarding all other nonlinear terms.

For plates ($d_{\alpha\beta} = 0$) this may be justified when the rotations in the middle-surface are small in comparison with the rotation of the normal, i.e., when the displacements are approximately perpendicular to the middle-surface. For shells in general an additional requirement is that the smallest principal radius of curvature R is considerably much larger than l, the characteristic length of the deformation pattern. Such an approximation is very reasonable for instance in the case of *shallow shells*.

Let us now derive the equations of equilibrium appropriate to the measures of strain (15.10) and bending (15.1) by applying the principle of virtual work. To do that, we substitute into (15.18) the following expressions,

$$\delta E_{\alpha\beta} = \tfrac{1}{2}(D_\alpha \delta v_\beta + D_\beta \delta v_\alpha) - d_{\alpha\beta}\, \delta w + \tfrac{1}{2}(w_{,\alpha} \delta w_{,\beta} + w_{,\beta} \delta w_{,\alpha}),$$

$$\delta K_{\alpha\beta} = D_\alpha D_\beta \delta w,$$

$$\delta r^\alpha = \varepsilon^{\alpha\beta}(d_{\beta\gamma} \delta v^\gamma + \delta w_{,\beta}).$$

Taking first $\delta v_\alpha = 0$ everywhere, equation (5.18) becomes

$$\iint_D (M^{\alpha\beta} D_\alpha \delta w_{,\beta} - d_{\alpha\beta} N^{\alpha\beta} \delta w + N^{\alpha\beta} w_{,\alpha} \delta w_{,\beta})\, dA$$

$$- \iint_D p\, \delta w\, dA - \oint_C (T\, \delta w + M_\alpha \varepsilon^{\alpha\beta} \delta w_{,\beta})\, ds = 0.$$

By repeated use of the divergence theorem, this may be transformed into

$$\iint_D [D_\alpha D_\beta M^{\alpha\beta} - d_{\alpha\beta} N^{\alpha\beta} - D_\beta(N^{\alpha\beta} w_{,\alpha}) - p]\, \delta w\, dA$$

$$+ \oint_C (M^{\alpha\beta} n_\alpha - M_\alpha \varepsilon^{\alpha\beta})\, \delta w_{,\beta}\, ds - \oint_C [T + (D_\alpha M^{\alpha\beta})n_\beta]\, \delta w\, ds = 0.$$

This must hold for arbitrary virtual displacements δw. Recalling the arguments used in Chapter 6, we find this to be possible only if all three integrands vanish. From the first we recover the equation of equilibrium in normal direction,

$$D_\alpha D_\beta M^{\alpha\beta} - d_{\alpha\beta} N^{\alpha\beta} - D_\beta(N^{\alpha\beta} w_{,\alpha}) - p = 0, \qquad (15.11)$$

which replaces the linear equation (15.3).

To get the equations of equilibrium in the tangent plane, we repeat the procedure, now taking $\delta w = 0$ everywhere and δv^α arbitrary. It may readily be verified that this gives

$$D_\alpha N^{\alpha\beta} + F^\beta = 0,$$

i.e., precisely the same equation as (15.2).

Using (15.1) and (15.2) we can rewrite (15.11) in the form

$$D_\alpha D_\beta M^{\alpha\beta} - (d_{\alpha\beta} + K_{\alpha\beta}) N^{\alpha\beta} - p^* = 0, \qquad (15.12)$$

where

$$p^* = p - F^\alpha w_{,\alpha} \tag{15.13}$$

is the load perpendicular to the deformed middle-surface.

Writing (15.3) in the deformed coordinate system, we get

$$D^*_\alpha D^*_\beta M^{\alpha\beta} - d^*_{\alpha\beta} N^{\alpha\beta} - p^* = 0,$$

where $D^*_\alpha T^{\beta\gamma\cdots}_{\delta\rho\cdots}$ differs from $D_\alpha T^{\beta\gamma\cdots}_{\delta\rho\cdots}$ according to (3.46) only by terms of relative order ε. Omitting such terms D^*_α may be replaced by D_α and we have precisely (15.12). Clearly, when applying the principle of virtual work, we can use the undeformed (original) coordinate system as our frame of reference, as we did here, instead of the deformed system like in Chapter 6. But, of course, the increments $\delta E_{\alpha\beta}$ (and $\delta K_{\alpha\beta}$) appearing in (15.18) must be interpreted accordingly. Thus the nonlinear terms of $E_{\alpha\beta}$ will contribute to $\delta E_{\alpha\beta}$ when we refer to the undeformed system, whereas there is no such contribution, when we use the deformed coordinate system as our frame of reference.

Using (15.1) and (15.4) we can write the equation of normal equilibrium (15.12) in the form

$$D\Delta^2 w - L[\Phi] - K[\Phi, w] - p^* = 0, \tag{15.14}$$

where L is the operator (15.7) and K is given by

$$K[\Phi, w] = \varepsilon^{\alpha\xi} \varepsilon^{\beta\eta} (D_\xi D_\eta \Phi)(D_\alpha D_\beta w). \tag{15.15}$$

The functions Φ and w must also satisfy the nonlinear equation of compatibility (3.48), which, omitting terms that introduce errors no larger than those already inherent in our theory, can be written as

$$D_\alpha D^\alpha E^\beta_\beta - D_\alpha D^\beta E^\alpha_\beta + d^\beta_\alpha K^\beta_\beta - d^\beta_\alpha K^\alpha_\beta = \tfrac{1}{2}[K^\alpha_\beta K^\beta_\alpha - K^\alpha_\alpha K^\beta_\beta]. \tag{15.16}$$

Using Hooke's law (7.11), (15.1) and (15.4) this equation can be written as

$$\frac{1}{Eh}\Delta^2 \Phi + L[w] + \tfrac{1}{2}K[w, w] = 0. \tag{15.17}$$

This completes the mathematical formulation of the theory.

4. Summary

The linear DMV theory is based on the simplified formula (15.1) for the bending tensor and leads to two coupled linear differential equations (15.6) and (15.9) each of fourth order. We have seen that the assumption that the influence of the normal displacement w predominates over the influence of the in-plane displacements v^α in the bending response of the shell is roughly equivalent to the assumption that the effect of the transverse shearing forces in the tangential equilibrium equations is negligible.

The nonlinear DMV theory is based on formulas (15.10) and (15.1) for the measures of strain and bending respectively. They lead to two coupled nonlinear fourth order differential equations (15.14) and (15.16). Many problems concerning stability of shells have been treated with this theory.

Bibliography

DONNELL, L.H. (1976), *Beams, Plates, and Shells*, McGraw-Hill, New York, pp. 325–358.

WLASSOW, W.S. (1958), *Allgemeine Schalentheorie und ihre Anwendung in der Technik*, Akademie-Verlag, Berlin, Chapter VII.

GOULD, P.L. (1977), *Static Analysis of Shells*, Lexington Books, Lexington, MA, pp. 362–367 and 372–373.

CHAPTER 16

NONLINEAR PLATE PROBLEMS

1. Introduction 365
2. Simply supported rectangular plate 366
3. Plates with geometrical imperfections 368
4. Clamped quadratic membrane with a uniform transverse load 369
5. Circular membrane 372
6. Large amplitude vibrations of a clamped, circular plate 375
   Bibliography 381

CHAPTER 16

NONLINEAR PLATE PROBLEMS

1. Introduction

The equations for large deflection of plates, obtained below as a special case of the nonlinear DMV theory, were in fact known long before the emergence of the DMV theory,[1] and had been generalized to *plates with a small initial curvature* (i.e., shallow shells) already in 1915.[2] When deriving them as a special case of the DMV theory we must remember that some of the assumptions on which this theory is based are *accurately* satisfied when $d_{\alpha\beta} = 0$.

We find from Chapter 15 for the measures of strain and bending

$$E_{\alpha\beta} = \tfrac{1}{2}(D_\alpha v_\beta + D_\beta v_\alpha + w_{,\alpha} w_{,\beta}), \tag{16.1}$$

$$K_{\alpha\beta} = D_\alpha D_\beta w, \tag{16.2}$$

respectively. We also find that the plate equations can be written either in the form

$$D_\alpha N^{\alpha\beta} + F^\beta = 0 \tag{16.3}$$

and

$$D\Delta^2 w = p^* + N^{\alpha\beta} D_\alpha D_\beta w \tag{16.4}$$

---

[1] KÁRMÁN, Th. VON (1910), *Encyklopädie der Mathematischen Wissenschaften*, Vol. 4, Teubner, Leipzig.

[2] TIMOSHENKO, S. (1915), *Memoirs of the Institute of Ways of Communication*, Vol. 89, Bull. Polytech. Inst. St. Petersburg.

or, using the stress function Φ, in the form

$$\frac{1}{Eh}\Delta^2\Phi + \tfrac{1}{2}K[w, w] = 0 \tag{16.5}$$

and

$$D\Delta^2 w = p^* + K[\Phi, w], \tag{16.6}$$

where the operator K is given by (15.15).

These are the VON KÁRMÁN equations for large deflection of plates. We shall now apply the theory to a few examples.

2. Simply supported rectangular plate

In this example we seek the deflection of a laterally loaded, simply supported plate, uniformly stressed in one direction (see Fig. 39).

In rectangular coordinates, equation (16.4) takes the form

$$\frac{\partial^4 w}{\partial x^4} + 2\frac{\partial^4 w}{\partial x^2 \partial y^2} + \frac{\partial^4 w}{\partial y^4} = \frac{1}{D}\left[p^* + N_x \frac{\partial^2 w}{\partial x^2} + 2N_{xy}\frac{\partial^2 w}{\partial x \partial y} + N_y \frac{\partial^2 w}{\partial y^2}\right], \tag{16.7}$$

where we shall assume the lateral load to be given in terms of a double Fourier series,

$$p^* = \sum_{m=1}^{\infty}\sum_{n=1}^{\infty} A_{mn} \sin\frac{m\pi x}{a} \sin\frac{n\pi y}{b}. \tag{16.8}$$

We seek the solution to (16.7) in the form

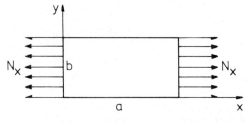

Fig. 39.

$$w = \sum_{m=1}^{\infty} \sum_{n=1}^{\infty} B_{mn} \sin \frac{m\pi x}{a} \sin \frac{n\pi y}{b} \tag{16.9}$$

and substitute into (16.7). Equating the resulting coefficients to zero, we find

$$B_{mn} = \frac{A_{mn}}{D\left[\left(\frac{m\pi}{a}\right)^2 + \left(\frac{n\pi}{b}\right)^2\right]^2 + N_x \left(\frac{m\pi}{a}\right)^2}. \tag{16.10}$$

Clearly, when $N_x > 0$ (tension), the deflection is diminished in comparison with the linear solution (9.21) and increased when $N_x < 0$ (compression).

Also, when the denominator approaches zero, the deflection tends to infinity. The smallest value of N_x for which the denominator becomes zero is the *limit of stability* of the plate. For loads at or above this limit the plate will *buckle*. We find this limit to be

$$N_x = -\underset{m,n}{\text{Min}} \left\{ \left(\frac{a}{m\pi}\right)^2 \left[\left(\frac{m\pi}{a}\right)^2 + \left(\frac{n\pi}{b}\right)^2\right]^2 D \right\}.$$

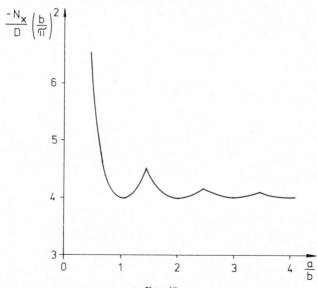

Fig. 40.

The minimum value is attained for $n = 1$ irrespective of m, and hence

$$-N_x = \operatorname*{Min}_{m} D\left(\frac{\pi}{b}\right)^2 \left[\frac{mb}{a} + \frac{a}{mb}\right]^2. \tag{16.11}$$

Figure 40 shows the critical load as a function of the side ratio a/b.

3. Plates with geometrical imperfections

If the middle-surface of the plate is not perfectly plane, but deviates from the plane by a lateral *initial deflection* $\tilde{w}$, we can describe the plate as a very shallow shell with curvature

$$\tilde{d}_{\alpha\beta} = D_\alpha D_\beta \tilde{w} \tag{16.12}$$

and, instead of (16.4) we have, according to (15.12),

$$D\Delta^2 w = p^* + N^{\alpha\beta}(\tilde{d}_{\alpha\beta} + D_\alpha D_\beta w). \tag{16.13}$$

Consider the plate in the example above, assuming that $p^* = 0$ and that it has an imperfection described by

$$\tilde{w} = A \sin\frac{\pi x}{a} \sin\frac{\pi y}{b}. \tag{16.14}$$

Looking for a solution in the form

$$w = B \sin\frac{\pi x}{a} \sin\frac{\pi y}{b}, \tag{16.15}$$

we find by substitution

$$D\left[\left(\frac{\pi}{a}\right)^2 + \left(\frac{\pi}{b}\right)^2\right]^2 B = -N_x \left(\frac{\pi}{a}\right)^2 (B + A). \tag{16.16}$$

Solving for B we find

$$B = -\frac{A\alpha}{1+\alpha}, \qquad \alpha = \frac{N_x}{N_c}, \qquad (16.17)$$

where N_c is the numerical value of the buckling load for $m = 1$ in the preceding example,

$$N_c = \left(\frac{a}{\pi}\right)^2 D\left[\left(\frac{\pi}{a}\right)^2 + \left(\frac{\pi}{b}\right)^2\right]^2. \qquad (16.18)$$

The result shows that when $\alpha \to -1$, i.e., when N_x approaches the buckling load, the deflection grows beyond any limit.

4. Clamped quadratic membrane with a uniform transverse load

The bending rigidity D of a membrane is taken to be zero, and since the boundary conditions are kinematic, the potential energy can be written as

$$U = \frac{1}{2}\frac{Eh}{1-\nu^2}\iint_{\mathscr{D}} \{(1-\nu)E^\beta_\alpha E^\alpha_\beta + \nu E^\alpha_\alpha E^\beta_\beta\}\, dA - \iint_{\mathscr{D}} pw\, dA \qquad (16.19)$$

or, in rectangular coordinates,

$$\begin{aligned}
U = \frac{1}{2}\frac{Eh}{1-\nu^2}\int_0^a\int_0^a &\left\{\left[\frac{\partial u}{\partial x}\right]^2 + \frac{\partial u}{\partial x}\left[\frac{\partial w}{\partial x}\right]^2 + \left[\frac{\partial v}{\partial y}\right]^2 + \frac{\partial v}{\partial y}\left[\frac{\partial w}{\partial y}\right]^2\right.\\
&+ \frac{1}{2}\left[\left[\frac{\partial w}{\partial x}\right]^2 + \left[\frac{\partial w}{\partial y}\right]^2\right]^2 + 2\nu\left[\frac{\partial u}{\partial x}\frac{\partial v}{\partial y} + \frac{1}{2}\frac{\partial v}{\partial y}\left[\frac{\partial w}{\partial x}\right]^2\right.\\
&\left.+ \frac{1}{2}\frac{\partial u}{\partial x}\left[\frac{\partial w}{\partial y}\right]^2\right] + \tfrac{1}{2}(1-\nu)\left[\left[\frac{\partial u}{\partial y}\right]^2 + 2\frac{\partial u}{\partial y}\frac{\partial v}{\partial x} + \left[\frac{\partial v}{\partial x}\right]^2\right.\\
&\left.\left.+ 2\frac{\partial u}{\partial y}\frac{\partial w}{\partial x}\frac{\partial w}{\partial y} + 2\frac{\partial v}{\partial x}\frac{\partial w}{\partial x}\frac{\partial w}{\partial y}\right]\right\}\, dx\, dy - \int_0^a\int_0^a pw\, dx\, dy, \qquad (16.20)
\end{aligned}$$

where (16.1) has been used. The side of the quadratic membrane is assumed to be a.

We shall seek an approximate solution to the problem using Ritz' method. Take the displacements to be

$$w = w_0 \sin \frac{\pi x}{a} \sin \frac{\pi y}{a},$$

$$u = c \sin \frac{2\pi x}{a} \sin \frac{\pi y}{a}, \qquad (16.21)$$

$$v = c \sin \frac{\pi x}{a} \sin \frac{2\pi y}{a},$$

where w_0 and c are coefficients to be determined. For w this is simply the first and leading term of the Fourier expansion and for u and v the first nonvanishing term of their Fourier expansions. Clearly, u is skew symmetrical about the middle $x = \frac{1}{2}a$, and similarly v is skew symmetrical about $y = \frac{1}{2}a$.

The displacements (16.21) satisfy the kinematic boundary conditions

$$w = u = v = 0$$

everywhere on the boundary of the membrane.

To determine the unknown coefficients w_0 and c we shall utilize the fact that U is a minimum in the state of equilibrium. Substituting (16.21) into (16.20) we find

$$U = \frac{Eh}{1-\nu^2} \left\{ \frac{5}{128} \frac{\pi^4}{a^2} w_0^4 + (\tfrac{5}{6} - \tfrac{1}{2}\nu) \frac{\pi^2}{a} cw_0^2 \right.$$

$$\left. + [\tfrac{1}{8}(9-\nu)\pi^2 + \tfrac{8}{9}(1+\nu)]c^2 \right\} - \frac{4pa^2}{\pi^2} w_0. \qquad (16.22)$$

The first condition, $\partial U/\partial c = 0$, yields

$$[9\pi^2(9-\nu) + 64(1+\nu)]c + \frac{\pi^2}{a}(30 - 18\nu)w_0^2 = 0$$

and hence

$$c = f(\nu)\frac{w_0^2}{a}, \tag{16.23}$$

where

$$f(\nu) = -\frac{6\pi^2(5-3\nu)}{9\pi^2(9-\nu)+64(1+\nu)}. \tag{16.24}$$

The second condition, $\partial U/\partial w_0 = 0$, yields

$$\frac{Eh}{1-\nu^2}\left[\frac{15\pi^2}{a}w_0^2 + 32(5-3\nu)c\right]\frac{\pi^2}{a}w_0 - 384\frac{pa^2}{\pi^2} = 0$$

and substituting c we can solve for w_0. We find

$$w_0 = g(\nu)a\sqrt[3]{\frac{pa}{Eh}}, \tag{16.25}$$

where

$$g(\nu) = \left[\frac{128(1-\nu^2)[9\pi^2(9-\nu)+64(1+\nu)]}{\pi^6[45\pi^2(9-\nu)-320(4-7\nu)-576\nu^2]}\right]^{1/3}. \tag{16.26}$$

For $\nu = 0.3$ we get

$$g(0.3) = 0.319, \qquad f(0.3) = -0.284.$$

We notice that the lateral displacement is proportional to the third root of the load

$$w_0 \sim p^{1/3}$$

and that the tangential displacements are proportional to the square of the lateral displacement,

$$c \sim w_0^2.$$

Hence, the membrane stresses are also proportional to the square of the lateral displacement, i.e., proportional to $p^{2/3}$.

We can now estimate the error introduced by neglecting the term $\frac{1}{2}D_\alpha v^\gamma D_\beta v_\gamma$ in the expression (3.10) for $E_{\alpha\beta}$. In this example the linear terms of $E_{\alpha\beta}$ are of order c/a and the quadratic term $w_{,\alpha}w_{,\beta}$ of order $(w_0/a)^2$, i.e., of the same order. The term $D_\alpha v^\gamma D_\beta v_\gamma$, however, is of order $(c/a)^2$ and hence, by omitting this term, we introduce errors no greater than of relative order $(w_0/a)^2$.

5. Circular membrane

In this example we shall determine the deformation of a circular membrane, loaded by a uniform lateral load p and a given tangential membrane stress $\mathcal{N}$ at the boundary (see Fig. 41).

For a membrane, the bending rigidity D vanishes, and the equation of normal equilibrium (16.6) takes the form

$$p + K[\Phi, w] = 0. \qquad (16.27)$$

In polar coordinates,

$$\Delta\Phi = \frac{1}{r}(r\Phi')',$$

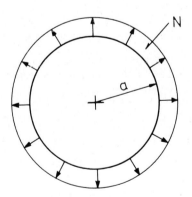

Fig. 41.

since Φ is a function of r only (see p. 165). Hence the equation of compatibility (16.5) can be written as

$$\frac{1}{r}\left\{r\left[\frac{1}{r}(r\Phi')'\right]'\right\}' + \frac{Eh}{2r}[(w')^2]' = 0.$$

Multiplying through by r and integrating we find

$$r\left[\frac{1}{r}(r\Phi')'\right]' + \tfrac{1}{2}Eh(w')^2 = C.$$

For $r = 0$, the slope $w' = 0$ and hence $C = 0$. In this equation as in (16.27), Φ does not appear itself, only the derivative Φ', which we shall denote by ψ. Hence we find the system

$$p + \frac{1}{r}(\psi w')' = 0, \qquad (16.28)$$

$$\left[\frac{1}{r}(r\psi)'\right]' + \frac{Eh}{2r}(w')^2 = 0. \qquad (16.29)$$

The first equation can be integrated to yield

$$\psi w' = -\tfrac{1}{2}pr^2, \qquad (16.30)$$

where the constant of integration must be zero since $w'(0) = 0$. Eliminating w' between these last two equations, we get the following second order nonlinear differential equation for ψ,

$$\psi^2\left[\frac{1}{r}(r\psi)'\right]' + \tfrac{1}{8}Ehp^2r^3 = 0. \qquad (16.31)$$

The membrane stresses are derived from the stress-function Φ and according to (15.4) we find

$$N_r = \frac{1}{r}\Phi' = \frac{1}{r}\psi \qquad (16.32)$$

and

$$N_\phi = \Phi'' = \psi' \tag{16.33}$$

for the radial and circumferential membrane stress respectively.

We cannot solve (16.31) in closed form, but it can of course be integrated numerically, using for instance Runge–Kutta's method. Another method is to expand ψ in a power series, substitute into (16.31) and equate the coefficients for each power of r to zero. Since N_r is an even function of r, ψ must be odd, i.e., only odd powers of r appear in Taylor expansion of ψ. Taking the first two nonvanishing terms we have

$$\psi = \psi_1 r + \psi_3 r^3 \tag{16.34}$$

and substitution into (16.31) yields

$$\psi_1^2 \psi_3 + \tfrac{1}{64} Ehp^2 = 0, \tag{16.35}$$

which indicates that ψ_3 must be negative.

From (16.30) we find

$$w' = -\frac{pr}{2(\psi_1 + \psi_3 r^2)},$$

which after integration yields the deflection

$$w = -\frac{p}{4\psi_3} \log \frac{\psi_1 + \psi_3 r^2}{\psi_1 + \psi_3 a^2}, \tag{16.36}$$

where $r = a$ is the radius of the membrane. The constant of integration has been determined from the condition $w(a) = 0$.

The condition that $N_r(a) = \mathcal{N}$ yields

$$\psi_1 + \psi_3 a^2 = \mathcal{N}, \tag{16.37}$$

which together with (16.35) determines ψ_1 and ψ_3. Thus ψ_1 is the solution of the following algebraic equation of third degree,

$$\psi_1^2(\mathcal{N}-\psi_1)+\tfrac{1}{64}Ehp^2a^2=0,\qquad(16.38)$$

and by solving it we have completed the solution. The result is shown in Fig. 42 where the deflection at the center δ is given as a function of the load p with $\mathcal{N}$ as a parameter. According to (16.33) and (16.34), $N_\phi = 0$ at $r = a$ when $\psi_1 + \psi_3 a^2 = 0$. In the region above the curve $N_\phi = 0$ (see Fig. 42), the solution requires compressive stresses, which cannot be supported by the membrane. In this region the membrane wrinkles.

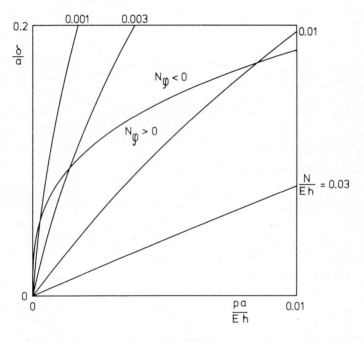

FIG. 42.

6. Large amplitude vibrations of a clamped, circular plate

Free vibrations of a clamped, circular plate were discussed in Chapter 9 using the linear theory. Of course, the result is valid for small amplitudes only. For larger amplitudes the stretching of the middle-surface may add considerably to the restoring force and thereby increase the frequency, which therefore is in fact dependent on the amplitude of the vibrations. Our subject here is to determine this dependence for amplitudes that are not too large. For simplicity we shall limit our analysis to axisymmetric modes.

Using VON KÁRMÁN's plate equations (16.5) and (16.6) we have in polar coordinates

$$\left[\frac{1}{r}(r\psi')\right]' + \frac{Eh}{2r}(w')^2 = 0, \qquad (16.39)$$

$$D\left[r\left[\frac{1}{r}(rw')'\right]'\right]' - \rho h r \frac{\partial^2 w}{\partial t^2} = (\psi w')', \qquad (16.40)$$

where $\psi = \Phi'$ as before.

Since the motion is periodic, we can expand w in a Fourier series,

$$w = \sum_n{}' G_n(r) \sin n\omega t, \qquad (16.41)$$

where, due to symmetry, only odd integers n appear. As an approximation we take only the first term of the Fourier series, and write

$$w = Ag(r) \sin \omega t, \qquad (16.42)$$

where we take $g(0)$ to be unity, so that A is the amplitude of the vibration.

From (16.39) we get

$$\psi = EhA^2 f(r) \sin^2 \omega t, \qquad (16.43)$$

where $f(r)$ is a solution to the following ordinary differential equation

$$\left[\frac{1}{r}(rf)'\right]' + \frac{1}{2r}(g')^2 = 0. \qquad (16.44)$$

Substituting (16.42) and (16.43) into (16.40), we find

$$DA\left[r\left[\frac{1}{r}(rg')'\right]'\right]' \sin \omega t = \rho h A r \omega^2 \sin \omega t + Eh(fg')' A^3 \sin^3 \omega t.$$

But

$$\sin^3 \omega t = \tfrac{3}{4} \sin \omega t - \tfrac{1}{4} \sin 3\omega t$$

and when this expression is introduced in the equation above, we find, by putting the coefficient for $\sin \omega t$ to zero,

$$\left[r \left[\frac{1}{r} (rg')' \right]' \right]' - \lambda rg - 9(1-\nu^2)\alpha (fg')' = 0, \qquad (16.45)$$

where

$$\lambda = \frac{\rho h}{D} \omega^2 \qquad (16.46)$$

and

$$\alpha = \left(\frac{A}{h}\right)^2. \qquad (16.47)$$

In addition to the system (16.44)–(16.45) we have the following boundary conditions,

$$w(a) = w'(a) = 0, \qquad w'(0) = w'''(0) = 0,$$

implying that

$$g(a) = g'(a) = 0, \qquad g'(0) = g'''(0) = 0. \qquad (16.48)$$

Also, the strain E_2^2 in circumferential direction at the boundary $r = a$ must vanish, since we assume the plate to be rigidly clamped there. Thus

$$E_2^2 = \frac{1}{Eh}(N_2^2 - \nu N_1^1) = \psi' - \nu \frac{\psi}{r} = 0 \quad \text{for } r = a$$

and hence

$$f'(a) - \nu \frac{f(a)}{a} = 0. \qquad (16.49)$$

But surely N_1^1 is bounded for $r \to 0$, and hence we have

$$f(0) = 0. \tag{16.50}$$

The dimensionless amplitude α appears as a parameter in the coupled nonlinear system (16.44)–(16.45), and the solution must therefore depend on α. Presumably this dependence is analytical, at least at $\alpha = 0$. We shall assume this to be the case and write

$$\begin{aligned} g &= g_0 + \alpha g_1 + \cdots, \\ f &= f_0 + \alpha f_1 + \cdots, \\ \lambda &= \lambda_0 + \alpha \lambda_1 + \cdots, \end{aligned} \tag{16.51}$$

with the aim of determining the functions g_0, g_1, f_0, f_1, and ultimately the numbers λ_0 and λ_1.

Substituting (16.51) into (16.45) and (16.44) and equating the coefficients of the zeroth and first power of α to zero, we get from (16.45)

$$\left[r \left[\frac{1}{r} (rg_0')' \right]' \right]' - \lambda_0 r g_0 = 0 \tag{16.52}$$

and

$$\left[r \left[\frac{1}{r} (rg_1')' \right]' \right]' - \lambda_0 r g_1 = \lambda_1 r g_0 + 9(1 - \nu^2)(f_0 g_0')' \tag{16.53}$$

and, from (16.44),

$$\left[\frac{1}{r} (rf_0)' \right]' + \frac{1}{2r} (g_0')^2 = 0 \tag{16.54}$$

and

$$\left[\frac{1}{r} (rf_1)' \right]' + \frac{1}{r} g_0' g_1' = 0. \tag{16.55}$$

Similarly, we find that the boundary conditions can be written as

$$g_i(a) = g_i'(a) = g_i'(0) = g_i'''(0) = 0, \quad i = 0, 1, \ldots, \tag{16.56}$$

and

$$f_i'(a) - \nu \frac{f_i(a)}{a} = f_i(0) = 0, \quad i = 0, 1, \ldots. \tag{16.57}$$

The solution to (16.52) is (see Chapter 9)

$$g_0(r) = A_1 J_0(kr) + A_2 J_0(ikr), \tag{16.58}$$

where

$$k^4 = \lambda_0. \tag{16.59}$$

The boundary conditions require that

$$A_1 J_0(\mu) + A_2 J_0(i\mu) = 0,$$
$$A_1 J_0'(\mu) + A_2 i J_0'(i\mu) = 0,$$

where $\mu = ka$. The roots of the determinant equation are given in Chapter 9, Table 3; the lowest being $\mu = 3.1962$.

The additional condition $g_0(0) = 1$ yields

$$A_1 + A_2 = 1.$$

With this condition the constants A_1 and A_2 are uniquely determined, and hence also g_0. With g_0 in hand we can integrate (16.54) and find f_0, which due to (16.49) will depend on Poisson's ratio ν. The functions $g_0(r)$, $g_0'(r)$ and $f_0(r)$ are shown in Fig. 43 for $\nu = 0.3$ together with $(1/r)f_0$, representing the radial membrane forces N_r.

The next approximation $g_1(r)$ is the solution of the nonhomogeneous equation (16.53). Since the operator on g_1 in (16.53) is identical with the operator on g_0 in (16.52), there is no solution g_1 unless the right-hand side of (16.53) is orthogonal to g_0.

Hence

$$\lambda_1 \int_0^a g_0^2 r \, dr + 9(1 - \nu^2) \int_0^a (fg_0')' g_0 \, dr = 0.$$

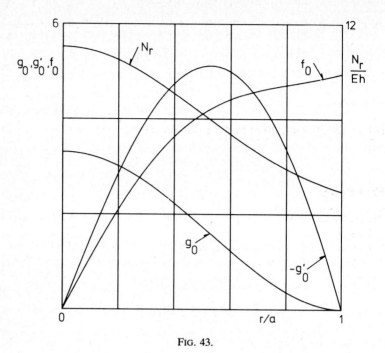

Fig. 43.

After integrating the second integral by parts and solving for λ_1, we find

$$\lambda_1 = 9(1-\nu^2)\int_0^a f_0(g_0')^2\,dr \Big/ \int_0^a g_0^2 r\,dr. \tag{16.60}$$

Numerical integration yields $\lambda_1 = 39.2/a^4$ and therefore

$$\omega^2 \frac{\rho h a^4}{D} = 3.1962^4 + 39.2\left(\frac{A}{h}\right)^2$$

or

$$\omega^2 = \omega_0^2\left[1 + 0.375\left(\frac{A}{h}\right)^2\right], \tag{16.61}$$

where

$$\omega_0 = 9.2963 \frac{h}{a^2} \sqrt{\frac{E}{\rho}} \qquad (16.62)$$

is the frequency when the amplitude is infinitely small. Formula (16.61) reveals that the nonlinear effect due to the membrane stresses is very strong.

Bibliography

TIMOSHENKO, S. (1959), *Theory of Plates and Shells*, 2nd ed., McGraw-Hill, New York, Chapters 12–13.

DONNELL, L.H. (1976), *Beams, Plates, and Shells*, McGraw-Hill, New York, pp. 159–225.

UGURAL, A.C. (1981), *Stresses in Plates and Shells*, McGraw-Hill, New York, pp. 153–172.

CHAPTER 17

BUCKLING OF PLATES AND SHELLS

1. Introduction 385
2. The classical model of elastic stability 385
3. Buckling of axially compressed cylindrical shells 388
4. Inadequacy of the classical model 394
5. References and literature on the nonlinear buckling theories 397
   Bibliography 398

CHAPTER 17

BUCKLING OF PLATES AND SHELLS

1. Introduction

In the linear theory of elasticity the deformation of a structure is assumed to be so small that equilibrium may be expressed in the undeformed instead of the deformed state with sufficient accuracy. This leads to a unique solution.

When equilibrium is expressed as a state of the deformed structure as in fact it should be, uniqueness of the solutions is no longer ensured, and there may be two or more states of equilibrium, stable or unstable. If we wish to study buckling of shells we must therefore go beyond the linear theory and take nonlinear effects into account.

2. The classical model of elastic stability

EULER's investigation of the elastic stability of slender bars in 1744[1] has been the model for the classical theory of buckling of shells. Prior to buckling, the deformation of the coordinate system[2] is assumed to be negligible and the equations of equilibrium in the undeformed state may be used to determine the state of stress in the shell. Also all changes of curvature are assumed to be negligible and hence the conditions for a membrane state (10.1)–(10.2) are present.

Let $S^{\alpha\beta}$ be a *unit membrane state* for the problem we wish to consider, corresponding to some unit external loading, F^α and p. The

---

[1] See Additamentum, "De curvis elasticis", in *Methodus Inveniendi Lineas Curvas Maximi Minimive Proprietate Gaudentes*, Lausanne, 1744.
[2] See Chapter 3, §1, Introduction.

membrane state is then uniquely determined from the following system of equations,

$$D_\alpha S^{\alpha\beta} + F^\beta = 0, \tag{17.1}$$

$$d_{\alpha\beta} S^{\alpha\beta} + p = 0 \tag{17.2}$$

and the boundary conditions on $S^{\alpha\beta}$. It should be noted that the existence of a solution to this system of equations with appropriate boundary conditions is necessary for the classical model of stability.

Let us assume that at some multiple λ of the unit membrane state buckling will occur. The membrane forces are then $\lambda S^{\alpha\beta}$. At the critical buckling load there exist an undeformed as well as deformed states of the shell in equilibrium. The resultant forces and moments in the buckled shell can be obtained by a superposition of the membrane state and a state of (small) additional forces $N^{\alpha\beta}$ and moments $M^{\alpha\beta}$. In other words, the resultant forces and moments after buckling will be written $\lambda S^{\alpha\beta} + N^{\alpha\beta}$ and $M^{\alpha\beta}$. This state has to satisfy the equations of equilibrium in the deformed coordinate system, i.e.,

$$D_\alpha^*(\lambda S^{\alpha\beta} + N^{\alpha\beta}) + 2d_\gamma^{*\beta} D_\alpha^* M^{\gamma\alpha} + M^{\gamma\alpha} D_\alpha^* d_\gamma^\beta + F^{*\beta} = 0, \tag{17.3}$$

$$D_\alpha^* D_\beta^* M^{\alpha\beta} - d_{\alpha\gamma}^* d_\beta^{*\gamma} M^{\alpha\beta} - d_{\alpha\beta}^*(\lambda S^{\alpha\beta} + N^{\alpha\beta}) - p^* = 0. \tag{17.4}$$

In order to express the equations of equilibrium in the undeformed coordinate system, we proceed to determine $D_\alpha^* S^{\alpha\beta}$ and $d_{\alpha\beta}^* S^{\alpha\beta}$,

$$D_\alpha^* S^{\alpha\beta} = D_\alpha S^{\alpha\beta} + E_{\alpha\gamma}^\beta S^{\alpha\gamma} + E_{\alpha\gamma}^\alpha S^{\beta\gamma},$$

$$d_{\alpha\beta}^* S^{\alpha\beta} = d_{\alpha\beta} S^{\alpha\beta} + K_{\alpha\beta} S^{\alpha\beta},$$

where (3.41) and (3.32) were used.

Substitution and linearization yields

$$D_\alpha N^{\alpha\beta} + 2d_\gamma^\beta D_\alpha M^{\gamma\alpha} + M^{\gamma\alpha} D_\alpha d_\gamma^\beta$$
$$+ \lambda (E_{\alpha\gamma}^\beta S^{\alpha\gamma} + E_{\alpha\gamma}^\alpha S^{\beta\gamma}) + (F^{*\beta} - \lambda F^\beta) = 0, \tag{17.5}$$

$$D_\alpha D_\beta M^{\alpha\beta} - d_{\alpha\beta} N^{\alpha\beta} - \lambda K_{\alpha\beta} S^{\alpha\beta} - (p^* - \lambda p) = 0. \tag{17.6}$$

It should be noted that this system of equations for the *additional* forces and moments may formally be regarded as a system of equations for the *total* state of resultant forces and moments; the external loads being the *membrane loads*, as a comparison with the system (5.22)–(5.23) clearly shows. In other words, the system of equations (17.5)–(17.6) is obtained from the system (5.22)–(5.23) with the external loading equal to the membrane loads

$$\hat{F}^\beta = \lambda(E^\beta_{\alpha\gamma}S^{\alpha\gamma} + E^\alpha_{\alpha\gamma}S^{\beta\gamma}) + F^{*\beta} - \lambda F^\beta, \qquad (17.7)$$

$$\hat{p} = \lambda K_{\alpha\beta}S^{\alpha\beta} - p^* + \lambda p, \qquad (17.8)$$

where the deviator $E^\beta_{\alpha\gamma}$ is given by (3.43) in terms of the strain tensor.

If the external load is *hydrostatic pressure*, it acts in the direction of the normal to the deformed middle-surface, and hence

$$F^{*\beta} = F^\beta = 0, \qquad p^* = \lambda p. \qquad (17.9)$$

The result with regards to (17.7)–(17.8) is precisely the same as if there were *no* external loads except boundary loads.

If on the other hand the shell is subject to *dead loading*, the cartesian components of the load in the fixed coordinate system remain unchanged during deformation. According to (4.14) we then have

$$\bar{F}^i = \lambda f^i_{,\alpha} F^\alpha + \lambda p X^i = f^{*i}_{,\alpha} F^{*\alpha} + p^* X^{*i}. \qquad (17.10)$$

Multiplication by $X^{*i} = X^i + \delta X^i$, considering (3.20) and (3.5), yields

$$p^* = \lambda p - \lambda a_{\alpha\delta}(a^{\delta\beta} w_{,\beta} + d^\delta_\gamma v^\gamma) F^\alpha$$

or

$$-p^* + \lambda p = \lambda(w_{,\alpha} + d_{\alpha\gamma}v^\gamma) F^\alpha, \qquad (17.11)$$

which should be substituted into (17.8).

Multiplying (17.10) by $f^{*i}_{,\gamma}$ instead, yields

$$F^*_\gamma = \lambda F_\gamma + \lambda(a_{\alpha\delta}D_\gamma v^\delta - d_{\alpha\gamma}w)F^\alpha + \lambda p(d_{\gamma\delta}v^\delta + w_{,\gamma})$$

and hence

$$F^{*\beta} - \lambda F^\beta = \lambda(D^\beta v_\alpha - d_\alpha^\beta w)F^\alpha + \lambda p(d_\delta^\beta v^\delta + D^\beta w), \quad (17.12)$$

which should be introduced in (17.7).

It is now clear that $\hat{F}^\beta$ and $\hat{p}$ are proportional to the *additional* deformation, and hence (17.5)–(17.6) becomes a linear and homogeneous system containing the unknown parameter λ. Together with a set of linear and homogeneous boundary conditions it constitutes an *eigenvalue problem*, analogous to the mathematical problem of free vibrations of shells. The mathematical theory of such problems is thoroughly treated in the literature, and several text-books are available in this field.[3]

In spite of the enormous literature on the stability of shells there is hardly any formulations of a general character to be found. The theory presented here is due to the present author.[4]

3. Buckling of axially compressed cylindrical shells

As an example, let us consider the case of a circular cylindrical shell (see Fig. 44) axially compressed by a force $\mathcal{N}$ per unit length of the circumference. Using the same cartesian coordinates as in Chapter 11, Fig. 23, the unit membrane state is conveniently taken to be

$$S^{11} = -\frac{Eh}{1-\nu^2}, \quad S^{12} = S^{22} = 0.$$

The compressive force at the boundaries is therefore

$$\mathcal{N} = \lambda \frac{Eh}{1-\nu^2}, \quad (17.13)$$

where λ is the dimensionless load parameter.

---

[3] See, for instance, COLLATZ, L. (1948), *Eigenwertprobleme und ihre Numerische Behandlung*, Chelsea Publishing, New York.

[4] NIORDSON, F.I.N. (1961), "On the linear theory of stability of thin elastic shells", *Bygn. Stat. Medd.* **32**, pp. 46–54; discussion by KOITER, W.T. and NIORDSON, F.I.N., *ibid.*, pp. 86–88.

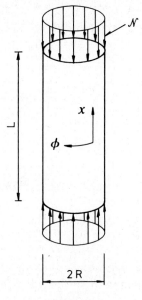

FIG. 44.

Since there are no external loads except at the boundaries, we have

$$F^{*\alpha} = F^{\alpha} = p^* = p = 0$$

and hence

$$\bar{F}^1 = \lambda(2E_{11}^1 + E_{12}^2)S^{11},$$
$$\bar{F}^2 = \lambda E_{11}^2 S^{11}, \qquad (17.14)$$
$$p = \lambda K_{11} S^{11}.$$

Using (3.43), (11.5) and (11.6) we get

$$\bar{F}^1 = -\lambda \frac{Eh}{1-\nu^2}\left(2\frac{\partial^2 u}{\partial x^2} + \frac{\partial^2 v}{\partial x \partial \phi} + \frac{1}{R}\frac{\partial w}{\partial x}\right),$$
$$\bar{F}^2 = -\lambda \frac{Eh}{1-\nu^2}\frac{\partial^2 v}{\partial x^2}, \qquad (17.15)$$
$$p = \lambda \frac{Eh}{1-\nu^2}\frac{\partial^2 w}{\partial x^2},$$

and introducing this into (11.12) we find the following system of linear and homogeneous differential equations for the displacements u, v and w,

$$\frac{\partial^2 u}{\partial x^2} + \tfrac{1}{2}(1-\nu)\frac{\partial u^2}{\partial \phi^2} + \tfrac{1}{2}(1+\nu)\frac{\partial^2 v}{\partial x \partial \phi} + \frac{\nu}{R}\frac{\partial w}{\partial x}$$

$$- \lambda \left[2\frac{\partial^2 u}{\partial x^2} + \frac{\partial^2 v}{\partial x \partial \phi} + \frac{1}{R}\frac{\partial w}{\partial x} \right] = 0,$$

$$\tfrac{1}{2}(1+\nu)\frac{\partial^2 u}{\partial x \partial \phi} + \tfrac{1}{2}(1-\nu)\frac{\partial^2 v}{\partial x^2} + \frac{\partial^2 v}{\partial \phi^2} + \frac{1}{R}\frac{\partial w}{\partial \phi} - \lambda \frac{\partial^2 v}{\partial x^2} = 0,$$

$$\frac{\nu}{R}\frac{\partial u}{\partial x} + \frac{1}{R}\frac{\partial v}{\partial \phi} + \frac{w}{R^2} + k\left(R\Delta + \frac{1}{R}\right)^2 w + \lambda \frac{\partial^2 w}{\partial x^2} = 0. \tag{17.16}$$

Like in the case of free vibrations of a cylindrical shell (Chapter 11), we assume the displacements to be the following,

$$u = A \cos \frac{mx}{R} \cos \frac{n\phi}{R},$$

$$v = B \sin \frac{mx}{R} \sin \frac{n\phi}{R}, \tag{17.17}$$

$$w = C \sin \frac{mx}{R} \cos \frac{n\phi}{R},$$

where n is an integer and

$$m = j\frac{\pi R}{L}, \quad j = 1, 2, \ldots, \tag{17.18}$$

where j is a natural number and L the length of the cylinder.

When (17.17) is substituted into (17.16), the trigonometric functions cancel and we are left with a set of linear homogeneous equations for A, B and C.

The condition for a nontrivial solution is that the determinant vanishes, i.e.,

BUCKLING OF AXIALLY COMPRESSED CYLINDRICAL SHELLS

$$\begin{vmatrix} -\tfrac{1}{2}(1-\nu)n^2 - m^2 + 2\lambda m^2 & \tfrac{1}{2}(1+\nu)mn - \lambda mn & \nu m - \lambda m \\ \tfrac{1}{2}(1+\nu)mn & -\tfrac{1}{2}(1-\nu)m^2 - n^2 + \lambda m^2 & -n \\ \nu m & -n & -1 - k(n^2 + m^2 - 1)^2 - \lambda m^2 \end{vmatrix} = 0.$$

(17.19)

When working this out, we find the following equation for λ,

$$a_3\lambda^3 + a_2\lambda^2 + a_1\lambda + a_0 = 0, \qquad (17.20)$$

where

$$\begin{aligned} a_0 &= \tfrac{1}{2}(1-\nu)[(1-\nu^2)m^4 + k(m^2+n^2)^2(m^2+n^2-1)^2], \\ a_1 &= -\tfrac{1}{2}(1-\nu)m^2[(m^2+n^2)^2 + (4+\nu)m^2 + n^2] \\ &\quad - k(2-\nu)m^2(m^2+n^2)(m^2+n^2-1)^2, \\ a_2 &= (2-\nu)m^4(m^2+n^2+1) + 2km^4(m^2+n^2-1)^2, \\ a_3 &= -2m^6. \end{aligned}$$

(17.21)

The critical load is given by the lowest root λ of this equation for all integer values of n ($n = 0, 1, 2, \ldots$) and all natural numbers j ($j = 1, 2, \ldots$).

The result, which is easily found numerically, is shown in Fig. 45. Here the critical load-factor λ is given as a function of L/R for different values of R/h. In Fig. 46, the corresponding modes, determined by the integers n and j, are indicated. For very long cylinders (large values of L/R) the shell buckles like an *Euler column*, i.e., in the mode $n = 1, j = 1$, shown in Fig. 45 as the line 1/1. Indeed, for very small values of m, the lowest root of (17.20) is obtained for $n = 1$ at

$$\lambda = \tfrac{1}{2}(1-\nu^2)m^2, \qquad (17.22)$$

which according to (17.13) and (17.18) is equivalent to a resultant axial force

$$P = 2\pi R \mathcal{N} = \pi^2 \frac{EI}{L^2}, \qquad (17.23)$$

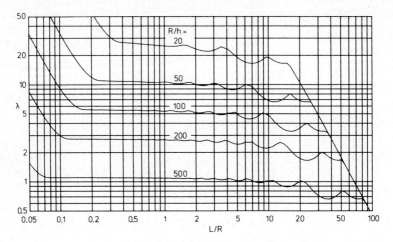

Fig. 45.

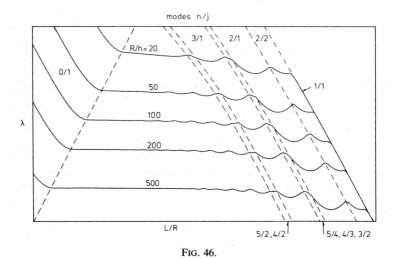

Fig. 46.

where $I = \pi R^3 h$ is the moment of inertia for a thin-walled circular section.

At the other end, for very large values of m the lowest root of (17.20) is found for $n = 0$ at

$$\lambda = km^2 \tag{17.24}$$

corresponding to

$$\mathcal{N} = \pi^2 \frac{D}{L^2}, \qquad (17.25)$$

where $D = Eh^3/12(1-\nu^2)$ is the bending rigidity per unit length of a plate of thickness h. In the region 0/1 of Fig. 46 the *wall* of the shell buckles like an Euler column.

To the left of the line 1/1 in Fig. 46 there is a region where the mode $n = 2$, $j = 2$ gives the lowest λ. Further to the left the lowest mode is $n = 2$, $j = 1$. Still further on we find a region with $n = 3$, $j = 1$. Between those two regions there is a narrow band with higher wave-numbers, either $n = 4$, $j = 3$ or $n = 3$, $j = 2$. Such narrow bands with higher wave-numbers interspace the main regions $j = 1$, $n = 2, 3, 4, \ldots$ as we approach the region of axisymmetrical buckling $n = 0$, $j = 1$.

In the main region of the diagram, between the line 1/1 and the region 0/1, $m^2 + n^2 \gg 1$ and we can approximate (17.20) by

$$-\tfrac{1}{2}(1-\nu)m^2(m^2+n^2)^2\lambda + \tfrac{1}{2}(1-\nu)[(1-\nu^2)m^4 + k(m^2+n^2)^4] = 0$$

and hence

$$\lambda = (1-\nu^2)\frac{m^2}{(m^2+n^2)^2} + k\frac{(m^2+n^2)^2}{m^2}. \qquad (17.26)$$

The minimum value of λ for *all* real numbers m and n is

$$\lambda_{\min} = 2\sqrt{(1-\nu^2)k} \qquad (17.27)$$

and the true critical value of λ, taking the integral properties of j and n into account, is only slightly larger. From this we can draw the conclusion that except for very large or very small values of L/R, all cylindrical shells buckle in axial compression at

$$\lambda = \frac{\mathcal{N}}{Eh}(1-\nu^2) = \frac{\sigma_c}{E}(1-\nu^2) = 2\sqrt{\frac{1-\nu^2}{12}}\frac{h}{R},$$

i.e., at the *critical stress*

$$\sigma_c \approx \frac{1}{\sqrt{3(1-\nu^2)}} \frac{Eh}{R} \approx 0.605 \frac{Eh}{R}. \tag{17.28}$$

Similar results for other loading cases as for instance hydrostatic pressure can be obtained in the same manner.

The linear problem of buckling of cylindrical shells has received much attention in the literature and results very similar to those presented here were obtained by FLÜGGE.[5]

4. Inadequacy of the classical model

The Euler formula for buckling of slender bars is still in use today as an engineering design formula due to the good agreement between the theoretically predicted buckling load and experiments. One had good reasons to expect that this model for elastic buckling would yield useful results also when applied to shells.

To most early investigators the violent disagreement between theoretical and experimental values of critical loads was therefore a baffling problem. *Knock-down factors* (representing the ratio between the lower bound of experimentally established buckling loads and the theoretical value) of the order of $\frac{1}{5}$ or even less were not uncommonly reported. Wide scatter in experimental results indicated that *geometric imperfections* might be responsible for the disagreement, and indeed, a number of investigations have shown that this is the case, although early attempts to explain the discrepancies quantitatively in this way failed.

The first progress along this road was due to VON KÁRMÁN and TSIEN[6] but the decisive step forward was taken by KOITER in his doctorate thesis of 1945,[7] where he derived a general theory is buckling and post-buckling behaviour. After a period of some 15 years the basic ideas of the theory started to become widely known and have by now been applied to a great number of problems.

It must be added that although the classical buckling theory as

---

[5] FLÜGGE, W. (1932), "Die Stabilität der Kreiszylinderschale", *Ing.-Arch.* **3**, pp. 463–506. See also FLÜGGE, W. (1960), *Stresses in Shells*, Springer, Berlin, Chapter 7.

[6] See §5, VON KÁRMÁN and TSIEN (1941).

[7] See §5, KOITER (1945).

presented above in the two preceding sections often leads to critical loads that are unrealistically high, the more realistic studies of *post-buckling behaviour* and *load-carrying capacity* (with or without imperfections) depend on this analysis, and therefore it is indispensible.

Figure 47 schematically shows the axial load P as a function of the axial compression δ in the case of an Euler column (a) and a cylindrical shell (b).

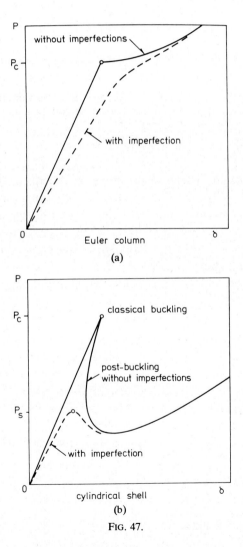

FIG. 47.

The solid lines represent the geometrically perfect structures and the dashed lines imperfect structures with some given small imperfection. The post-buckling behaviour of the two perfect structures is entirely different and it is easy to see why the shell (b) is so much more imperfection sensitive than the column (a).

A careful analysis of the stability of an axially compressed cylindrical shell with an axisymmetric initial imperfection of the type

$$\tilde{w} = \delta \cos \beta x$$

shows a considerable lowering of the critical stress (see Fig. 48).[8] In the diagram the theoretically computed stability limit for the stress σ_s is given as a function of the initial imperfection for the worst possible value of β. We see that an imperfection of the type of only $\frac{1}{10}$ of the shell-thickness h reduces the buckling load to half the classical value (17.28). The effect of other types of imperfections is normally less drastic.

Mode interaction is another important and often dangerous phenomenon in structures with equal or nearly equal critical loads for overall and local buckling modes. This interaction has always a detrimental effect on the load carrying capacity of the structure, in particular in the presence of unavoidable imperfections.

Post-buckling behaviour, imperfection sensitivity and mode interaction has been a subject of intense research in the past 25 years. The

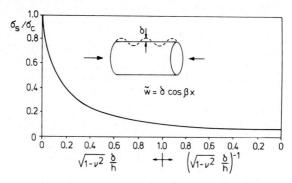

FIG. 48.

[8] See §5, PEDERSEN, P.T. (1973).

number of research papers in this field is very high and there are several survey and review papers available, some of them listed below. For a first introduction to the subject, BUDIANSKY (1974), is recommended.

5. References and literature on the nonlinear buckling theories

KÁRMÁN, Th. VON and H.S. TSIEN (1941), "The buckling of thin cylindrical shells under axial compression", *J. Aero. Sci.* **8**, p. 303.

KOITER, W.T. (1945), *On the Stability of Elastic Equilibrium*, Thesis, Techn. Univ. of Delft (in Dutch with English summary), H.J. Paris, Amsterdam.

KOITER, W.T. (1963), "The effect of axisymmetric imperfections on the buckling of cylindrical shells under axial compression", *Proc. Kon. Ned. Akad. Wet.* **B66**, p. 265.

KOITER, W.T. (1963), "Elastic stability and post-buckling behavior", *Proc. Symp. Nonlinear Problems*, R.E. Langer, Ed., University of Wisconsin Press, pp. 257–275.

BUDIANSKY, B. and J.W. HUTCHINSON (1966), "A survey of some buckling problems", *AIAA J.* **4**, pp. 1505–1510.

KOITER, W.T. (1966), "General equations of elastic stability for thin shells", *Proc. Symp. in Honor of Lloyd H. Donnell*, D. Muster, Ed., University of Houston Press, pp. 187–227.

HUTCHINSON, J.W. and W.T. KOITER (1970), "Postbuckling theory", *Appl. Mech. Rev.* **23**, pp. 1353–1366.

BUDIANSKY, B. and J.W. HUTCHINSON (1972), "Buckling of circular cylindrical shells under axial compression", in: *Contributions to the Theory of Aircraft Structures (van der Neut Anniversary Volume)*, Delft University Press, pp. 239–260.

BUDIANSKY, B. (1974), "Theory of buckling and post-buckling behavior of elastic structures", in: *Advances in Applied Mechanics, Vol. 14*, Academic Press, New York, pp. 1–65.

KOITER, W.T. (1976), "General theory of mode interaction in stiffened plate and shell structures", *Rept. No. 590*, Lab. of Eng. Mech., Delft (*WTHD Nr. 91*).

KOITER, W.T. and M. PIGNATARO (1976), "An alternative approach to the interaction between local and overall buckling in stiffened panels", *Proc. IUTAM Symp. in Buckling of Structures, 1974*, B. Budiansky, Ed., Springer, Berlin, pp. 133–148.

KOITER, W.T. (1976), "Current trends in the theory of buckling", *Proc. IUTAM Symp. on Buckling of Structures, 1974*, B. Budiansky, Ed., Springer, Berlin, pp. 1–16.

TVERGAARD, V., (1976), "Buckling behavior of plate and shell structures", *Proc. 14th IUTAM Congress*, North-Holland, Amsterdam, pp. 233–247.

BUDIANSKY, B. and J.W. HUTCHINSON (1979), "Buckling: Progress and challenge", *Trends in Solid Mechanics*, J.F. Besseling and A.M.A. van der Heijden, Eds., Delft University Press.

PEDERSEN, P.T. (1973), "Buckling of unstiffened and ring stiffened cylindrical shells under axial compression", *Internat. J. Solids & Structures* **9**, pp. 671–691.

Bibliography

TIMOSHENKO, S. (1936), *Theory of Elastic Stability*, McGraw-Hill, New York, pp. 439–491.

FLÜGGE, W. (1960). *Stresses in Shells*, Springer, Berlin, pp. 404–472.

WLASSOW, W.S. (1958), *Allgemeine Schalentheorie und ihre Anwendungen in der Technik*, Akademie-Verlag, Berlin, pp. 379–391.

DONNELL, L.H. (1976), *Beams, Plates, and Shells*, McGraw-Hill, New York, pp. 386–416.

BUDIANSKY, B., Ed. (1976), *Buckling of Structures, Proc. IUTAM Symp.*, Springer, Berlin.

SUBJECT INDEX

absolute change	22, 80
admissible functions	153
AIRY, G.B.	138
Airy's stress function	138
d'Alembert forces	151
alternating symbol in three dimensions	36
– symbol in two dimensions	62
– tensor	62
alternative measures of bending	127, 266, 290
analogy, static-geometric	128
angle between coordinate lines	20
– between vectors	20
apex of a dome, conditions at	192
approximate solutions	283, 355
area element	47, 60
ARON, H.	355
associated Legendre functions	303
– tensors	19
assumption of small strains	115
axial component of membrane stresses	268
– displacements	271, 342
– membrane forces	343
axisymmetrical bending, circular plates	166
– bending, conical shells	272
– bending, cylindrical shells	242
– bending, shells	265
– bending, spherical shells	281
– bending, strains	266
– bending, toroidal shells	282
– membranes, displacements	218
– membranes, strains	218
– membranes, wind loading	201
– ring-loads	244
– surfaces, equations of	182, 266
base of a dome, conditions at	193
beam-states of cylindrical shells	241
bending	64
– due to axisymmetrical ring-loads	244
– energy	115
– moment	82
– moment in polar coordinates	165
– moment in rectangular coordinates	145
– of axisymmetrical shells	265, 339
– of conical shells	272
– of cylindrical panels	261
– of cylindrical shells	225
– of spherical shells	281, 289
– of toroidal shells	282, 349
– tensor	64
– tensor, axisymmetrical shells	266
– tensor, cylindrical shells	227
– tensor, DMV theory	356
– tensor, plates	144
bending tensor, spherical shells	282
– tensors, alternative	127, 266, 290
– theory of plates	144
bending, alternative measures of	127, 266, 290
– pure	120
– skew	146
Bessel functions	168

biharmonic functions	138
boundary conditions	103
– conditions, axisymmetrical shells	341
– conditions, cylindrical shells	231
– conditions, kinematic	153
– conditions, membranes	185
– conditions, spherical shells	297
– curve	75
– moment	104
buckling behaviour	395
– equations, cylindrical shells	390
– of cylindrical shells	388
– of plates	367, 385
– of shells	385
buckling, critical stresses	394
BUDIANSKY, B.	133, 397
cartesian coordinates	7
Cauchy stress tensor	106
CAUCHY, A.	2
Cauchy–Riemann equations	204, 314
Cauchy's integral formula	209
characteristic equation	173, 237
– length of deformation	119, 244
characteristics	186
Christoffel deviator	68, 386
– symbols, normal coordinates	40
– symbols, polar coordinates	165
– symbols, definition	22
– symbols, spherical coordinates	299
circular membrane under transverse load	372
– plates	166
– plates, bending of	166
– plates, vibrations of	168, 375
circumferential membrane force	192
clamped edge of plates	145
– rectangular plates	160
closed cylindrical shells	237
co-latitude	298
Codazzi's equations	46
COHEN, G.A.	348

compatibility	67
compatibility, equations of	69, 70
– plates	138
– spherical shells	293
complex form of shell equations	131
– functions, plates	139
– functions, spherical membranes	204
compliance	154
components of tensors	7, 8, 11
components of vectors	7, 9, 10
concentrated forces at corners	146
– forces, spherical membranes	209
conditions at the apex of domes	192
– at the base of domes	193
conical coordinates	220
– membranes	220
– shells, axisymmetrical bending	272
– shells, example	276
– surfaces	219
– surfaces, equations of	220
constitutive equations	121, 228
contraction	14
contravariant derivative	28
– law	9
– vectors	10
coordinate transformations	8
coordinates, cartesian	7
– conical	220
– cylindrical	225
– deformed	56
– Eulerian	55
– geographical	203
– isometric	202
– Lagrangian	55
– locally cartesian	17
– normal	35
– orthogonal	21
– of principal curvature	49
– rectangular	7
– skew	182
– spherical	202, 298
– toroidal	213
– transformations between	8

SUBJECT INDEX

corners, behaviour of plates at	145, 172
– obtuse, stresses at	175
covariant derivative	23
– derivative of metric tensors	24
– derivative of product	24
– derivative of scalar	24
– derivatives, order of	22, 291, 357
– differentiation	21
– law	10
– vectors	9
critical stresses at buckling	394
curvature of surfaces	42
– tensor	39
curvature, axisymmetrical surfaces	183
– conical surfaces	220
– cylindrical surfaces	226
– Gaussian	43, 185
– mean	43
– principal	42
– radius of	41
– spherical surfaces	289, 299
cylindrical coordinates	225
– membranes	219
– panels, bending of	261
– shells	225
– shells, beam state	241
cylindrical shells, buckling of	388
– shells, closed	237
– shells, critical stresses	394
– shells, elementary cases	246
– shells, ring-stiffened	248
– shells, sensitivity	396
– shells, strains	226
– shells, vibrations of	259
– surfaces, equations of	219, 225
dead loading	387
deformation, analysis of	55
deformed coordinate system	56
– state	56
delta function, Dirac's	149
delta, Kronecker's	11
derivative, contravariant	28
– covariant	23
determinant, Jacobian	8, 49
– metric tensor	18, 35
developable surfaces	44, 219
dimensional reduction	2
Dirac delta function	149
displacement gradient	57
– vector	55
displacements	55
displacements, axisymmetrical membranes	197
– conical membranes	220
– cylindrical membranes	219
– developable membranes	218
– normal	56
– shells of revolution	342
– tangential	56
– virtual	91
distributive rule	24
divergence theorem	77
DMV theory	356
DMV theory, nonlinear plate problems	365
domes	190
– of given circumferential stresses	196
– of uniform strength	194
– of uniform thickness, spherical	193
domes, conditions at apex	192
– conditions at base	193
DONNELL, L.H.	230, 355
DONNELL–MUSHTARI–VLASOV theory	355
drop container	187
dummy indices	9
effective forces and moments	91
– loads	86
– membrane forces	109
– membrane stresses	96
effective moment tensor	96
– shear-force	108

- shear-force in polar coordinates 165
- shear-force in rectangular coordinates 145
- width 247
eigenvalue problem by buckling 388
Einstein's summation convention 9
elastic energy 115
elasticity, plane 139
element of area 47, 60
- of shell 75
elementary cases for cylindrical shells 246
elliptic differential equations 185
elliptical plates 150
embedding of surfaces 33, 38
energy methods 152
energy, elastic 115
equations for axisymmetrical shells 343
- for circular plates 376
- for cylindrical shells 229, 236
- for plates, VON KARMAN's 366, 376
- for spherical shells 295, 315
- of Codazzi 46
- of compatibility 69, 70
- of equilibrium 84, 99
- of Gauss 41, 46
equations, constitutive 121
equilibrium of axisymmetrical membranes 184
- of axisymmetrical shells 270, 341
- of conical membranes 220
- of cylindrical membranes 219
- of domes 191
- of membranes 179, 181
- of shells 83, 99
- of spherical membranes 203, 204
equilibrium, axisymmetrical shells 270, 341
- cylindrical shells 227, 229
- deformed state 359, 361, 386
- equations of 84, 99
- plates, nonlinear 365
- simplified equations of 356

- spherical shells 292
errors, relative 120, 152, 229, 372
Euclidean space 17
Euler column, buckling of 395
EULER, L. 3, 385
Eulerian coordinates 55
EVANS, T.H. 161
external forces 103
- forces due to gravity 190, 387
- loads 82, 387

FEDERHOFER, K. 336
field functions 347
- method 346
first fundamental form of surfaces 33
first order theory 110
flat shells 137
FLÜGGE, W. 230
force, physical concept of 20
forces, concentrated at corners 146
Fourier series 148, 199, 237, 262, 299, 366, 399
free edges of plates 145
- indices 9
frequencies of circular plates 168, 170, 381
- of cylindrical shells 260
- of skew plates 162
- of spherical shells 321, 330
fundamental form, first 33
- form, second 37
- tensor of space 16
- tensor of surfaces, first 34
- tensor, second 39

GAUSS, C.F. 33
Gauss' equations 41, 46
Gaussian curvature 43
- curvature, axisymmetrical surfaces 185
GECKELER, J.W. 283
generators 217
geographical coordinates 203
geometrical imperfections 368, 396
geometry of surfaces 33

SUBJECT INDEX

– of surfaces, intrinsic	38
GOURSAT, E.	140
gradient	10, 23
– of displacements	57
Greek indices	34
Hamilton–Cayley's theorem	50, 116
harmonic functions	204
hemispherical shell, example	311
Hooke's law	115
l'Hospital's rule	189
hydrostatic pressure	387
hyperbolic differential equations	185
hypergeometric functions	302
imperfection sensitivity	395
in-plane loaded plates	138
indices, dummy	9
– free	9
– Greek	34
– Latin	8
– lower	10
– lowering of	19
– raising of	19
– upper	10
inextensional deformations	294, 329
initial state	56
inner product	14
integrability conditions	46
intrinsic geometry of surfaces	38
invariance	28
invariant	14
– operator	28
invariants of curvature	41
– of strain	59
inverse problem for surfaces	44
– transformations	9
isometric coordinates	202
– surfaces	37
Jacobian determinant	8, 49
JELLETT, J.H.	301, 323
JENSEN, J.	122
JORDAN, P.F.	348

KARMAN, Th. VON	365, 394
Kelvin functions	273
– functions, asymptotic formulas	278
kinematic boundary conditions	153
kinematics of shells	55
KIRCHHOFF, G.	2
Kirchhoff's assumptions	2, 93, 151
KOITER, W.T.	111, 230, 388, 394
KÖNIG, H.	189
Kronecker's delta	11
Lagrangian coordinates	55
LAMB, H.	323
Laplacian, general form	28
– polar coordinates	165
– rectangular coordinates	235
– skew coordinates	162
– spherical coordinates	299
large amplitude vibrations	375
– deflections of plates	366
Latin indices	8
Laurent series	209
Legendre functions	302
– functions, associated	303
– polynomials	303
Legendre's equation	302
LEIPINS, A.	216
limit of stability for plates	367
linearly elastic shells	116
Liouville's theorem	205
load-carrying capacity of shells	395
loads, dead	387
– effective	86
local deformation	67
– measures of deformation	56
locally cartesian coordinates	17
longitude	298
longitudinal membrane force	218
LOVE, A.E.H.	2, 230, 355
lowering indices	19, 56
LURIE, A.I.	133
magnitude of vectors	20

SUBJECT INDEX

mathematical problem of shell theory	122
Maxwell's reciprocity theorem	280
mean curvature	43
measures of bending	64, 66, 127
Meissner operator	267
– operator, conical shells	272
– operator, spherical shells	281
– operator, toroidal shells	282
MEISSNER, E.	265
membrane	179
– force vector	97
– force, effective	109
– loads	387
– state of shells	179
– state, axisymmetrical loading	201
– state, axisymmetrical shells	181
– state, conical shells	220
– state, developable shells	216
– state, general solution	199
– state, orthogonal coordinates	180
– state, pre-buckling	385
– state, spherical shells	202, 294
– state, toroidal shells	213
– state, wind loading	201
– stress tensor	81
– stress tensor, effective	96
– stresses, effects of	381
membranes, circular, transverse load	372
– negative Gaussian curvature	187
– rectangular, transverse load	369
– transversely loaded	369, 372
– zero Gaussian curvature	216
Mercator projection	202
meridian	182
meridional membrane force	192
metric tensor, axisymmetrical surfaces	183
– tensor, conical surfaces	220
– tensor, cylindrical surfaces	219, 225
– tensor, space	17
– tensor, spherical surfaces	206, 299
– tensor, surfaces	34
middle-surface	33

MILLER, R.E.	348
minimum potential energy, principle of	152
mixed fundamental tensor	19
mixed system	46
– tensors	11
mode interaction	396
moment equilibrium	84
– tensor	82
– tensor, effective	96
– vector	82
MORLEY, L.S.D.	230
Morley–Koiter equations	229
MUSHTARI, Kh.M.	355
MUSKHELISHVILI, N.I.	144
NADAI, A.	147
NAVIER, C.	148
NEUT, A. VAN DER	336
NIORDSON, F.I.	122, 152, 228, 336, 388
non-axisymmetrical bending	250
– bending of shells	339
non-Euclidean surfaces	35
nonlinear DMV theory	359, 365
– effects	381
– plate problems	365
– strains	359
normal coordinates	35
– displacement	56
– to boundary curve	76
– to deformed surface	61
– to middle-surface	35
numerical solutions of shell problems	346
obtuse corners, stresses at	175
operator method	234, 294
operator, Laplacian	28, 162, 165, 299
– Meissner's	267, 272, 281, 282
– substitutional	11
optimal container	187
order of covariant derivatives	22, 27, 291, 357

– of Legendre functions	303	pressure, hydrostatic	387
orthogonal coordinates	21	principal curvatures	42
– trajectories	217	– radii of curvature	43
outer product	13, 24	– strains	59, 117
		principle of minimum potential energy	152
panels, cylindrical, bending of	261		
parabolic differential equations	185	– of Saint-Venant	219
parallel circle	182	– of virtual work	91, 297, 359
– vector field	79	product, inner	14
parametric equations of surfaces	33	– outer	13, 24
Pedersen, P.T.	396	– scalar	14
peripheral displacements	342	proper boundary conditions	98
Piola–Kirchhoff stress tensor	106	– element of shell	75
plane elasticity	139	pure bending	120
plates	137, 366, 376	– shear	119
– in polar coordinates	165	– tension	119
– in rectangular coordinates	144, 150	– twist	119
– in skew coordinates	161	Pythagoras' theorem	16
– with geometrical imperfections	368		
		quotient law of tensors	14
– with initial curvature	365	quotient, Rayleigh's	158
plates, axisymmetric bending of circular	166		
		radial component of membrane stresses	268
– behaviour at corners	172		
– bending theory	144	– displacements	271, 342
– buckling of	367	– membrane forces	343
– circular	166	radially loaded spherical shells	334
– circular, vibrations of	168, 375	radius of curvature	41
– elliptical	150	raising indices	19, 56
– large deflections of	366	Rayleigh quotient	158
– loaded in the plane	138	Rayleigh, J.W.S.	323, 329
– nonlinear problems of	365	Rayleigh–Ritz method	152
– rectangular	148, 160, 367	rectangular plates	148, 160, 367
– skew	161	– plates, clamped	160
– square	149	regular boundaries	346
– stressed in the plane	366	– transformations	8
– von Karman equations	366, 376	Reissner, H.	265
Poisson, S.D.	2	relative errors	120, 152, 229, 372
Poisson's ratio	115	Riemann–Christoffel tensor	25, 35, 37, 40
polar coordinates, plates in	165		
positive definite	17	– tensor, spheres	291
post-buckling behaviour of shells	395	Riemannian geometry	17
potential function, cylindrical shells	237	rigid body displacements	56
		rigorous quotient theorem	14
– function, spherical shells	296	ring-stiffened cylindrical shells	248

ring-stiffeners for domes	196	– spherical	202, 289, 294
Ritz' method	370	shells, toroidal	213, 281, 349
Rodriques' formula	303	– vibrations of	259, 312
rotation	57, 60	– zero Gaussian curvature	216
– of a vector	63	SHELLY, P.E.	348
– of an element	60	simply supported edge	145
– of boundary edge	109	– supported plates	145, 148
– vector	60	singular boundaries	346
rotation, average	63	– points	8
rotational inertia	152	skew deformation of plates	146
RUNGE, C.	189	– plates, vibrations of	161
Runge–Kutta method	189, 194	small strains, assumption of	115
		solution by operator method	234, 294
		space, Euclidean	17
		– Riemannian	27
Saint-Venant's principle	219	spherical coordinates	202, 298
SANDERS, J.L.	133, 216	– domes of uniform thickness	193
scalar	7	– harmonics	317
– function	10	– membranes	202
– product	14	– membranes, concentrated	
– stress function	132	forces	209
second fundamental form of		– shells, axisymmetrical	
surfaces	37	bending	281
– fundamental tensor	39	– shells, radially loaded	334
sectoral harmonics	317	– shells, vibrations of	312
self-consistency, tensor law	10	– surfaces, equations of	202
separation of plate problem	137	square plate, simply supported	149
shallow shells	359, 365	stability: see buckling	
shear, pure	119	static-geometric analogy	128
shear-force	98	statically determined state	179
– vector	82	– equivalent forces	79
shear-force, effective	108	statics of shells	75
– supplemented	98	stereographic projection	202
shell-thickness	33	strain invariants	59
shells, approximate solutions of	283	– tensor	58
– axisymmetrical	181, 265, 339	strain-energy density	116
– buckling of	385	strains in axisymmetrical	
– conical	220, 272, 276	membranes	198
– cylindrical	219, 225	– in cylindrical shells	226
– flat	137	– in spherical shells	290
– membrane state	179	strains, assumption of small	115
– nonpositive Gaussian		– principal	59
curvature	187	– virtual	99
– shallow	359	stress	115

– concentrations	175	tensors, associated	19
– distribution	109, 143	– dimension or type of	12
– function, Airy's	138	– mixed	11
– functions for shells	131, 237, 294, 357	– outer product of	24
		– rank or order of	8, 12
– tensor	79	tesseral harmonics	317
stress–strain relations	115	theorem of Liouville	205
stress-trajectories	211	TIMOSHENKO, S.	365
stresses at obtuse corners	175	toroidal coordinates	213
stresses, Cauchy's	106	– membranes	216, 221
– critical	394	– shells, axisymmetrical bending	282
– Piola–Kirchhoff's	106		
stretching of middle-surface	56	– shells, membrane state	213
substitutional operator	11	– surfaces, equations of	213
summation convention	9	transformation matrix	10
supplemented shear-force	98	– of coordinates	8
surface coordinates	33	translation	60
– element	47	transverse membrane force	217
– harmonics, spherical	317	TSIEN, H.S.	394
– tension	187	twist, pure	119
– tractions	79	twisting moment	82
surfaces, axisymmetrical	182		
– conical	220		
– cylindrical	219	uncoupled stress–strain equations	121
– first fundamental form of	34	uniform strength, domes of	194
– geometry of	33	unit membrane state	385
– intrinsic geometry of	38	upper indices	10
– parametric equations of	33		
– second fundamental form of	39	vector fields	21
– spherical	202	– stress-function for shells	132
– toroidal	213	vectors	8
symbolic manipulation	122	vectors, angle between	20
		– contravariant	10
tangent plane	35	– covariant	9
– surfaces	219	– inner product of	14
– vector to boundary curve	76	– magnitude of	20
– vector to coordinate line	20	– scalar product of	14
tangential displacements	56	vibrations of plates	151
tension, pure	119	– of circular plates	168, 375
tensor algebra	13	– of cylindrical shells	259
– analysis	7	– of plates, large amplitude	375
– equation	16	– of skew plates	161
tensor, alternating	62	– of spherical shells	312
tensors	8, 11	virtual displacements	91

– strains	99	wind loaded axisymmetrical membranes	201
– work, principle of	91, 297, 359	Wissler, H.	283
Vlasov, V.Z.	213, 355	work, principle of virtual	91, 297, 359
volume element	47		
– forces	83, 91		
von Karman plate equations	366, 376	Young's modulus	115